FA

Richard Holmes is Professor of Military and
Security Studies at Cranfield University and the
Defence Academy of the UK. He has written over
twenty books, including the acclaimed *Redcoat*
and *Tommy*, and has presented seven TV series'
for the BBC. In addition to being President of the
British Commission for Military History, he is
Patron of the Guild of Battlefield Guides and
Chairman of Project Hougoumont, which seeks
to restore the farm complex on Wellington's right
centre at Waterloo

RICHARD HOLMES

Fatal Avenue

A Traveller's History of the Battlefields of
Northern France and Flanders 1346–1945

VINTAGE BOOKS
London

Published by Vintage 2008

2 4 6 8 10 9 7 5 3 1

First published in Great Britain in 1992
by Jonathan Cape
First published in paperback in 1993 by Pimlico

Random House, 20 Vauxhall Bridge Road,
London SW1V 2SA

www.vintage-books.co.uk

Addresses for companies within The Random House Group Limited
can be found at: www.randomhouse.co.uk/offices.htm

The Random House Group Limited Reg. No. 954009

A CIP catalogue record for this book
is available from the British Library

ISBN 9781844139385

The Random House Group Limited supports The Forest Stewardship
Council (FSC), the leading international forest certification
organisation. All our titles that are printed on Greenpeace approved
FSC certified paper carry the FSC logo. Our paper procurement
policy can be found at www.rbooks.co.uk/environment

Typeset by SX Composing DTP, Rayleigh, Essex
Printed in the UK by CPI Bookmarque, Croydon, CR0 4TD

Contents

For
Jessie and Corinna
with a father's love

List of Illustrations

1. Monkey statue, Mons
2. Main Gate, Lille citadel
3. Valmy windmill
4. Von Bredow's Charge, Rezonville
5. Smashed cupola, Fort Loncin
6. Fort de Leveau, Maubeuge
7. Command post, Bois des caures, Verdun
8. Somme skyline
9. First World War wire pickets: Beaumont Hamel
10. Rusty rifle and shrapnel shells, the Somme, 2007
11. Lochnager crater
12. Trenches: Newfoundland Park, Beaumont Hamel
13. 38th (Welsh) Division Memorial, Mametz Wood
14. Indian Memorial, Neuve Chapelle
15. Fortress ditch and Resistance memorial: Arras
16. Flying Services Memorial, Arras

All photographs copyright © Michael St Maur Sheil

Maps

1. France (page 2)
2. The cockpit of Europe (pages 14–15)
3. Flanders, Artois and Picardy (page 103)
4. Ypres (page 154)
5. The Somme (page 183)
6. France 1940 (page 261)

The fortification diagram (page 369) is after Christopher Duffy, *The Fortress in the Age of Vauban and Frederick the Great* (Routledge and Kegan Paul 1985)

Just as a portrait suggests the sitter's destiny, so the map of France tells our fortune. The main body of our country has at its centre a citadel, a rugged mass of ancient mountains, flanked by the plateaux of Languedoc, the Limousin and Burgundy; all around stretches a broad glacis, for the most part difficult of access for any invader, protected by the trenches of the Saône, the Rhône and the Garonne, barred by the walls of the Jura, the Alps and the Pyrenees, and plunging, in the distance, down to the Channel, the Atlantic and the Mediterranean. But in the north-east there is a terrible breach that links German territory to the crucial basins of the Seine and the Loire. The Rhine was given by nature to the Gauls for boundary and protection, but scarcely has it touched France than it swings away, leaving her exposed.

It is true that the Vosges throw up a wide rampart, but this can be outflanked through the Belfort gap or the Salt Marshes. It is true that the Moselle and the Meuse heights, resting on the Lorraine plateau at one end and the Ardennes at the other, form appreciable obstacles, but they are shallow and vulnerable to loss through blunder, surprise or neglect, and are exposed by the first withdrawal in Hainault or Flanders. In these low-lying plains there is no wall or ditch to which the defence might cling; no lines of dominant heights, no rivers running parallel to the front. Even worse, geography conspires with the invader by offering multiple penetrating routes, the valleys of the Meuse, the Sambre, the Scheldt, the Scarpe and the Lys, along which rivers, roads and rails guide the enemy.

The north-east frontier looks disturbing in relief, and is no less so in outline. The adversary who attacks simultaneously in Flanders, the Ardennes, Lorraine, Alsace and Burgundy strikes concentric blows. Victor at any one point, he crumbles the entire French defence system. The first steps forward lead him to the Seine, the Aube, the Marne, the Aise or the Oise, where he has only to follow the easiest path to stab at their confluence – Paris, the heart of France.

This breach in its ramparts is France's age-old weakness. Through it Roman Gaul saw the barbarians rush in on its riches. It was there that the monarchy struggled with difficulty against the power of the Holy Roman Empire. There Louis XIV defended his power against the European coalition. The Revolution almost perished there. Napoleon succumbed there. In 1870 disaster and disgrace took no other road. In this fatal avenue we have just buried one-third of our youth.

Charles de Gaulle
Vers l'armée de métier, 1934

Introduction

The Fatal Avenue

Charles de Gaulle called it a 'fatal avenue'. And well he might, for his own military career was rooted in that bloody slab of territory lying north and east of Paris. He was born in the fortress-city of Lille and educated at St-Cyr before going off as a subaltern to Colonel Pétain's 33rd of the Line at St-Omer. Wounded and captured as a captain at Verdun in 1916, he next saw action as a colonel commanding the part-formed 4th Armoured Division near Laon in the desperate days of May 1940. He landed in Normandy in July 1944 and made his triumphant entry into Paris in August the same year.

These events took place in the area bounded to the west by the Channel coast and the east by the valley of the Moselle. To north and east lie the classic routes used by invaders across the centuries, and the names of the battlefields upon them read like a dictionary of military history: Agincourt and Arras, Béthune and Bapaume, Calais and Crécy . . . Fortresses speckle the landscape. Two thousand years of military architecture rear earth, stone and concrete above field and street. The towns of the area have fortifications circling them like the rings of an onion, from medieval walls and Vauban citadel in the centre, past the ravelins and lunettes of the seventeenth to early nineteenth centuries, to the concrete and cupolas of the twentieth. Isolated forts stand guard over river crossing or railway line, while defensive belts, from Vauban to Maginot, use artifice to strengthen or supplant nature.

Evidence of a martial past is never far away, but much of it is invisible to the unprompted eye. The traveller who sees northern France only as the glacis of the south, to be crossed as quickly as the *gendarmerie* will allow, hurtles along the A26 autoroute north of Arras, watching for the junction that will swing him southwards. Just short of the intersection he passes between Notre-Dame-de-Lorette on his right, with its lighthouse tower and whaleback chapel, and the gentle reverse slope of Vimy Ridge on his left, with the Canadian memorial, visible through rides cut through the pines, crowning its crest. He may be forgiven for missing either feature, for by now the white-on-brown signboards at the roadside have lost their novelty value, and in any case the bald statements *Notre-Dame-de-Lorette* and *Monument Canadien* say little to the uninitiated. The military significance of the Lorette spur and Vimy Ridge, which between them dominate the Douai plain and the northern approaches to the ancient route centre of Arras, will probably escape him, for what is high ground in this part of France might pass without comment elsewhere. He may easily fail to notice the many Commonwealth War Graves Commission cemeteries, for the combination of roadside embankment and growing crops means that often only the cross of sacrifice, Sir Reginald Blomfield's elegantly proportioned white stone monument which stands in all the larger cemeteries, can be glimpsed above maize or turnips.

Few British travellers stray to the French National Memorial and Cemetery of Notre-Dame-de-Lorette, with its 20,000 individual graves and its ossuary containing the remains of at least another 20,000. More swing north from the autoroute to visit the Canadian Memorial, commemorating the 66,000 Canadians who died in France and Belgium during the First World War, and to pick their way amongst preserved trenches, their orderly concrete-filled sandbags and prim armoured shields saying little about the grimy terror of trench warfare. Even if the traveller does pause long enough to visit Vimy Ridge, and perhaps one of the cluttered little museums nearby, he is still unlikely to

feel the full pulse of history that beats through this
downland, a history which indissolubly links the English-
speaking world to France.

The future King James II served here as a lieutenant-
general in the French service in 1654, when Marshal
Turenne forced the Spaniards to abandon the siege of Arras.
In 1711 the Duke of Marlborough faced Marshal Villars in
the lines of *Non Plus Ultra*, which ran from the River
Canche on the Channel coast to Namur on the Sambre,
incorporating the Rivers Gy, Scarpe and Sensée in the Arras
sector. On the night of 4–5 August Marlborough
manoeuvred around Vimy Ridge – part of his route
remarkably similar to that of the A26 – to pierce Villars's
lines near Arleux, itself a dozen kilometres from the First
World War battlefield of Bourlon Wood. The area saw fierce
fighting throughout the First World War, with the Canadian
capture of Vimy Ridge in April 1917 as a remarkable feat of
arms. In the Second, it witnessed the Arras counter-attack
by the improvised Frankforce, which hooked down from
Vimy Ridge on 21 May to cause the redoubtable Rommel
serious concern. It is typical of the ironies of history that the
First World War memorial in the forecourt of Arras station
is splashed by splinters of the German bombs that destroyed
the station in 1940, and that the commanding officer of one
of the Durham Light Infantry battalions taking part in the
counter-attack had fought as a subaltern in the Durhams
over the same ground in 1917.

South of Arras lie the First World War battlefields of the
Somme. In 1923 Charles Carrington, author of the evocative
memoirs *A Subaltern's War* and *Soldier from the Wars
Returning*, revisited part of the battlefield near Péronne. 'I
found a trench still full of the flotsam and jetsam of war,' he
wrote. 'I dug an old gun out of the mud and found to my
surprise that it was not a modern rifle but a Brown Bess
musket, dropped there by some British soldier during
Wellington's last action against a French rearguard in 1815.'
Péronne itself was devastated when the Germans withdrew
to the Hindenburg Line in 1917 – 'Don't be angry, only
wonder' announced a German placard amongst the ruins –

but the town's main gates survive, covered by a hornwork taken by the Guards Light Companies on 26 June 1815.

Two other examples, one from each flank of the fatal avenue, show that the repetitive pattern of war is not confined to Arras and the Somme, but marks the entire area. In Normandy, the memorial opposite the Bayeux War Cemetery bears a Latin inscription on its frieze, which translates as: 'We, once conquered by William, have now set free the Conqueror's native land.' William's castle in Caen, begun by the Conqueror in 1060, was badly damaged by Allied bombing in 1944. The battlefield of Brémule, where Henry I of England made good his claim to Normandy in 1119, saw American armour break out from Normandy in 1944.

Lorraine, on the eastern flank of the fatal avenue, is every bit as bloody. The German assault on the fortress of Verdun in 1916 was only the most recent example of an attack on this important crossing point over the Meuse. The battlefield of Gravelotte–St-Privat, just west of Metz, was the scene of an expensive German victory on 18 August 1870. In September 1944 infantry and armour of Patton's 3rd Army found themselves attacking over exactly the same ground but with less happy results. Two of the farms used as strongpoints on this battlefield of two wars – held by the French in 1870 and the Germans in 1944 – had been owned by veterans of Napoleon's army. They took their names from Napoleonic battles – Moscou and Leipzig.

Time and Place

There are undoubtedly more attractive parts of France than the downlands of Artois and the post-industrial debris around Lens and Lille to their north. But for me the chilly north always exercised a stronger pull than the warmer south. In part this reflects my own upbringing, for my father grew up under the dark shadow of the Great War, which had turned jovial uncles to stark names in newsprint, and began to visit its battlefields in the 1920s. Amongst his photographs are Tyne Cot cemetery, wooden crosses rising

to a bleak skyline, and the Ploegsteert Memorial, with
H. Charlton Bradshaw's circular colonnade still in the
course of construction.

Somehow the dark attraction of the north went beyond
the simple desire to follow in my father's footsteps, for I was
not drawn exclusively to the British battlefields of two
world wars. Perhaps it was a natural affinity with lost causes
that made the Franco-Prussian War appeal to me. My
abiding memory of my first visit to Paris, over thirty years
ago, is of neither the Eiffel Tower nor Notre-Dame, but
dusty cuirassiers in the *Salle Detaille* of the then
unmodernised *Musée de l'Armée*; huge figures on big horses,
romantic and anachronistic in a way so often epitomised by
French military uniform. So when I embarked upon research
for my doctorate I chose, not a safely British topic that might
have immured me in the Public Record Office at Kew, but a
decidedly unsafe French one: the army of the Second Empire
on the eve of the Franco-Prussian War. Instead of Kew, it
was the *Service historique de l'Armée* in the Château de
Vincennes – where Henry V of England, the Duc d'Enghien
and Mata Hari had all died.

The stylised copperplate of official military handwriting
presented a barrier that was soon overcome. But it took me
some time to penetrate the bureaucratic armour, which
insisted that only five boxes of documents could be taken
out in any one day, a hard law where cataloguing was often
sketchy. Many a scarce request was wasted when I forgot to
add that extra stroke, which turned an English seven into its
Continental cousin: how often carton MR 1891 appeared
instead of MR 7897. It took me even longer to overcome the
only just unspoken suspicion that French history was
somehow only a proper topic for French historians.
Although, with my dark hair, moustache and *méridional*
features, I could look the part, there was no danger (then or
now) of my ever sounding it.

But eventually I crossed some invisible divide. My sevens
were no longer misunderstood and rules about numbers of
boxes became only a basis for negotiation. I discovered the
considerable delights of the French railway system and

escaped from the city at weekends to spend my time scrambling amongst the hollows of Cazal, where General Margueritte's *Chasseurs d'Afrique* thundered to destruction at Sedan in 1870, or picking my way amongst the German regimental memorials on the field of Rezonville. If my accent got no better, my vocabulary improved, and eventually I reached the summit of happiness – and was mistaken for a Spaniard.

All this had its limits. I cannot share Richard Cobb's claim to 'a second identity', and I still find myself back in the infant class when it comes to explaining to flinty-faced *adjutants* of *gendarmerie* why I was speeding on the Vire-Falaise road. But it has helped give me a sense of place as profound as that which I feel about anywhere in England, even my own comfortable corner of Hampshire. And perhaps the undertow runs deeper still, for I am Huguenot on my mother's side. No distant noble ancestors here, no Gascon *hobereaux* or chevaliers from the Beauce, only the stolid Jacqueses and Boulangers, the Jameses and Bakers of French bourgeois life. Yet France, for all its irritations and inconsistencies, pomposities and perversities, is in my blood.

This is neither a guidebook in the normal sense of the word, nor a personal account of an expedition, still less a formal historical essay. It is a military history of northern France and southern Belgium – de Gaulle's fatal avenue at its most liberal interpretation – which attempts to convey something of the spirit of the place and to suggest how readers might most profitably savour it for themselves. It starts with the most general overview of the area and the events that have taken place upon it, putting wars into their political context and relating the campaigns which unrolled across the fatal avenue to the weapons and tactics in use at the time. With my scaffolding solidly bolted together, I then examine individual sections of terrain, considering the events on them across the centuries, rather than pursuing a detailed chronology. Thus the reader can either work his way through the book as a whole, or use a specific chapter as a guide to, say, Normandy or Champagne.

The first edition of this book appeared in 1992, and when offered the opportunity to amend it for a fresh edition I was uncomfortably aware that the usual Baldric-like cunning plan adopted by authors in such circumstances, simply writing a new foreword, would not work. In the first place, the intervening fifteen years have revealed errors of fact and interpretation, which cannot be allowed to stand uncorrected. Moreover, there are things that seemed important to me in the late 1980s and seem less so now, and vice versa. In the second, a good deal has happened to the landscape I describe. New roads have appeared and old ones have been modernised: new museums have opened and old ones have closed. This edition embodies, therefore, a significant reworking of the earlier version, the new information gleaned partly from my own visits to the Fatal Avenue. For example, I rode my trusty grey horse Thatch from Mons to Fontainebleau not long after this book first appeared: I visit the Somme about twice a year, and Normandy perhaps as often. Part of the fresh information comes from two reconnaissances, made in early 2007, by Dr Jim Storr, a former doctoral pupil and infantry officer who brings a little of both his personae to the field of battle, and who checked my original words against the modern landscape and road system, and found me wanting more often than I might have wished.

None of the maps and photographs in the original edition have survived into this version. The new maps are the work of my old friend and fellow military historian Hugh Bicheno, and the new photographs were taken by Mike Sheil, a professional photographer who came, late in his career, to the interpretation of battlefields, and with whom I continue to enjoy the most fruitful relationship.

The maps in this book are intended simply to enable the reader to pilot his way through the text. For visits to the battlefields themselves good maps are indispensable. The yellow-jacketed Michelin 1/150,000 *local* series is adequate for most navigation, but the *Institute Géographique National* 1:100,000 M 663 series adds contours and other detail not available on the *local* series. For really detailed work it is impossible to beat the admirable *IGN Série Bleue* in 1:25,000, but the British sector of the 1916 Somme battlefield alone requires four of these sheets, so the series is perhaps best confined to the specialist rather than the more casual visitor. For those with a specific interest in the First World War, there are now at least two commercial software programmes that superimpose trench maps on to a modern GPS.

This is not a book for the reader interested in one specific war. I am not in competition with the late Rose Coombs's magisterial *Before Endeavours Fade*, her guide to the battlefields of the First World War, and the reader interested only in Second World War Normandy would do better to read Tonie and Valmai Holts's *Visitor's Guide to Normandy Landing Beaches* than this. Indeed, since the first edition of this book appeared there has been a proliferation of guidebooks to battlefields, with the Pen and Sword *Battleground Europe* series offering a very wide selection, the best of them very good indeed. Martin and Mary Middlebrook's magisterial *The Somme Battlefields* must be required reading for those with a serious interest in that uplifting but haunted landscape.

Professional battlefield guides have grown in both number and professional expertise since the first edition of this book appeared, and it is always a pleasure to meet my colleagues of the Guild of Battlefield Guides plying their trade in rain and shine, on stubble and on turf. Most professional guides plan their tours on the basis of specific 'stands', places on a battlefield where they will discuss specific episodes of the action, either to develop its history or perhaps to illustrate specific themes. Individual stands require working up, ideally, in the case of the British army in the world wars, from official histories, unit war diaries in the National Archives at Kew, or, for other conflicts, from the best sources available. They must be planned after careful consultation of the relevant maps, with due consideration given to parking (increasingly tricky now that thieves target cars with British number plates parked in isolated spots) and access. The facts that a turnip field was fought over, a farmhouse served as a battalion headquarters, or an especially attractive bunker lurks in an orchard do not necessarily mean that their owners welcome the sudden and unexpected ingress of enthusiastic visitors. Do not walk through growing crops or over sown fields, ask if in any doubt and remember that a soft answer, even in franglais, often turneth away wrath.

Picking up objects on battlefields can be plain dangerous (I lost a good friend to the nosecap of a Second World War shell) or simply insensitive. The French police increasingly suspect unauthorised folk who visit battlefields with shovels and metal detectors of seeking at best to despoil historical sites or at worst to rob war graves: caution and common sense will avoid embarrassing interviews. There are many places that sell safe and legally acquired memorabilia: I think, for example, of *The Shell Hole* at Ypres and the incomparable Avril Williams at Auchonvillers on the Somme.

Some battlefields now groan beneath the weight of the memorials and cemeteries built upon them, and it is perfectly possible to drift from statue to obelisk and from memorial panel to headstone, forgetting, in the process, that objects like this post-date the battle they commemorate and would make little sense to the men who fought there. I would not expect a visitor not to pause in silence beneath the soaring arch at Thiepval, which bears the names of more than 74,000 officers and men, 'the Missing of the Somme', but it would be a pity to do so without, say, walking along the track past Flatiron Copse Cemetery in the shadow of Mametz Wood, which shows too clearly why this evil spot was so important in July 1916.

I make no attempt to advise the reader where to stay or what to eat: there are other capable authors in that particular business. In any event I have mixed feelings about inciting invasion of my favourite haunts, though that prince of *routier* stops, *Le Poppy*, on the 1 July 1916 front line where the Albert–Bapaume road dips in front of La Boisselle, will be able to cope, in its brisk and benevolent way, with more visitors than this book will inspire. If you are in luck there will be skate wings with capers on the menu.

Remember that in France and Belgium quick lunches often extend to two lost (if delightful) hours: the presence of snowy napery and elegant cutlery tends to indicate an establishment where a swift *steak frites* might be hard to come by. Similarly, I do not indicate museum opening times, although it is worth noting that Continental museums, open over the weekend, often close, by way of compensation, on Mondays. The internet makes checking such things so much easier than was the case fifteen years ago. When I wanted to find out about the new Malplaquet museum in Bavay, a little light Googling produced details of the excellent *Musée du 11 Septembre 1709*, which really is worth a visit.

The internet is invaluable for checking details of British and dominion servicemen killed in the two world wars. The Commonwealth War Graves Commission's website has a search facility that enables one either to check an individual name or details of a cemetery. Where I refer to a burial in the pages that follow I use the Commission's standard form of reference, with the plot number in Roman numerals, followed by the letter designating the row and the Arabic numeral for the plot itself. A bronze-doored cupboard near the entrance to each cemetery contains, amongst other things, a plan of the cemetery's layout, which makes finding a specific plot comparatively easy.

There will be times when my narrative will do violence to the pedantry of lines on maps. My definition of Champagne is rather wider than a French civil servant might think wise, and the chapter dealing with the area around the confluence of Sambre and Meuse flits repeatedly across what is now the Franco-Belgian border. In short, I have chosen blocs of territory that make geographical and historical sense, at the risk of offending administrative sensibilities. I cannot promise that my spelling will please both Walloons and Flemings, but it would be bewildering to call Ypres by its widely used modern name of Ieper, and I have used Marlborough's Oudenarde rather than the current Oudenaarde or Audenarde.

So much for place. As for time, of the many possible start

dates I have chosen 1337, the beginning of the Hundred Years War. Although going back earlier has superficial attractions, it is difficult enough to relate even battles of the Hundred Years War to the ground and extraordinarily hard – save perhaps in the case of sieges – to do so with much certainty for earlier conflicts. Including the Hundred Years War appreciably widens the book's focus, but it would be sad to tramp through France without discussing Agincourt as well as Arras, Crécy as well as Cambrai. It seemed a broad canvas when I set about making my first hesitant daubs all those years ago and it has certainly become no smaller since. Let us set about repainting it.

I

The Cockpit of Europe

E VERY OTHER CHAPTER of this book is concerned with military operations in a given area: the ground remains the constant factor and what changes are the events that unrolled across it. But these make little sense out of context. Why was Verdun important long before the first German shell smashed into the bishop's palace in February 1916 and why did the Somme meander through British folk memory long before the assaulting waves plodded to catastrophe on 1 July that year? This chapter is, therefore, a history of northern France from the beginning of the Hundred Years War to the present day and it provides a framework for what follows. Its emphasis is not exclusively military, for wars do not take place in a political vacuum, and at least part of what happens on the battlefield is comprehensible only in broader terms. English-speaking readers may find it francocentric, and so it is – quite properly, for the majority of the events described took place on French soil and in most cases French soldiers participated in them.

The Hundred Years War

In the darkness of an October night in 1330 the eighteen-year-old King Edward III of England entered Nottingham Castle through an underground passage and seized Roger Mortimer, Earl of March, lover of his mother Isabella and in effect ruler of England. Isabella was sent off to dignified captivity, Mortimer suffered a traitor's death and the young Edward declared that he had taken up the reins of government.

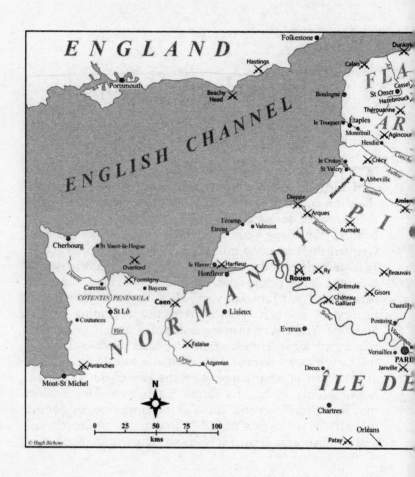

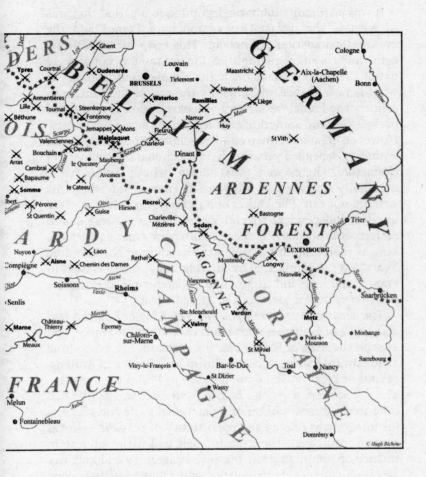

© Hugh Bicheno

It was a suitably dramatic beginning to a period that was to witness the devastation of much of France and the eventual exhaustion of England. This crisp use of national definitions is an over-simplification, for modern concepts of nationalism would have made little sense to inhabitants of what is now France, although by the end of the thirteenth century the French monarchy was based upon consolidated possessions and something approaching a sense of country. However, great noblemen enjoyed immense power and royal authority depended very much on the monarch's strength of character. There were vast areas whose rulers were nominally liegemen of the King of France, but where his writ did not run. The Dukes of Burgundy, for instance, were in effect independent sovereigns, with huge possessions in eastern and northern France, as often in arms against the king as marching in his support. The story of the Hundred Years War is complicated by the ebbs and flows of faction in France and England alike, with noblemen jockeying for position against a backcloth of sporadic war.

The seeds of war were dynastic. The King of England was also Duke of Guienne, with lands in south-west France, and as such the King of France was his suzerain. Indeed, in June 1329 Edward had knelt before Philip VI of France in Amiens, capital of Picardy, and done homage to Philip for Guienne. The relationship soon became strained. Edward had inherited a throne weakened by factional strife and successful foreign war offered the opportunity of reasserting royal control. Moreover, the king himself had little interest in finance or administration, but loved pageantry and knightly accomplishments: both politics and temperament pushed him towards war.

Nor was the status quo more satisfactory for Philip. He was the first Valois king and his predecessors, the Capetians, had increased royal power: it was hard for him to tolerate English authority over a large tract of France. Yet the English could not withdraw willingly, for their trade with Gascony was important and Edward's position at home would have been weakened by concessions. And there were further difficulties. The great towns of Flanders – Lille,

Ypres and Bruges amongst them – were the best customers for English wool. They were in conflict with France and looked towards England for support. Edward tried to find allies in the Rhineland: Philip, for his part, looked to Scotland, England's traditional enemy.

It was on the Scots that Edward first stropped his martial edge. Robert Bruce had died in 1329 when his son David II was only five. Edward supported a group of disinherited Scottish noblemen under Edward Balliol, who duly defeated David and was crowned King of Scots. However, the substantial territorial concessions made by Balliol as the price of English support alienated most Scots and it was only continuing English military assistance that kept Balliol on the throne. Although the Scottish cause seemed hopeless, the disunity of Balliol's party and the sheer fervour of Scottish resistance prolonged the war despite English victories at Halidon Hill (1333) and Neville's Cross (1346). King David was captured at Neville's Cross and kept a prisoner in England for eleven years. After his release David agreed that Edward or one of his sons should succeed him, but found himself shackled by a baronial opposition. When he died in 1371 England was too deeply embroiled in France to prevent the accession of Robert Stewart. The monarchy was to remain at the mercy of over-mighty subjects and the Borders were perennially lawless, but Scotland had emerged from the maelstrom with her territory intact and her national identity vigorous.

By that time the Scottish war had lit two long-burning fuses. The first was political. Edward visited France in 1331, ostensibly to seek agreement with Philip: there was talk of a marriage between the royal houses and Edward spoke of an Anglo-French expedition to the Holy Land. But instead of mounting a crusade, Edward moved north against the Scots, and Philip then threatened to support David and moved his crusading fleet from Marseilles to the Channel ports. This persuaded Edward to withdraw from Scotland in the autumn of 1336. Instead of meekly joining the crusade, he asked Parliament for subsidies for war with France.

The second fuse was military. Under Edward I the

longbow had begun its rise to favour, and at the Battle of
Falkirk (1298) the combination of archers and armoured
cavalry had proved fatal to Scottish spearmen. At Halidon
Hill, in July 1333, the example of Falkirk was repeated. The
Scots lost seventy lords and 500 knights and squires: the
English, one knight, one man-at-arms and twelve archers.

It took some time for Edward to employ his archers on the
Continent. War broke out in 1337, but fighting in Flanders
and Picardy was inconclusive, though in a naval battle off
Sluys in 1340 most of the French fleet was captured or
destroyed. That year Edward laid claim to the throne of
France on the grounds that his mother, Isabella, was a
daughter of the house of Capet. Although Salic Law banned
women from succeeding to the throne, Edward argued that
his own claim descended through her. It was a flimsy
pretext, but gave some comfort to Edward's Flemish allies,
now able to deny that they were in rebellion against their
lawful king.

A truce in 1340 was soon imperilled by a disputed succession to the duchy of Britanny, with England and France
supporting rival claimants. In 1345 Edward denounced the
truce, and Henry Earl of Lancaster raided from Guienne into
French territory. The following year Edward led an expedition of his own. One contemporary estimated his force at
20,000, but it is unlikely that Edward actually had more
than 12,000 men. Theoretically Philip enjoyed a considerable advantage. He was the strongest sovereign in
Christendom: his feudal vassals owed him knight-service,
peasant *communes* could be called out under arms, and
there were mercenaries available for hire. In practice he was
less fortunate. As Sir Charles Oman wrote:

> The strength of the armies of Philip and John of Valois
> was composed of a fiery and undisciplined aristocracy
> which imagined itself to be the most efficient military
> force in the world but was in reality little removed
> from an armed mob. A system which reproduced on
> the battlefield the distinctions of feudal society was
> considered by the French noble to represent the ideal

form of warlike organization ... Pride goes before a fall, and the French noble was now to meet infantry of a quality such as he had never supposed to exist.

Archers were the mainstay of Edward's infantry. Most were volunteers, often with experience of the Scottish wars, and followed a lord, knight or gentleman who had contracted to supply the king with soldiers, though there were still pressed men to be found in the English ranks. In 1346 one Robert White was released from prison after his committal for 'homicides, felonies, robberies, rapes of women and trespasses' to serve at the siege of Calais. Archers were paid at the high rate of sixpence a day and had little in common with the rabble of footmen that trailed behind the French knights. Many were recruited from Lancashire, Cheshire and Wales itself. Good though Edward's archers were, they were an adjunct to, not a substitute for, armoured knights and men-at-arms: and neither Edward nor his son the Black Prince deliberately sought battle against a numerically superior French army. They preferred to mount destructive raids into enemy territory, (*chevauchées* in the terminology of the day) and usually fought pitched battles only when cornered.

Edward landed on 12 July at St-Vaast-la-Hougue on the Cotentin, set off for Rouen with the intention of joining his Flemish allies and took Caen on the 25th. Philip already lay between him and the Flemings, but the English managed to ford the Somme at Blanchetaque, between Abbeville and St-Valéry, on 24 August, narrowly escaping being caught with the river at their backs. On Friday 25 August Edward offered battle near the village of Crécy, fifteen kilometres north-east of the ford. The French began their attack late in the afternoon and until midnight wave after wave of knights struggled into the arrow storm. Their army began to disintegrate shortly afterwards, having lost, by one estimate, 1,500 knights and 10,000 common soldiers. There were only forty English dead. Edward marched to Calais to re-embark his army, but found it resolutely defended, and not until July 1347 did the town surrender. It was to be held for another two centuries and became the last English possession in France.

Defeat at Crécy and Calais tarnished Philip's prestige and, to make matters worse, his Scottish ally David had been defeated and captured at Neville's Cross in October 1346. But French and English alike faced an enemy more deadly than arrow or broadsword. The virulent plague known as the Black Death appeared in southern France in 1347 and the following year it spread like wildfire through the land, reaching England, probably in a ship that had taken supplies to the siege of Calais. The immense social and economic damage done by the Black Death made the continuation of military operations almost impossible, and there were successive truces. In 1354 Philip's successor, John II, agreed to a costly treaty, giving up western France in return for Edward's renunciation of his claim to the French throne. In the event, the Treaty of Guines was never ratified, and John withdrew from the bargain and prepared for war.

In 1355 Edward unleashed his son, Edward the Black Prince, in a destructive raid from Bordeaux into Armagnac and on into Languedoc, devastating town and country alike. Edward himself had landed at Calais, but soon withdrew to face the resurgent Scots, while his cousin Henry, Duke of Lancaster, raiding in Normandy, prudently retired when John approached with a large army. For the 1356 campaign Edward reinforced his son with more archers and in early August the Prince set off with an Anglo-Gascon force of some 7,000 men. On 20 September he was brought to battle by John with a much larger army just south of Poitiers. Once again the French were utterly defeated, and King John and his younger son Philip were both captured.

When John agreed to a lavish ransom and extensive territorial concessions his eldest son, de facto ruler of France, repudiated the settlement, but another English raid, this time across the north, compelled him to agree to the Peace of Bretigny, signed in May 1340. Edward made huge gains, including the duchy of Aquitaine, the counties of Poitou and the Limousin, and the northern district of Ponthieu: in addition, John was to pay a large ransom. Only a third of this had been raised when John died in London, leaving his weak and divided kingdom to his sickly son Charles V.

Before the war was resumed the Black Prince, now Duke of Aquitaine, with his capital at Bordeaux, had marched to the assistance of King Pedro the Cruel, ousted from the throne of Castile by his half-brother Henry of Trastamare. The Black Prince duly defeated Henry at Najera, but Pedro failed to pay the war expenses and was soon murdered by Henry. Aquitaine was heavily taxed to pay for the war and the Black Prince, who had caught fever in Spain, fell prey to declining health, which worsened his already ugly temper. The taxes imposed on Aquitaine to pay for the Spanish campaign encouraged two of Prince Edward's vassals to appeal to Charles of France. Although, strictly speaking, he lacked jurisdiction over Aquitaine – for the Peace of Bretigny had granted it to King Edward without fealty to the French king – the protest gave Charles the pretext he needed to declare Aquitaine confiscated. From 1369 to 1396 a sporadic conflict – Alfred Burne's term a 'sub-war' is a fair description – bubbled around north and west France. It was marked by deliberate French refusal to offer battle to English *chevauchées*, which scorched their way across the country-side. At the same time the French made steady progress, first in improving the royal army, then in using it to grab a town here and a castle there, steadily eroding English dominions in France. This new policy was largely the work of Bertrand Duguesclin, an ugly Breton professional soldier who was probably the first commoner to be appointed Constable of France. The conflict was ended in 1396 by the Peace of Paris, a truce which recognised the status quo but left the question of Aquitaine – English possession as of right, or French territory held only by English homage – unresolved.

It was not until 1415 that war flared up once more. By this time there had been dynastic disputes in France and England alike. The Black Prince's son, Richard II, had behaved moderately until 1396 when he married Isabella, daughter of Charles VI of France, and revealed a taste for French manners and absolutism. He alienated many of his powerful

subjects, and was imprisoned and murdered: Henry
Bolingbroke, grandson of Edward III, claimed the throne on
the grounds of descent, conquest and the need for better
rule. Henry IV, first of the Lancastrian kings, ruled till 1413,
when he was succeeded by his son Henry V, a self-confident
young man of twenty-five. In France there was full-scale
civil war between the Orleanists, supporters of Louis, Duke
of Orléans, the brother of Charles VI, and adherents of John,
Duke of Burgundy. The unfortunate Charles teetered on the
edge of insanity, his divided kingdom ever more vulnerable
to external pressures.

Henry saw France as a suitable sphere for his nobility to
exercise their martial talents and, as a devout man, believed
in the divine justice of his claim to its throne. He spun out
negotiations until he had persuaded Parliament to grant
subsidies for war and the Burgundians to agree to his descent
upon Normandy. The Exchequer Rolls show how Henry
raised his army, laying down pay scales and drawing up
indentures with the nobles and gentlemen who were to
furnish contingents, usually at the rate of one man-at-arms
for three archers. At one extreme the Duke of Clarence,
Henry's brother, provided the largest contingent of 240 men-
at-arms and 720 archers, and at the other at least 122 men
agreed to provide less than ten men apiece. The muster rolls
give us the names of over 5,000 of the archers, with
Welshmen like Mered ap Jenan and Gwillym ap Griffuth
serving alongside Thomas Armygrove and John of Kent.

Henry assembled an expeditionary force of something
around 9,000 men in Southampton and Portsmouth in July
1415, and landed near Harfleur on 14 August. His opponent,
presiding over a realm torn between Orleanists and
Burgundians, appointed his eldest son, the Dauphin Louis,
to command his army, with the assistance of Charles
d'Albret, the Constable.

Harfleur fell on 22 September after a difficult siege. Henry
had originally intended to march on Paris, but eventually
decided on a *chevauchée* across Normandy and up to Calais.
The French held the north bank of the Somme against him
and it was only after hard marching that the English were

able to cross near Péronne on 19 October. Instead of blocking Henry's path immediately the French fell back to Bapaume, allowing him to follow the direct route for Calais. On 24 October, with Henry nearing his objective, scouts reported that the way was blocked by a mighty host. On the next day the armies fought between the villages of Agincourt and Tramecourt in a battle that demonstrated yet again the power of the longbow over the armoured knight. French losses were appalling and Henry marched on to Calais unhindered, paused briefly there and returned to England. He then busied himself concocting an anti-French coalition with Sigismund, the Holy Roman Emperor, and John the Fearless, Duke of Burgundy. In the meantime there was further fighting in Normandy: in March 1416 the Earl of Dorset, raiding from his garrison of Harfleur, fought a running battle with the Count of Armagnac along the coast between Valmont and Harfleur, emerging victorious against all the odds. However, it proved far less easy for the English to sustain Harfleur than Calais and in August the Duke of Bedford led a fleet to Harfleur's relief. The French met him in the Seine estuary and on 15 August Bedford won a costly victory.

Henry decided to increase his pressure on the French in an effort to secure the duchies of Aquitaine and Normandy, and in 1417 he led an expedition to Normandy, storming Caen and going on to hammer Falaise into submission. In the spring of 1418 the Duke of Gloucester subdued the Cotentin, the Earl of Huntingdon took Coutances and Avranches, and in 1419 the Duke of Clarence cleared the Seine as far as Rouen. Despite a vigorous defence – and a political volte-face which brought the Duke of Burgundy into the war against the English – Rouen fell on 20 July 1418. By the end of February 1419 most of Normandy was in English hands and of the surviving castles all save Mont St-Michel were eventually captured. After tortuous negotiations the Treaty of Troyes was signed in the spring of 1420: after the death of the insane Charles VI the kingdoms of England and France would be united in the person of Henry V or his successor, although France would retain its own language and customs. Henry married the French Princess Catherine shortly afterwards and expressed the hope that 'perpetual peace' was assured.

His confidence was misplaced, for the dispossessed Dauphin, holding court at Bourges, was not prepared to accept the verdict of Troyes, and over the next months the English, French and Burgundians took Sens, Montereau and Melun from him. Henry entered Paris in triumph with Charles VI at his side and returned to England early in 1421. But the war went on. The Dauphinists not only held much of southern France, including the strategically placed city of Orléans, but occupied Champagne and Picardy, between Burgundy proper and the Burgundian possessions of Artois and Flanders. Thomas, Duke of Clarence, Henry's brother, was defeated and killed by a Franco-Scottish force (the auld alliance was active yet again) at Bauge, between Angers and Tours. Henry mounted his last expedition to France in the summer of 1421, landing at Calais and campaigning on the Loire before marching north once more to besiege the Dauphinist stronghold of Meaux, which fell in May 1422. He was on the march southwards from Paris in July when he fell ill and was taken back to Vincennes to die. He bequeathed the regency to his eldest brother John, Duke of Bedford, and died on 31 August.

For the next few years the war twanged on in the minor key, with raids, skirmishes and sieges. In April 1423 Bedford succeeded in bringing about the Treaty of Amiens, a tripartite agreement between England, Burgundy and Brittany, and in July 1423 an Anglo-Burgundian force routed the Dauphinists at Cravant, on the Yonne south of Auxerre. In August the following year Bedford won the major battle of Verneuil, not without some difficulty, for the French charged while his archers were still hammering in the stakes they used to strengthen their position. Bedford followed this impressive victory by overrunning most of Maine, Anjou and Picardy.

There were already cracks in the alliance between England and Burgundy, and it required all Bedford's skill to patch up a quarrel that arose in 1424 when Humphrey, Duke of Gloucester landed in Flanders in an effort to regain the lands of his wife Jacqueline, Countess of Hainault and Holland in her own right, and former wife of the Duke of Brabant, cousin to the Duke of Burgundy. Although the Burgundians theoretically remained loyal to the Treaty of Amiens, their active collaboration in the war dwindled.

Nor was Gloucester's influence on the home front much more benign, for he was engaged in a long-running quarrel with his uncle, the powerful Cardinal Beaufort, and this forced Bedford to remain in England in 1426. On his return to France he resumed the business of siege warfare, a process enlightened by Lord Talbot's sparkling relief of Le Mans in the spring of 1428.

In the summer of 1428 the arrival in Paris of a detachment of fresh troops under the Earl of Salisbury encouraged the English leaders to advance against Orléans, the Dauphinist stronghold midway between Paris and Bourges, the rival capitals. The siege began well, but Salisbury was mortally wounded by a cannon shot and his successors – a triumvirate of the Earl of Suffolk, Lord Talbot and Lord Scales – settled down for a lengthy stay. The Dauphin sent the Count of Clermont to the city's relief, but he stumbled into the redoubtable Sir John Fastolf, bringing a convoy of herrings and other supplies to the besiegers, near Janville in mid-February. The day began well for Clermont, whose light guns peppered the English from a safe distance, but an impetuous advance by the Scottish contingent gave Fastolf the opportunity to counter-attack and he won a stunning victory over a much-superior force in what became known as 'The Day of the Herrings'.

This unlooked-for English triumph was followed by a defeat no less remarkable. Joan of Arc, a peasant girl from the hamlet of Domrémy, on the Lorraine-Champagne border, had heard the voices of saints urging her to rescue France from English domination. She met the Dauphin at Chinon and persuaded him to send another force to relieve Orléans. She accompanied it clad in armour and her presence gave an unprecedented boost to French morale. The relief force entered Orléans on 3 May and the English drew off in good order on the 8th. Suffolk unwisely dispersed his army, and was himself soon defeated and captured at Jargeau. The French went on to take Meung and Beaugency, and on 18 June they roundly defeated Talbot at Patay, north-east of Orléans. The hesitant Dauphin was at last persuaded to make his way to Rheims, where, on 17 July, he was anointed and crowned as Charles VII.

Coronation did little for Charles's resolve. He declined to face Bedford north of Paris, then mounted a half-hearted attack on the capital in which Joan was wounded by a crossbow bolt. Worse was to come. News that a fresh English army, accompanied by the boy-king Henry VI, was to land at Calais may have persuaded the Duke of Burgundy to resume active operations against Charles. In any event, the Burgundians were back in the war with a vengeance and in May 1430 Joan was captured at Compiègne by a Burgundian force with a strong English element. Sold to the English, she was tried by a French ecclesiastical court for heresy and witchcraft, and burnt at Rouen on 30 May 1431.

English recovery began promisingly, but the defection of the Duke of Burgundy, after failed negotiations between French, English and Burgundian representatives at Arras in the summer of 1435, struck the death blow to English hopes. The collapse of the Burgundian alliance hastened the death of old Bedford, who lies buried in Rouen cathedral. In April 1436 the English evacuated Paris and it was only the superb skill of Lord Talbot, captured at Patay but later ransomed, that enabled the English to hang on. In January 1437 he beat the two capable French commanders La Hire and Poton de Xantrailles at Ry, near Rouen, and went on to take Pontoise by a clever *coup de main*. In the summer he defeated a Burgundian force that was besieging Le Crotoy, north of the Somme estuary, using the ford at Blanchetaque to cross the river.

The inconclusive character of the war encouraged the combatants to negotiate, but discussions, held between Calais and Gravelines in mid-1439, came to nothing. The English were unable to renounce their king's claims to the throne of France and without this no French sovereign could make peace. Talbot routed the Constable Richemont at Avranches in December 1439, but the English failed to capitalise on the factional strife within France, where the Dauphin was at odds with his father. In 1441 the Duke of York and Talbot fought a fast-moving campaign between Seine and Oise, but could not prevent the fall of Pontoise. Two years later the Duke of Somerset, nephew of Cardinal Beaufort, led an aimless expedition around Maine. Further peace negotiations foundered – this time the sticking point was the old question of English homage for Guienne and Normandy – but a two-year truce was concluded and Henry VI married Margaret of Anjou.

The death of Somerset and the retirement of Cardinal Beaufort left Beaufort power in the hands of the Earl of Suffolk, whose mismanagement looms large in what follows. Humphrey Duke of Gloucester was accused of treason and died under suspicious circumstances, and Suffolk ruled the kingdom with the connivance of Queen Margaret. Suffolk had agreed to hand over Maine to France in return for an extension of the truce and when he failed to do so Charles resumed the war. It was a one-sided contest. The French army had been thoroughly reformed, with a substantial professional element replacing the old rabble of gentility, while English forces in France, though numerous, were tied down in garrisons.

English grip over Normandy was loosened in 1449–50. Rouen fell in October 1449, Harfleur the next month and Honfleur in January 1450. Suffolk sent out an army under Sir Thomas Kyriel, who landed at Cherbourg and marched to the relief of Caen, only to be decisively defeated at Formigny on the Bayeux–Carentan road on 15 April. Kyriel was captured, but most of his men perished. With the destruction of the last English field army, the end in Normandy could not be long delayed: Caen fell in June and on 11 August Cherbourg, bravely held by Thomas Gower, at last opened its gates.

With Normandy secured, Charles turned his attention to Gascony and French troops entered Bordeaux in June 1451. They were less than popular, for Guienne had been linked to the English crown for a century, and trade between England and Gascony cemented the relationship. The burgesses of Bordeaux begged Henry VI to support them and in response he sent out old Talbot, now Earl of Shrewsbury. He landed in October 1452: Bordeaux threw out its French garrison and Shrewsbury was soon in possession of the Bordelais. The French counter-attack developed slowly and it was not until the summer of 1453 that Charles sent three armies against the Earl. Shrewsbury planned to remain in Bordeaux until the nearest was within striking distance, but when Charles's centre army besieged the small town of Castillon, Shrewsbury intervened. He moved with his accustomed speed, and early on 17 July cleared the French from an outpost and fell upon their well-fortified camp. The assault was already going badly when a French detachment took the attackers in the flank. Shrewsbury's horse was killed by a cannon ball and while he was pinned beneath it a Frenchman brained him with a battleaxe. Thus ended the English dominion in Gascony and with it the Hundred Years War.

The Wars of Religion

Charles VII died in 1461 and was succeeded by his son Louis XI. During his long reign – which lasted till 1483 – Louis strengthened the royal hand. He withstood the pressure of the feudal leagues and, after a long struggle against Burgundy, brought the duchy to the French crown by a dynastic marriage. Brittany soon followed: in 1491 its young duchess Anne married Louis's son Charles VIII and the duchy was united to the crown when Anne's daughter Claude married the future Francis I. Charles's foreign policy imperilled domestic gains. In 1494 he invaded Italy, beginning a series of wars that brought France into repeated conflict with the Holy Roman Empire. He died without a son in 1499 and his uncle Louis of Orléans succeeded him as Louis XII, bringing his fief of Orléans to the crown. Louis maintained the French presence in Italy, only to see his armies defeated by the Swiss at Novara in 1513. In the same year an English and Imperialist force, besieging Thérouanne in Flanders, drove off a relieving army at such pace that the clash, at Guinegate, became known as the Battle of the Spurs.

The reign of Francis I, Louis's nephew and successor, saw the continuation of the Italian wars. Following his failure to secure election as Holy Roman Emperor, Francis became entangled in a long war with the successful candidate, Charles V. Despite a lavishly orchestrated meeting with Henry VIII of England at the Field of the Cloth of Gold, near Guines on the Channel coast in 1520, he was unable to prevent an English invasion of Picardy. Worse still, in 1525 he was defeated and captured at Pavia in northern Italy, though it was not until 1529 that he renounced his Italian claims. Even then there was no lasting peace, and further fighting on the borders of France resulted in yet another English invasion and the capture of Boulogne, which Henry VIII agreed to restore to France by the Treaty of Ardres in 1546.

Francis died the following year. Despite his frequent military failures, he had held France together and had done much to encourage the spirit of the Renaissance. But his reign also witnessed the first signs of the rift that was to divide France in the second half of the sixteenth century. Protestantism gained ground in the 1520s, despite official disapproval, and the situation worsened under Henry II. Much influenced by the Duke of Guise, he renewed the conflict with the Emperor, occupying the three strategically important Lorraine bishoprics of Metz, Toul and Verdun, and holding Metz against Charles V in the winter of 1552-3. The Imperialists captured Thérouanne and Hesdin before a short-lived peace was concluded. When war was resumed the French found themselves fighting a reconstituted enemy, for Charles had resigned his Spanish throne in favour of his son Philip, and later went on to renounce the imperial title, to which his brother Ferdinand was elected. This dynastic redeployment resulted in the French fighting the Spaniards in Picardy, a less bizarre state of affairs than one might think, for Philip had succeeded to his father's domains in the Low Countries and for the next century the Spanish presence in the north was a very real one. In 1557 the French were soundly defeated by the Spaniards at St-Quentin, although the Duke of Guise went a long way towards restoring French fortunes when he captured Calais from the English early the following year. The war was ended by the Treaty of Cateau-Cambrésis, signed on 3 April 1559, under whose terms France retained Calais and the three bishoprics, but renounced her claims in Italy.

Henry had been encouraged to make peace so that he could turn his undivided attention to the extirpation of Protestantism in France, but he was accidentally killed at a tournament in Paris in July 1559. Historians disagree over the extent to which religion became a convenient cloak for political ambition and it is certainly the case that the rival camps drew adherents whose motives were more political than religious. Some Huguenots were more influenced by the desire to weaken royal authority than by purely religious motives, and many of the lesser nobility who contributed so much to Huguenot strength hoped for court favours and church lands, or saw war as a means of repairing the damage done to their incomes by spiralling inflation. Equally, we can neither disregard the genuine religious enthusiasm that gripped many actors in the drama, nor ignore the powerful bonding effect of persecution on the Huguenots.

After Henry's death the Guise faction tightened its grip, identifying itself with the ultra-Catholic cause. Most of the rival house of Montmorency remained in the Catholic camp, but one of its scions, Gaspard de Coligny – hero of the defence of St-Quentin in 1557 – was to become a notable Huguenot leader. The most implacable opponent of the Guises was Louis de Bourbon, Prince of Condé, veteran of the siege of Metz and the capture of Calais. Catherine de' Medici, mother of the young King Francis II, encouraged the Huguenots as a counterweight to the Guises, but massacres of Huguenots at Vassy and Sens in the spring of 1562 led to the outbreak of civil war. The Catholics captured Rouen in October 1562, and Condé's elder brother, Anthony of Navarre, died of wounds received in the siege. In a confused battle at Dreux in December Condé was defeated and captured, though not before his men had taken the Constable Montmorency, the Catholic commander. Coligny, meanwhile, enjoyed considerable success in seizing Catholic-held towns in Normandy, while the Duke of Guise besieged Orléans. The assassination of Guise in February 1563 made possible the Peace of Amboise, which gave the Huguenots a measure of religious toleration. Both sides immediately joined to eject the English from Le Havre (the new port of Havre de Grâce, a few kilometres from Harfleur), to which they had been admitted as price for their support for the Huguenots during the war.

A meeting between Francis II, the Queen Mother, the Queen of Spain and the Duke of Alva at Bayonne in 1565 fuelled Huguenot fears, and their attempt to seize the King and remove him from Catholic influence failed. The second War of Religion, which followed, was brief and there was a bitter but indecisive battle at St-Denis, where Montmorency was killed. The Edict of Longjumeau (March 1568) confirmed the terms of the Treaty of Amboise, but brought peace for only a few months. Guise's brother Charles, Cardinal of Lorraine, planned to seize the Huguenot leaders, but the scheme misfired and the third war broke out. Catherine now sided firmly with the Catholics. The edicts of toleration were revoked and when Condé was captured in the scrambling little Battle of Jarnac (13 March 1569) he was shot shortly afterwards. Coligny, his natural successor, was soundly defeated at Moncontour, near Poitiers, but the Huguenots continued to hold several major cities, including their great stronghold of La Rochelle, and Catherine was persuaded to make peace: the Treaty of St-Germain, concluded in August 1570, not only granted the Huguenots religious toleration, but permitted them to hold church services in two towns in each province.

Coligny won the confidence of Charles IX, brother and successor of Francis II, and encouraged him to support the Netherlands Protestants in their struggle against Spain. In April 1572 an Anglo-French defensive alliance was concluded: English volunteers landed in the Low Countries and

Louis of Nassau, with a Huguenot force, captured Mons. These successes proved temporary and in the wake of Spanish military victories the Queen Mother's attempt to remove Coligny precipitated the Massacre of Saint Bartholomew on 22 August 1572, when Coligny and many of his kinsmen and confederates were butchered in Paris.

The massacre was followed by a fourth War of Religion, ended by the Treaty of La Rochelle on terms similar to those of previous peaces, and a fifth war, concluded in May 1576 by the Peace of Monsieur. This time the terms reflected a strengthening of the Huguenots' position through their alliance with the *Politiques*, the moderate Catholic party. The Huguenot leader Henry of Navarre, son of Anthony, who had fallen at Rouen in 1562, had married Margaret of Valois, sister of Charles IX, just before the massacre and had escaped with his life on condition of accepting Catholicism. He resumed both his old religion and command of the Huguenot army as soon as he gained his freedom, and obtained the governorship of Guienne under the terms of the Peace of Monsieur. His co-religionist Condé, son of the leader shot after Jarnac, became governor of Picardy.

Catholic resentment at Huguenot success led to the formation of leagues, first local and then national, with the establishment of the Catholic League under Henry of Guise. The unsteady Henry III had succeeded to the throne in 1574. He had no heir and the death of his only surviving brother Anjou in 1584 left Henry of Navarre heir presumptive to the throne. This strengthened ties between the Huguenots and the *Politiques* in defence of the hereditary monarchy, and at the same time drove the League to oppose the crown. Henry III failed in his attempt to create a third force around his person, turned briefly to the League, but then had Guise assassinated. This provoked the League into an outright break with the crown: it declared that Guise's brother, the Duke of Mayenne, was custodian of royal authority, driving Henry into alliance with Henry of Navarre. Their joint armies were approaching Paris in July 1589 when the king was murdered and Henry of Navarre succeeded to the throne.

For the next nine years war rippled across France as Henry IV fought the League and its Spanish allies. On 21 September 1589 he beat the much-superior Catholic army under Mayenne at Arques, near Dieppe. The following year Henry met Mayenne again, this time at Ivry, on the Eure near Dreux, and combined the fire of his arquebusiers and the shock of his cavalry to win against the odds. He marched on to besiege Paris, but it held fast against him and the Duke of Parma, Spanish commander in the Low Countries, moved south to raise the siege. In 1591 Parma relieved Rouen and beat Henry at Aumale, but the Duke died the following year and Henry was not thereafter faced by a commander of real quality. In 1593 Henry announced that he was prepared to accept Catholicism and the alliance against him crumbled rapidly: he entered Paris in March 1594 to scenes of wild enthusiasm. He continued the struggle against Spain, winning Fontaine-Française (1595) and going on to take Amiens, principal Spanish garrison in northern France. In 1598 he made peace with Spain and published the terms of the Edict of Nantes, which settled the religious question in France, at least for the time being. The Huguenots were allowed liberty of worship and of conscience, and kept eight towns and their own armed force.

Henry embarked upon a policy of national reconstruction, aided by his old comrade-in-arms, the Duke of Sully, who recast the administration and put fleet and arsenals into good order. By the time he was murdered by a Catholic fanatic, François Ravaillac, in 1610, Henry had secured the Bourbon dynasty. Yet the crown's authority was still circumscribed by that of the great nobles; the power of the Huguenots, especially in the south-west; and the activities of the Parisian and provincial *parlements*, law courts with wide police powers and the duty of registering royal decrees in their areas. During the first years of the reign of Louis XIII, Henry's son, power lay in the hands of the Queen Mother, Marie de' Medici. Armed conflict between the Guise and Condé factions was averted only with difficulty and in 1614 the States-General – an unwieldy national assembly of nobles, clergy and 'Third Estate' – met without useful result

and for the last time till 1789. More political squabbles led,
in 1622, to a brief royal campaign against the Huguenots,
ended by the Treaty of Montpellier, which confirmed the
Edict of Nantes and left the Huguenots with two *places de
sûreté*, Montauban and La Rochelle.

The Thirty Years War

The next two decades were dominated by Armand Jean
Duplessis, Duke of Richelieu, who had abandoned the
profession of arms to accept the family bishopric of Luçon.
Appointed cardinal in 1622, in 1624 Richelieu became the
King's minister of state. He quickly concluded an alliance
with England, strengthened by the marriage of the King's
sister Henrietta Maria to Charles I. The Thirty Years War
had already broken out, and Richelieu recognised that before
France could play an active part in it she must first put her
own house in order. In 1627-9 he broke the power of the
Huguenots, after the long siege of La Rochelle, in which an
English force under the Duke of Buckingham vainly
intervened. The Huguenots retained their freedom of
worship, but were no longer a real political force. Appointed
chief minister in 1629, he continued his vigorous policy at
home and abroad, thwarting the Queen Mother's attempt to
overthrow him, and by a mixture of military and diplomatic
pressure he obtained the fortress of Pinerolo, between Turin
and Briançon, as well as the Duchy of Lorraine.

Richelieu had initially given financial support to the
Protestants in their struggle against the Empire, but in 1635
he entered the war directly, in alliance with the Protestant
Dutch and Swedes, and in opposition first to the Spanish
alone and, after 1638, to the Empire too. He succeeded in
carrying out a partial reform of the army and putting huge
forces into the field – in July 1635 the King's nine armies had
a paper strength of 134,000 foot and 26,000 horse. Although
there had been no decisive victory by the time Richelieu
died in 1642, it was already clear that the Franco-Swedish
alliance had the edge over its enemies. The capture of
Breisach in 1638 cut the land link between Spain and the

embattled Spanish Netherlands, and a Dutch naval victory off Dover deprived the Spaniards of control of the sea route. In August 1640 the French took Arras, and in their hands the town covered the line of the Somme and exposed the Spanish Netherlands to attack. Richelieu's last years were dogged by plots: in 1641 the King's favourite, the Marquis of Cinq-Mars, made a secret pact with Spain, but Richelieu sniffed it out and had him executed.

Richelieu died only months before his master, Louis XIII. The new King, Louis XIV, was a four-year-old child and his mother, Anne of Austria, became regent. Although Spanish by birth, Anne did not reverse French policy, largely because she fell under the influence of Cardinal Mazarin, Richelieu's successor, who adhered to Richelieu's line. He was rewarded by success. In 1642 Don Francisco de Melo, governor of the Spanish Netherlands, planned a full-scale invasion of France with the intention of taking Paris. To clear his path he had first to breach the fortress line on the upper Oise and sat down before Rocroi. Here the young Duke of Enghien, son of the Prince of Condé, met him on 19 May 1643 and in the ensuing battle most of Melo's Spanish infantry were killed or captured. There could be no replacing them from a depopulated and impoverished Spain, and Rocroi signalled a permanent decline in Spanish fortunes. In southern Germany Enghien and the Viscount of Turenne cleared the Rhine valley, and eventually pushed on as far as the Inn, posing a threat which helped persuade the Emperor to make peace.

The Treaty of Westphalia, which ended the war in 1648, had wide-ranging implications. It marked the end of the alliance between the Austrian and Spanish Hapsburgs, a point of great significance for France. French sovereignty over Metz, Toul and Verdun was recognised, and France gained most of Alsace, albeit by terms so vague that subsequent difficulties were inevitable.

Frondeurs and Spaniards

Civil strife had erupted even before peace terms were agreed. The long war had left France exhausted, and the Paris

Parlement protested vigorously at what it saw as 'abuses which had crept into the State'. The arrest of some of its leaders provoked the first parliamentary *Fronde* – a coalition of popular and parliamentary leaders who attacked Mazarin and the Queen. There was fighting around Paris in 1649, with royal troops under the command of Condé (the victor of Rocroi, who had succeeded to his father's title in 1646). A short-lived treaty increased the *Parlement*'s influence but did not exile Mazarin (one of the demands of the *Frondeurs*) and a new *Fronde* soon formed, this time focused on the Princes. Condé and his brother-in-law, Conti, were arrested, and the royal army beat Turenne at Rethel. Mazarin, however, found the *Parlement* firmly in favour of the arrested Princes, and the combination proved too much for him, persuading him to spend much of 1651 in Germany.

In the interim the political merry-go-round spun again. Condé entered into negotiations with Spain, while Turenne returned to the royal fold. There was a confused battle between the two in the Faubourg St-Antoine on 2 July 1652 and although Condé had the worst of it, he was able to enter the capital. He remained supreme there till October, when the King, who had now attained his majority, returned. Mazarin himself re-entered the capital in February 1653 and with the submission of Bordeaux later that year the *Frondes* ended.

The Spanish war went on. The Spaniards had capitalised upon the dislocation caused by the *Frondes* to improve their position in the north and in September 1652 they took Dunkirk. Mazarin hesitated before making an alliance with England, but in March 1657 he agreed to the Treaty of Paris, under whose terms Cromwell was to give military assistance to France in return for Dunkirk and Mardyke. In 1658 the Spaniards were beaten by an Anglo-French force at the battle of the Dunes, just outside Dunkirk, which fell shortly afterwards. The war was ended in November 1659 by the Peace of the Pyrenees, which gave France territory to north and south, including the whole of Artois. The strain of war had been colossal, because of its devastation and the burden of taxation required to pay for it, but France emerged from her long rivalry with Spain indisputably the victor.

Wars of the Sun King

When Mazarin died in 1661 the twenty-two-year-old Louis XIV made it plain that he would conduct the affairs of state himself. The country soon felt the smack of personal rule. Fouquet, the Finance Minister, was accused of corruption and imprisoned in a fortress for life: the capable Colbert replaced him, and instituted reforms in aspects as varied as financial administration, road transport and agriculture. The *parlements* lost their rights of remonstrance and moves against the Huguenots culminated in the Revocation of the Edict of Nantes in 1685, resulting in large-scale emigration of these industrious folk. Periodic rebellions – the most serious in Brittany in 1675 – were sharply put down by royal troops. By the ennoblement of useful commoners and the emasculation of the great nobility Louis emphasised his own status as the fount of all honour and authority, a role underscored by the elaborate ceremony of his own life. The stupendous palace of Versailles reflected the glory of the Sun King to Frenchmen and foreigners alike.

The securing of defensible frontiers was to remain one of Louis's major preoccupations. He quickly closed the Dunkirk gate to his kingdom by buying it back off the English, and went on to give the Duke of Lorraine full pockets and senior rank in the French peerage in return for agreeing to French sovereignty over his duchy. The death of his father-in-law, Philip IV of Spain, enabled Louis to defend his wife's alleged right to a large part of the Spanish Netherlands and in 1666 he invaded these provinces with a well-found army, making so much progress that England and Holland, until recently at one another's throats, allied to resist him. Nevertheless, the Treaty of Aix-la-Chapelle (1668) gave France useful chunks of territory around Dunkirk and Lille, as well as Oudenarde, Ath, Tournai and Charleroi.

Success in the War of Devolution encouraged Louis to

mount a direct assault on the Dutch. His War Minister, the
Marquis of Louvois, built upon the firm foundations laid by
his father Michel le Tellier to produce a well-equipped army,
while the engineer Vauban built fortresses in an attempt to
guarantee the new frontiers of the kingdom against
invasion. Louis went to war in 1672, marching up the Rhine
almost as he pleased. Although the Dutch opened their
dykes to make French progress more difficult, the French
took several major towns, but failed to turn military victory
into diplomatic success by offering the Dutch acceptable
peace terms. A popular revolution brought the Prince of
Orange to power in Holland and the Dutch fought on. In
1673 the French took the fortress of Maastricht after a siege
conducted with great precision by Vauban. This feat of arms
brought victory no nearer, for Louis refused peace terms that
would deprive him of his prize, despite the fact that both
Spain and the Empire prepared to enter the war against him.

The brilliance of Condé and Turenne, coupled with the
meticulous war administration of Louvois, enabled the
French to conduct a measured retreat from Holland and to
drive the Imperialists from Alsace. In 1675 Louis took his
army through the Spanish Netherlands, marching from
Cateau-Cambrésis to Mons (the same route, albeit in a
different direction, as that pursued by the British Expeditionary Force in August 1914) and on towards Namur,
strengthening his grip on Maastricht. But a chance cannon
ball killed Turenne near Strasbourg and the situation in
Alsace deteriorated rapidly. Louis invaded the Spanish
Netherlands again in 1676, capturing Bouchain, and the
following year the French took St-Omer, Cambrai and
Valenciennes, beating the Dutch at Cassel for good measure.
Early in 1678 Louis jabbed up to Ghent, encouraging the
Dutch to make peace on generous terms. The Treaty of
Nijmegen left France with a much improved frontier to her
north, and with gains to the east, including Franche-Comté.
True, Louis had been forced to draw back from the high-
water mark of his success, but the war's results were
anything but derisory. Moreover, during the next nine years
Louis used local pretexts to 'reunite' to France several

strategically important areas on the German frontier, and by securing Strasbourg, Saarbrücken, Zweibrücken, Luxembourg and Trier, he greatly strengthened his position. The Emperor, preoccupied with the threat posed by the Turks, was in no position to interfere and in the Truce of Ratisbon he recognised Louis's recent gains.

In 1686 the Dutch leader William of Orange assembled the League of Augsburg, an anti-French coalition that included the Empire, Sweden, Spain, Bavaria, Saxony and the Palatinate. A disputed succession in the Palatinate and a disputed election to the archbishopric of Cologne gave rise, in 1688, to the War of the League of Augsburg. William's succession to the English throne in 1689 made him an even more formidable adversary, but Louis hoped for a quick and decisive victory before a long war had time to tear the fabric of the state. The war began well: Louis took Philippsburg on the Rhine and went on to ravage the Palatinate to prevent its fortresses being held against him. This devastation roused a powerful anti-French sentiment and helped make it impossible for Louis to secure his gains by a quick peace. By 1690 Louis found almost all his neighbours in arms against him: William had brought England into the war when Louis gave military support to the deposed James II. Nevertheless, his armies fought doggedly. In July 1690 the Duke of Luxembourg beat Waldeck at Fleurus, north of Charleroi. An unlikely naval victory followed, when Tourville defeated the Anglo-Dutch fleet off Beachy Head, while in the south Catinat checked an Imperial and Savoyard army at Staffarde. The following year the King himself commanded the army that took Mons, but it was a victory which hardly equalled the loss suffered by Louis when his War Minister Louvois died suddenly shortly afterwards. The powerful fortress of Namur fell in June 1692 and when William tried to recapture the place he was roughly handled at Steenkirke.

For the 1693 campaign Louis led a huge army in Flanders, with Marshal Luxembourg 'commanding under his orders'. He sent the Dauphin off to the Rhine to press the Germans, and Luxembourg beat William at Neerwinden and went on to take Charleroi. Catinat trounced the Savoyards and

Noailles held his ground in Spain. Yet the victories of 1693 brought peace no nearer and a succession of poor harvests undermined an already shaky economy. Louis found himself on the defensive and the Electoral Prince of Bavaria took Namur. But Louis managed to make peace with Savoy, whose defection helped William to conclude that a nego- tiated peace was in his interests too. The Treaty of Ryswick (1697) represented a real humiliation for Louis, who lost most of the territory acquired since the Treaty of Nijmegen.

Events in Spain made the peace all too short. The death of Charles II in November 1700 left three claimants to the Spanish throne: Louis's grandson, the Emperor Leopold and the Electoral Prince of Bavaria. The status of two of the claimants meant that the balance of power in Europe was at stake, and the ensuing War of Spanish Succession saw Louis, the Spanish Bourbons and the Bavarians ranged against William's Grand Alliance of England, Holland, the Empire and many of the smaller German states. The war was scarcely under way when William died, to be succeeded by his sister-in-law, Anne.

Louis soon discovered that, although he had men of talent amongst his own marshals, he was facing opponents of real genius. In September 1701 the Imperialist general Prince Eugène of Savoy beat Villeroi at Chiari in Lombardy, and in the north the Earl of Marlborough snatched a number of French-held fortresses, including Liège. In 1703 the French capitalised on an early success, when Villars took Kehl, across the Rhine from Strasbourg, to produce a plan for a grand attack on the Empire. Squabbling amongst French commanders prevented its development, although the Duke of Burgundy, his hand firmly held by Vauban, took Breisach. Villars's strained relations with the Elector of Bavaria led to his replacement by the milder Marshal Marsin and Villars himself was dealing with a Huguenot revolt in the south when the fateful campaign of 1704 opened.

This time the Anglo-Dutch took the initiative as Marlborough marched for the Danube, moving at speed and safely joining his allies the Margrave of Baden and Prince Eugène. They stormed the Schellenberg, key to Donauworth, on 2 July, but despite the ravaging of his territory, the Elector of Bavaria refused to fight or negotiate and awaited the appearance of Marshal Tallard's army from the Rhine. Tallard duly arrived and on 13 August Marlborough routed the Franco-Bavarian army at Blenheim. The Allies went on to occupy most of Bavaria and the Archduke Charles landed in Spain to make good his claim to its throne.

In 1705 Louis assessed that the greatest danger lay in the north and deployed three armies there. Villeroi and the Elector of Bavaria were based on Namur and Antwerp; Villars had a smaller force at Metz, while Marsin covered Alsace. Marlborough found his hands tied by inter-Allied wrangling and was assembling troops in the Palatinate, prior to campaigning along the Moselle, when Villars pounced. He took Huy on 10 June and went on to capture the city of Liège though he failed to take its citadel. Villeroi fell back as the Duke approached and Marlborough, after briefly considering returning to the Moselle, determined on a campaign in Flanders. The Duke left the Dutch General Overkirk to besiege Huy and in mid-July manoeuvred Villeroi out of the Lines of Brabant – a belt of fortifications running from Antwerp, through Aerschot, Diest and Merdorp, to the Meuse – to occupy Tirlemont. In August he again outmanoeuvred Villeroi and only difficulties with the Dutch prevented him from giving battle near what was later to be the field of Waterloo.

The decisive events of the 1706 campaigning season seemed likely to occur in northern Italy, where the advantage lay with the French, and in Spain, where the Allies were making steady progress. Indeed, Allied successes across the Pyrenees compelled Louis to lend ever-increasing support to his grandson. Marlborough hoped to march south to Italy and redress the balance there, but was thwarted by French offensives in Alsace and Italy. Early success elsewhere encouraged Louis to attack in Flanders too and Villeroi set off from Louvain to besiege Leau, while Marsin moved up to Metz to support him if necessary. Marlborough was concerned that the French would enjoy overwhelming superiority if Villeroi and Marsin joined forces. Although his own army was as yet incomplete, he advanced from his camp near Maastricht through a demolished section of the Lines of Brabant between Diest and the Meuse in the hope of bringing Villeroi to battle. Villeroi, increasingly confident and spurred on by orders from Versailles, was equally anxious for a general action and marched south from Diest to face the Allies. The armies met at Ramillies on 23 May and Villeroi was utterly defeated. Ramillies opened the Spanish Netherlands to the Allies. Louvain was occupied on 25 May, Brussels, the capital, fell three days later and throughout the early summer Marlborough pressed his advantage, taking a score of fortresses including Antwerp and Ostend. Menin and Dendermonde, with robust garrisons and staunch governors, stood long sieges but they too fell, and in September Marlborough closed the campaigning season by taking Ath. Elsewhere the Allies had mixed fortunes. Eugène had restored the situation in north Italy by a brilliant victory at Turin, but in Spain the Allies had proved unable to hold Madrid.

In 1707 the Allies failed to take advantage of the previous year's promise. The centrifugal tendencies of the Grand Alliance threatened to pull it apart and Marlborough's footing on the heeling raft of English politics was unsteady. His plan for the year – a thrust, under Eugène, from north Italy to the French naval base of Toulon, combined with an offensive of his own in Flanders – went terribly wrong. In Spain, the Duke of Berwick, natural son of the dispossessed Stuart, James II of England, won the decisive battle of Almanza, dashing Allied hopes in the peninsula. On the Rhine, things went almost as badly: Villars surprised the Margrave of Bayreuth, took Rastadt and entered Stuttgart. Marlborough campaigned inconclusively against Vendôme, Villars's successor, in the Netherlands, then established a fortified camp at Meldert, south-east of Louvain, to await news of the Toulon project. Eugène was unable to set out till 30 June and it was not until late July that he reached Toulon. The French sent reinforcements south and Eugène had no stomach for a formal siege: he abandoned the operation on 21 August. By the year's end the French had repaired much of the damage done in 1706 and the political ripples spread by failure in 1707 were to make Marlborough's life increasingly difficult. However, peace negotiations foundered when it became clear that the Dutch required not only the removal of Philip V from the throne of Spain, but also the cession of Ypres, Menin, Condé and Maubeuge to buttress the barrier between Holland and France.

Louis resolved to persevere in 1708, despite France's financial exhaustion. His main effort was to take place in the north, where his grandson, the Duke of Burgundy, shared command with Marshal Vendôme. After a failed attempt to land the Chevalier de St-George, pretender to the British throne, in Scotland, Vendôme and Burgundy concentrated on their task of luring Marlborough away from French Flanders and bringing about a pitched battle on favourable terms. Burgundy and Vendôme got on badly, and marching and countermarching between Soignies and Brussels in June produced no useful result. But on the Moselle Eugène was shadowed by the able Berwick and he was unable to join Marlborough as quickly as planned. Worse still, hearing that the citizens of Bruges and Ghent were dissatisfied with Anglo-Dutch rule, the French planned to push two flying columns into them, covering the operation with the main field army.

The French took Bruges and Ghent on 5 July, and Vendôme and Burgundy followed up, crossing the Dender and breaking down its bridges behind them. Although Marlborough moved fast, he failed to intercept the French and this neat coup left the French in possession of much of Spanish Flanders. However, Marlborough was helped by a dispute between Burgundy and Vendôme, which enabled him to reinforce the garrison of Oudenarde, to prepare a week's supply of bread and strip his army to the bare essentials for a rapid march. He jabbed down to Lessines, where he secured the crossings of the Dender, causing the French to break away to the north. Marlborough concentrated south of Oudenarde while the French camped north of the town and on 11 July the Allies won the battle of Oudenarde, seizing the initiative in Flanders at a stroke.

Marlborough put forward a bold plan for an advance along the Channel coast, supported by the fleet. Neither the Dutch nor Eugène much liked it and the Allies instead laid siege to the enormously strong fortress of Lille. The operation progressed slowly, for the governor, Marshal Boufflers, was energetic and his garrison strong. Boufflers eventually gave up the town on 22 October and withdrew into the citadel, where he held out till 9 December. Bruges and Ghent were both recovered before the Allies went into winter quarters.

The winter of 1708–9 was so severe that horses and men froze to death in camp, and the sufferings endured in France, with its failed harvests and collapsed economy, were truly terrible. The disasters of 1708 encouraged Louis to seek peace, but Allied overconfidence ensured that the price was too high. French armies took the field in Spain; in the Dauphiné, facing the Imperialists; opposing the Elector of Hanover on the Rhine; and, under Villars, meeting Marlborough and Eugène in the north. Marlborough entertained a scheme similar to his initial project for the previous year, and hoped to break the French lines near Dunkirk and move down to the Somme, supported by the fleet. This failed to win Allied approval and he instead drew his siege train up to Menin, as if to threaten Béthune or Ypres, then swung quickly south-east to besiege Tournai, which fell on 5 September after a bitter defence. The Allies went on to besiege Mons, while Villars, now authorised to offer battle on favourable terms, closed up behind them. On 8–9 September he took up a commanding position around Malplaquet, between Mons and Bavay, and the Allies attacked him there on the 11th. The French drew off in good order after a bloody battle, which dented even Marlborough's reputation. Mons fell in late October and Marlborough disposed his army in winter quarters at the end of the month.

Malplaquet raised French spirits, and its long butcher's bill caused recriminations in Holland and in England, where Marlborough's political position grew almost daily weaker. The 1710 campaign in Flanders focused on the Allied sieges of Douai and Béthune, which fell in late June and late

August respectively. Marlborough again suggested an advance along the Channel coast and was again overruled, and the Allies took the fortresses of Aire and St-Venant on the Lys. Elsewhere they had little to show for their efforts: French armies had held their own on other fronts, while in Spain Marshal Vendôme had administered the *coup de grâce* to Charles III's hopes of securing his kingdom.

When he returned to Flanders after a dispiriting winter amongst the politicians, Marlborough found that Villars had thrown up a long and well-sited fortified barrier from the River Canche on the Channel coast to Namur on the Sambre. The Duke was deprived of the trusted assistance of Eugène, called back to face a French offensive on the upper Rhine. And he campaigned in an atmosphere of growing uncertainty, for the Archduke Charles (Allied claimant to the throne of Spain) had become Holy Roman Emperor: his candidacy for the Spanish throne was no longer in the Anglo-Dutch interest, for a union between Spain and the Empire would distort the balance of power. Marlborough nevertheless set about manoeuvring Villars out of his lines of *Non Plus Ultra*, as the confident marshal called them, and in early August he triumphantly outgeneralled Villars and went on to take Bouchain.

The capture of Bouchain was the last act of the Grand Alliance, for Britain and France signed preliminaries of peace three weeks later. Formal treaties were not agreed till 1713, when England, Holland, Savoy and Prussia made peace with France at Utrecht, while the Empire agreed to terms at Rastadt in 1714. After British withdrawal Villars had handled Eugène very roughly indeed, beating him at Denain and going on to take Douai, Le Quesnoy and Bouchain, strengthening the hand of French negotiators. France emerged from the war without substantial territorial loss, although she agreed to the demolition of the fortifications of Dunkirk and the cession of some overseas possessions. Philip V was recognised as King of Spain, but renounced his rights to the French throne. Britain came off best, with territorial gains in the Mediterranean and the New World, and French agreement to the Hanoverian succession. The Dutch obtained the Spanish Netherlands as a much sought-after barrier between themselves and the French, and duly ceded them to the Austrians, believing that this would serve as the best insurance against future French expansion. Austria was given territory in Italy, and both Savoy and Prussia – its monarch's royal title now recognised – made useful gains.

The Age of Limited War

Louis XIV died on 1 September 1715. He was succeeded by his five-year-old grandson, with the old King's raffish nephew Philip of Orléans as regent. The arrangement failed to commend itself to Philip V of Spain who still hankered after uniting the Bourbon thrones despite the Treaty of Utrecht, driving Orléans into a brief alliance with England. The regent attempted to inspire the economic recovery of France, but his cure proved almost as bad as the disease and the collapse of the Mississippi Company ruined thousands of speculators. Orléans died in 1723 and the next prince of the blood, the unappealing Duke of Bourbon, duly took his place. He organised the marriage of Louis XV to Marie Leczinska, daughter of the ex-King of Poland, and fell from power shortly afterwards when he tried to use his influence with the Queen to remove Bishop Fleury, the King's tutor.

In 1726 Louis announced that he would take the reins of government personally, but in practice Fleury – now elevated to the cardinalate – was chief minister till 1743. Fleury strove to continue Orléans's policy of peace and prosperity, stabilising the currency and developing foreign trade. There was little he could do about the underlying weaknesses of French government. The king enjoyed absolute power, transmitted through a hierarchy of officials, most of whom held their offices by payment and were difficult to dismiss. Although in theory the king exercised authority through a series of councils – the Council of State was the most influential – in practice the secretaries of state and the controller-general enjoyed immense power. Co-ordination required the energetic personal effort of the king or his first minister. Fleury never formally took this title and from his death till 1787 there was no first minister at all.

Fleury was right to recognise that above all France needed peace. Unfortunately, in 1733 the death of Augustus, Elector of Saxony and elective King of Poland, presented him with a crisis he could not circumvent. Louis XV's father-in-law set off for Warsaw and was proclaimed King. An anti-French faction at once proclaimed his rival, Augustus III, son of the previous monarch, and the Russians and Austrians invaded to make good his claim. This gave such powerful impetus to the war party in France that there was nothing Fleury could do to prevent war. However, he secured British neutrality and ensured that there were no military operations in the Low Countries, where they would have aroused British fears. The French occupied Lorraine, crossed the Rhine and took Philippsburg, and old Villars gained some ground in northern Italy.

These successes enabled Fleury to make peace on favourable terms. The peace settled the destiny of that shuttlecock duchy, Lorraine, so often in the French grasp but as frequently relinquished. Stanislaus was made King of Lorraine, establishing his capital at Nancy. His French chancellor administered the little kingdom almost as a part of France and when Stanislaus died in 1766 it was absorbed into France.

Fleury's wise policy did much to colour the landscape of the fatal avenue. Already it bore the heavy footprint of that master of military engineering, Marshal Sebastien le Prestre de Vauban. No traveller in northern France can avoid an encounter with the old fellow sooner or later. Sometimes the meeting is long expected: the majestic citadel of Montmédy scrapes the skyline long before the town itself is visible. At other times earth and ashlar come as a surprise, like the westerly bastion at Arras, jumping up by the traffic lights on the Rue Roger Salengro, or the east gate of Verdun, ambushing the traveller as he drops down the last stretch of the Metz road. Modern townscape often gives a clue to Vauban's proximity: watch for an unexpected park between

ancient centre and modern suburbs, and in it you have an
excellent chance of finding redan, ravelin and a classical
gateway surmounted by the lilies of France or the sunburst
of *le Roi Soleil*. Sometimes, where the threat had visibly
disappeared and the municipal coffers were deep enough to
undo what Vauban and his colleagues had done, roads were
driven along the line of old fortifications, providing a useful
ring road around the town centre. The word *boulevard*
actually means bulwark, an indication of the route followed
by these broad and convenient roads.

Fleury's contribution is almost as pervasive. Overseas
trade and domestic economy had suffered during the wars of
Louis XIV, and the peace which followed them left its mark
on the land. Burgeoning overseas trade saw the great ports
grow in size and importance: in the north, Le Havre and
Dunkirk flourished, and the spacious houses that survive in
both towns testify to the wealth flowing from Atlantic and
Baltic respectively. In large provincial centres, lavish
buildings reflected the fact that each major town sought to
duplicate Paris, with its palaces for royal *intendants* (the
powerful provincial governors established by Henry IV),
bishops, and nobility of sword and robe, and scarcely less
lavish houses for officials, merchants and financiers. A few
still house one family, but more are offices, hotels, or
apartments with an architectural hierarchy that takes
dwellers from the double doors and marble chimneypieces
of the lower floors to the romantic inconvenience of dormer
windows and low ceilings in the old servants' quarters at the
top. The clearing of medieval walls made redundant by
gunpowder gave wide scope for formal public gardens. The
new *place royale*, that cynosure of provincial elegance,
might find itself, four-square in high-windowed formality,
on the fringes of the town where land was cheaper, or in the
centre where reeking alleys and cobbled courtyards were
demolished to make way for it.

Industry prospered. In a curious way it resembled the
monarchy itself, part modern and part ancient. Local guilds
proliferated, growing in number during the eighteenth
century, and in most crafts there were scores of small enter-

prises, each with a master and a handful of journeymen. Many trades became, in effect, hereditary with huge fees payable for entry as a master. The state siphoned money from the guilds by selling meaningless offices, which enabled their holders to collect fees, or, indeed, provoke the guilds to buy them out. Government regulations spawned a shoal of officials who enforced them and drew fees for their pains.

Yet there were signs of real progress. Factories sprang up, the most important, like Gobelin tapestries and Sèvres porcelain, state supported. Textile factories were set up in some areas – the Jacobite refugee John Holker founded a successful factory in a Rouen suburb and almost half the Rheims cloth workers were concentrated in factories. In the north, however, it was more usual for textile workers to work on their own in villages or suburbs, provided with raw materials by town-based merchants who collected and sold the finished product. Mining encouraged large-scale enterprise and the eighteenth century saw the establishment of some of the mines whose slag heaps pimple the face of the north. Yet overall, a high degree of state control, the continuing strength of the guilds and frequent shortage of capital meant that French industry was not as innovative or well-developed as that in England.

As far as transport was concerned, however, France enjoyed a clear lead over England. Sully had done his best to improve the wretched roads and initiated the practice of planting poplar trees at their verges, something recently reversed because of fears that collisions with roadside trees increase fatalities amongst drivers and their passengers. In the eighteenth century the strong central influence, which industry found so stifling, was applied to roads and it worked wonders. In 1736 roads were put under the *Contrôle Général*, and eleven years later the Department of Highways and Bridges was created, with a huge staff of engineers and officials. Labour was provided by *corvée*, the tax of forced labour, formalised in the 1730s. Roads ran from Paris like the spokes of a wheel, with broad, straight, tree-lined highways speeding the passage of goods, merchandise and armies. Wide use was made of navigable rivers, connected by a growing network of canals. The first of the great canals, connecting the Seine and the Loire, had been dug in the reign of Henry IV and the Canal du Midi, linking the Atlantic and Mediterranean coasts, built under Colbert's sponsorship, was completed in 1681.

Fleury's last years were dogged by conflict. The King found himself at loggerheads with the *parlements*, which helped spread the notion of opposition to royal authority. There were waves of Huguenot unrest, which harsh laws failed to suppress, and the church was riven by the dispute between Jansenists and Jesuits. Perhaps most serious was the struggle for financial reform, which ran squarely into the vested interests of clergy and nobility alike.

Crises at home were matched by failure abroad. With Fleury in failing health, the ambitious Count of Belle-Isle hustled along a forward foreign policy, trying to secure the election of the Elector of Bavaria to the throne of the Holy Roman Empire, vacant following the death of Charles VI in 1740. Frederick II of Prussia invaded the Austrian province of Silesia while Belle-Isle was hard at his diplomacy, and Belle-Isle then brought about a Franco-Prussian alliance, invaded Austrian territory and duly secured the Elector's selection as Holy Roman Emperor.

Belle-Isle's triumph soon turned sour as the Austrians counter-attacked into Bavaria. The picture darkened further with a change in British policy in favour of intervention on the Continent. Frederick made peace with Austria, retaining Silesia. He re-entered the war two years later, in 1744, but by this time France and her ally Spain were locked in a struggle not only with Austria, but also with the increasing maritime strength of Britain.

In the fighting that followed France fared badly at sea, and in 1747 her navy was crippled by battles off Cape Finisterre in May and at Belle-Isle in October. The sheer logic of geography had long suggested that the French army should be strong even if the navy was weak and the reverse was true for Britain. France was able to field over 150,000 regular infantry, backed by another 100,000 irregulars of varying quality. At their head marched the Household Regiments, the French Guards and the Swiss Guards, then followed the regiments of the line. There were several foreign regiments, mostly Swiss and Irish, but including the *Royal-Ecossais*, raised in 1744 and commanded by Lord John Drummond. The cavalry numbered over 40,000 and again the Household troops took precedence, followed by the regiments of line cavalry and fifteen regiments of dragoons, still essentially mounted infantry. The artillery, strangely, ranked forty-sixth amongst the infantry regiments as the *Royal-Artillerie*, but had been reorganised in the 1730s in an effort to standardise the weights and calibres of its pieces.

Regular troops were volunteers, serving for ten or twelve years, but often re-enlisting at the conclusion of their first stretch with the colours. Louvois had introduced proper uniforms into the infantry in the 1660s: French regiments generally wore white or grey, while red was worn by the Swiss and Irish, and blue by irregulars. Officers were by no means universally aristocratic, and there were a good many ex-rankers amongst the lieutenants and captains, and the occasional commoner made his way even higher. François de Chevert, Verdun's local hero, was a poor orphan but rose to the rank of lieutenant-general. In common with his comrades in other European armies, the French soldier did not spend all his time uniformed and under arms. Provided he returned to his barracks by the appointed hour, he could follow any trade he chose and the harvest season saw scores of soldiers, in civilian clothes, at work in the fields.

His harvester's smock replaced by long-skirted coat on the day of battle, the French infantryman shouldered a muzzle-loading flintlock musket and stood in line to deliver close-range volleys to the packed ranks opposite. Cavalry hovered on the flanks, to charge shaken infantry, sweep away opposing horsemen or cut in to turn retreat into rout. Heavy guns, awkward to move off roads, would be placed in battery on commanding ground, while lighter pieces would be manhandled into the gaps between infantry units.

In 1743 the French army found itself engaged against an Allied army with British, Hessian, Hanoverian and Austrian contingents, under the personal command of King George II. In June the Allies won an improbable victory at Dettingen on the River Main, but dissensions in their ranks prevented proper exploitation. The following spring the Allies concentrated at Bruges and Ghent, and a larger French force, under Marshal Noailles and Maurice de Saxe set about the fortresses in the Austrian Netherlands, taking Menin, Ypres, Courtrai and Maubeuge. French gains in Flanders were counterbalanced by reverses in Alsace, where the Austrian general Charles of Lorraine took Lauterbourg and Wissembourg. Louis XV set off from Flanders with a strong force to deal with him, giving the Allies a chance to attack in the north in his absence. This window of opportunity slid shut quickly, for Frederick of Prussia, alarmed at Austrian successes, had re-entered the war and Saxe, facing the now-superior Allied army in Flanders, knew that it was only a matter of time before the threat posed by Frederick drew the Austrians from Alsace and enabled the French to reinforce Flanders.

The Allies threatened Lille but lacked the resources to besiege this powerful fortress, so fell back on Tournai and thence to Oudenarde, with Saxe, emboldened by news of Charles of Lorraine's withdrawal from Alsace, pressing them vigorously. In October the Allies went into winter quarters around Ghent after a campaign of wasted opportunities. Worse was to come. Saxe laid siege to Tournai in April 1745 and the Allied army, under George II's son William Augustus, Duke of Cumberland, marched on the town to raise the siege. Saxe divided his army, leaving part to contain the garrison while he took the remainder to prepare a strong position astride Cumberland's line of advance. When the Duke attacked on 11 May his infantry assault foundered on Saxe's redoubts and the Battle of Fontenoy was a bloody Allied defeat. Tournai surrendered soon afterwards and Ghent followed suit in mid-July. Bruges fell without a fight and Oudenarde lasted only two days.

English misfortunes in Flanders encouraged Prince Charles Edward, 'the Young Pretender', to press for French support, but little was forthcoming. The prince reached

Scotland with a handful of followers and raised the royal standard at Glenfinnan. The situation in Flanders continued to deteriorate and Saxe took Ostend in mid-August. The gains made in Scotland by Prince Charles persuaded the British government to withdraw most of its troops from Flanders and with Cumberland at its head the King's army in Scotland duly defeated Charles Edward at Culloden on 16 April 1746.

The French profited from Cumberland's absence to take Brussels, and even the return of troops from Britain failed to prevent Saxe from taking Mons and Namur. In October 1746 he defeated Charles of Lorraine at Rocourt, north of Liège. In 1747 things were no better for the Allies, although French declaration of war on Holland turned the Dutch from auxiliaries to combatants in their own right. Saxe advanced on Maastricht in late June and on 2 July he beat the Allies at Laffeldt: they were fortunate to be able to retire across the Meuse and under the guns of Maastricht without serious loss. Saxe did not feel able to attack Maastricht and turned his attention instead to Bergen-op-Zoom, at the neck of the Walcheren peninsula. Hearing that Cumberland was on his way to the town's relief, the French stormed the place before dawn on 13 September 1747. Peace negotiations were already in progress at Aix-la-Chapelle when Saxe turned his attention to Maastricht, which surrendered on 10 May 1748 as part of the armistice preceding a general peace.

By the terms of the Treaty of Aix-la-Chapelle, France evacuated the Austrian Netherlands and Madras, while Britain returned Louisbourg and Cape Breton Island in North America. The Hanoverian succession to the British throne was guaranteed and a commission was set up to consider Anglo-French frontier disputes in North America. Austria was the chief loser, for Silesia remained in Prussia's hands and some Austrian possessions in Italy were lost. The peace settled nothing on a permanent basis, for it was clear that Anglo-French colonial rivalry in North America and India would persist regardless of the ebb and flow of European diplomacy. And while the French army would exercise an important influence on events in Europe, the

British navy, which had emerged from the war with considerable credit, would exercise a decisive influence on
overseas affairs.

Friction in the colonies, culminating in an undeclared war
that broke out in North America in 1754, led to the next
round of European war. This time the alliances of the War of
Austrian Succession were dramatically reversed: Prussia
sided with Britain and Austria with France. Frederick's
invasion of Saxony in the summer of 1756 precipitated a
general conflict. Frederick forced the Saxons to surrender
near Dresden, then beat a relieving Austrian army.
Cumberland took the field again, this time at the head of a
force of Hanoverians, Hessians and Brunswickers defending
Hanover. Defeated at Hastenbeck, Cumberland was forced
to capitulate shortly afterwards. Although Frederick was not
uniformly successful, his victories at Rossbach and Leuthen
encouraged the British to send an even larger subsidy to
Prussia, and to resume operations on the mainland, putting
an Anglo-Hanoverian force into the field under Prince
Ferdinand of Brunswick. In 1758 Frederick fared badly,
winning a ruinously expensive victory against the Russians
at Zorndorf, but losing to the Austrians at Hochkirch.
Brunswick, in contrast, expelled the French from Hanover
and Brunswick, and went on to clear Hesse and Westphalia.
When the French counter-attacked in 1759 they recovered
some of the lost ground, only to be beaten at Minden.
Frederick, meanwhile, suffered further reverses, but was
saved from disaster by his enemies' inability to combine and
bring the war to a conclusion. He was ultimately rescued by
'the miracle of the House of Brandenburg'. The Empress
Elizabeth of Russia died and was succeeded by her nephew,
Paul III, who had long admired Frederick. Paul took Russia
out of the war, enabling Frederick to expel the Austrians
from Silesia. Paul was speedily deposed and murdered, but
the damage was done.

The Duke of Choiseul, France's new Foreign Minister,
eventually brought Spain into the war so as to put more
pressure on Britain. His plans came to naught, for Spain's
entry into the conflict simply gave Britain the opportunity

to seize some of her territories overseas. Both France and Austria, under heavy pressure, were forced to make peace in 1763. The treaties – Paris for most combatants and Hubertusburg for Austria and Prussia – brought gains for Britain and Prussia. British primacy was established in North America and India. In addition to relinquishing overseas possessions, France gave up Prussian territory on the Rhine, which she occupied during the war. Prussia retained Silesia: although Frederick's fortunes had hung by a thread, he had succeeded in his principal war aim. The same could not be said of France. She had lost her first colonial empire and made no compensating gains in Europe. Her government had been revealed as vacillating and incompetent, and the strain of war had done serious damage to the French economy.

Choiseul implemented far-reaching reforms, building up the navy and improving the army, and cementing the Austrian alliance by securing the marriage of Maria Theresa's daughter, Marie-Antoinette, to the Dauphin. But his support for the *parlements* drew the opposition of those who championed the King's rights and he fell in 1770. The triumvirate that succeeded him curbed the *parlements* and tried to put the country's finances on a firmer footing, but the death of Louis XV in 1774 ushered in a wave of reaction, with the *parlements* regaining their old powers. An attempt at large-scale fiscal reform failed with the dismissal of Controller-General Turgot in 1776, but some of his ministerial colleagues left more enduring marks. St-Germain improved the organisation and discipline of the army, and enabled Gribeauval to bring his reform of the artillery, begun under Choiseul, to a successful conclusion. Like Turgot, St-Germain stirred up hostility, which swept him from office and as part of the aristocratic reaction a decree of 1781 effectively restricted promotion above the rank of captain to noblemen.

Nevertheless, France gained a measure of revenge for the humiliations of the Seven Years War by helping the American colonists in their War of Independence, and the foundations laid by Choiseul and St-Germain were to show their strength in the years that followed. Not least amongst the products of St-Germain's reforms were twelve provincial military schools and it was to the Royal School at Brienne that a young Corsican, Napoleone di Buonaparte, reported in 1779.

The Revolution and Napoleon

Napoleon dated the Revolution from a damaging scandal of 1785, when the Cardinal de Rohan was tried for using the Queen's name to obtain an expensive diamond necklace without paying for it. The affair gravely damaged the prestige of the monarchy. Necker, Finance Minister from 1777 to 1781, failed to repair the mischief done to the exchequer by the American war and the wiser schemes of Calonne also foundered when he sought to persuade the nobility to accept taxation, from most of which they were exempt. Lomenie de Brienne became first minister in 1787, but his team was weak and he himself was not the man for such a time of peril. The *parlements* played a key role in fomenting the crisis. They had identified themselves as guardians of an unwritten contract between monarch and people, arguing that fundamental laws should be expressions of the general will. When Brienne unwisely referred his taxation proposals to the *parlements* they threw them out, and when he used the procedure of *lit de justice* to register the measures despite this, the *parlements* declared them invalid. After protracted negotiations Brienne agreed to withdraw the edicts, but the crisis deepened. In May 1788 the *parlements* were suspended and replaced by new courts, but there was a nationwide revolt of nobility and magistracy alike. The States-General were summoned to meet on 1 May 1789: Brienne fell and was replaced by Necker.

The States-General had not met since 1614 and the *Parlement* decided that they should be constituted as they were then, with three estates – nobility, clergy and Third Estate. The latter included the lawyers, bureaucrats and merchants who had supported the *parlements*, but now feared that they would be outvoted by the nobility and clergy. Simply doubling the representation of the Third Estate solved nothing, since each estate was to sit and vote separately. The Third Estate declared itself a National Assembly and invited the other estates to join it: on 20 June it met in an indoor tennis court and swore not to be dissolved until the constitutional crisis had been resolved. At a joint royal session three days later the King declared that the estates should continue to meet separately, but many of the nobility and clergy had joined the Third Estate, and the King, fearful of the growing unrest, ordered the remainder to do so. Necker was dismissed and his successor, Breteuil, headed a government that sought to reassert royal control, with the assistance, if need be, of the growing numbers of troops around Paris.

This was easier said than done. The capital was in ferment, with rioting mobs seeking arms. On 14 July the mob – with the assistance of a mutinous detachment of *Gardes Françaises* – took the Bastille, the fortress dominating the Faubourg St-Antoine on the capital's eastern edge. The King vacillated. Some of his advisers counselled withdrawing from Paris and reasserting order by force: Louis himself considered retiring to Metz, but Broglie could not undertake to furnish a loyal escort. Eventually the King dismissed Breteuil and recalled Necker, before going to the Hôtel de Ville to receive the new national cockade of red, white and blue. Breteuil and the hardliners departed abroad, amongst the first of a growing number of emigrants.

The detail of the political events of the next few years lies beyond the scope of this book. But in them lay the origins of the wars that followed, so the briefest outline must be sketched. The Third Estate's 'revolution' of the summer of 1789 was followed by widespread popular unrest in the provinces, as peasants sought to free themselves from seigneurial dues. The National Assembly abolished most of these, and went on to adopt a Declaration of the Rights of Man and of the Citizen. On 6 October the King was taken from Versailles to Paris by an armed mob of Parisians and National Guard. The subsequent move towards a constitutional monarchy was imperilled by a rift with the papacy

over the status of the Church and by the activities of the
émigrés. In June 1791 the royal family attempted to flee the
country and reached the little town of Varennes, so near to
safety, in the Argonne before being stopped. The Assembly
duly produced a constitution, whose elected body, the
Legislative Assembly, met on 1 October 1791. It did so in an
atmosphere of growing tension, for in August the Emperor of
Austria and the King of Prussia had announced, in the
Declaration of Pilnitz, that they favoured the restoration of
Louis to full monarchical powers. In fact the declaration was
toothless, but it did not seem so in France, where fear of
invasion increased. Political manoeuvrings within an
unstable government encouraged France to declare war on
Austria in April 1792.

It was a rash decision, for the French army was in a state
of turmoil. In January 1791 its old regimental titles had been
replaced by numbers and a decree of 1793 was to take the
process still further, abolishing infantry regiments altogether and replacing them with *demi-brigades*, each
consisting of one regular battalion and two battalions of the
new volunteers. Morale was decidedly volatile and there had
been a serious mutiny at Nancy in 1790, put down with at
least 200 casualties. By 1793 dismissal and emigration had
resulted in the disappearance of seventy per cent of
infantry and cavalry officers. The artillery and engineers,
with their smaller proportion of noble officers, had been less
hard hit.

The opening moves of the 1792 campaign bore witness to
the army's wretched state. A four-pronged offensive into the
Austrian Netherlands failed miserably, and the unedifying
episode encouraged the dominant Girondin party to accuse
the King and Queen of duplicity, while the extremist
Jacobin faction used it as a stick with which to beat the
Girondins. The Duke of Brunswick's declaration that any
action against the royal family would be met with sharp
reprisals only worsened the situation and in August the
Paris mob stormed the Tuileries. Louis had already sought
the Assembly's protection, but he was suspended from his
functions and a National Convention was summoned to

produce a constitution, which would replace the failed monarchical version.

The growing military threat to France accelerated the drift towards extremism. Brunswick crossed the frontier between Sedan and Metz with a large army in mid-July, and plodded steadily forward to take Longwy and Verdun. His approach helped provoke an outburst of popular fury in Paris and in early September over 1,000 prisoners – priests, Swiss guards, political prisoners and ordinary criminals – were massacred. The Marquis of Lafayette, commanding the army facing Brunswick, had fled to the Austrians and his place was taken by Dumouriez. He had previously favoured attacking into the Netherlands, but, recognising the seriousness of Brunswick's threat, he decided to concentrate in 'the Thermopylae of France', the hilly, wooded country of the Argonne through which the road from Verdun to Paris wound its way. He marched down from Sedan, disposing his army in the passages of the Argonne, and when Brunswick turned his northern flank was able to concentrate around St-Menehould, where he was joined by Kellermann, commander of the Army of the Centre, previously based on Metz.

On 20 September the French offered battle. The fight consisted largely of a duel between Kellermann's guns on the high ground around Valmy windmill and Brunswick's more numerous but older pieces. Although some contemporaries, looking at the few casualties inflicted, thought that the cannonade of Valmy was much noise about nothing, the day's message was clear: the Prussian infantry could not risk an assault against such powerful artillery. The weather was atrocious and the Allied army, ravaged by sickness, had little option but to fall back. Brunswick relinquished Verdun and Longwy, and slunk off across the frontier.

Dumouriez returned to Paris a popular hero and, with the support of Danton, now the dominant figure in the government, was given fresh volunteers for his offensive into the Netherlands. On 6 November he beat the Austrians at Jemappes and went on to enter Brussels. In the east, meanwhile, Custine had taken Spires, Worms and Mainz, and went on to seize Frankfurt. These victories encouraged the Convention to make far-reaching decrees offering support to all peoples who wished to recover their liberty. Within France, the monarchy was abolished and the dawn of the new era was signalled by the dating of decrees from Year 1 of the Republic. In an atmosphere of sharp political conflict the Convention voted for the King's death and on 21 January 1793 Louis was guillotined in what is now the Place de la Concorde.

In bellicose confidence the Convention widened its aims, annexing Savoy, declaring the Sheldt open to all shipping, and going on to declare war on Britain, Holland and Spain. Pride went before a fall. The French failed before Maastricht, and in March Dumouriez was beaten at Neerwinden and evacuated Brussels. He then embarked upon a devious plan to march on Paris and restore a monarchical constitution, but the scheme collapsed and he fled to the Austrians on 5 April. Custine, too, had failed to retain the gains of 1792 and the Convention's perils were increased by the outbreak of a serious revolt in the Vendée, provoked largely by attempts to enforce conscription. The economic strains of the war generated inflation and with it a new food crisis. This, coupled with the shock of defeat, helped inspire draconian legislation. Representatives on mission carried arbitrary authority into the provinces and the armies, and 'rebels' taken in the act were to be summarily executed.

The new Committee of Public Safety floundered in a sea of troubles. Valenciennes fell, the port of Toulon was occupied by the English and revolts multiplied. The Revolutionary Tribunal redoubled its efforts, and the guillotine claimed aristocrat and revolutionary, defeated general and failed moderate. The army, swollen by a *levée en masse*, which sought to propel all young men into its ranks, began to win again. In October Jourdan attacked Coburg's Allied force besieging Maubeuge, defeated it at Wattignies and raised the siege. Toulon was recaptured, with the active assistance of Captain di Buonaparte, who was promoted to brigadier-general for his pains and soon took to spelling his name in the French style as Napoleon Bonaparte. Lazare

Hoche was victorious in Alsace, Kellermann cleared Savoy and the Vendeans were beaten – though not permanently pacified – in a welter of atrocity and counter-atrocity. In 1794, with a fresh outburst of guillotining in Paris, Jourdan went on to beat Coburg at Fleurus and take Liège and Antwerp, bringing the whole of the Austrian Netherlands under French domination. The execution of Robespierre in July signalled the end of the worst excesses and in May 1795 the army helped the moderates to dismantle the structure that had made the Reign of Terror possible. The Committee of Public Safety was placed firmly under the Convention's control and provincial revolutionary committees were abolished.

The reaction of 1794–5 left the constitutional question unsolved. Plans for a revival of the monarchy were thwarted by the death of the young Louis XVII, and the declaration by the émigré Count of Provence – *de jure* Louis XVIII – that restoration would be accompanied by the root and branch re-establishment of the old regime. A new constitution established the Directory – a five-man executive – and a new Legislative Assembly, heavily based on the Convention. When the mob marched on the Tuileries, seat of the Convention, on 5 October 1795, it was met by Brigadier-General Bonaparte – temporarily struck off the active list because of his association with the fallen members of the Committee of Public Safety. Bonaparte was summoned to the Convention's support and sited his guns to command the streets leading to the Tuileries. The insurgents were met with 'a whiff of grapeshot'. The episode demonstrated that the mob no longer held the balance in the capital and accelerated Napoleon's rise: he was rapidly promoted general of division and appointed commander-in-chief of the Army of the Interior. Before long he departed to be commander-in-chief of the Army of Italy and in 1796–7 he set the seal on his reputation in a stunning campaign, which brought the Austrians to the peace table. They recognised French annexation of the Austrian Netherlands and the left bank of the Rhine, and agreed to the establishment of the Cisalpine Republic in northern Italy.

While Napoleon was in Italy the army again moved into the political arena, helping Barras and two other Directors to remove their colleagues and the royalists in the legislature. The Directory continued the war against Britain, but the Royal Navy's command of the sea made invasion impossible. In 1798 Napoleon took an expeditionary force to Egypt, only to have his fleet destroyed by Nelson in Aboukir Bay. In his absence the Second Coalition united Turkey – alarmed by French irruption into the Middle East – with Britain, Russia and Austria. News of Austrian successes encouraged Napoleon to return to France, leaving his army behind under the command of Kléber.

Napoleon returned to discover that defective Allied strategy had enabled Masséna to beat the Russians in Switzerland, while defeat in Holland was soon to persuade the British to withdraw their expeditionary force. Only in northern Italy, where Joubert was defeated and killed at Novi, did the war go badly for the French. Joubert's death deprived Sièyes, one of the Directors, of a charismatic military backer for his attempt to stage a coup in order to forestall an anticipated Jacobin attempt. Despite personal antipathy towards Sièyes, Napoleon spent late October and early November rallying prominent soldiers to his cause, and on 9–10 November, in the *coup d'état de Brumaire*, a three-man Consulate replaced the Directory and the Council of Five Hundred was expelled by troops. The Consulate was speedily modified: Sièyes and Roger-Ducos departed, and although two new consuls were appointed, real power lay in the hands of the First Consul – Napoleon himself.

While he consolidated his position at home, Napoleon strove to bring the war to a victorious conclusion. He led an expedition into northern Italy, winning an untidy victory at Marengo, and Moreau occupied Munich, beat the Austrians at Hohenlinden and approached Vienna. The Treaty of Lunéville, signed in February 1801, confirmed French possession of the left bank of the Rhine and recognised the French client states – the Batavian Republic in Holland, the Cisalpine and Ligurian Republics in northern Italy, and the Helvetic Republic in Switzerland. It was not until March 1802 that, isolated diplomatically, Britain signed the Treaty of Amiens.

These successes enabled Napoleon to tighten his grip on France. Plotters were executed or exiled and, in an act of judicial murder, the young Duke of Enghien, last of the Condés, who had commanded an émigré detachment in 1796–9, was seized in Baden, tried and shot in the fortress-ditch at Vincennes. Napoleon was already Consul for life and in 1804 the Senate proclaimed him Emperor, a decision confirmed by plebiscite. He used the brief period of peace following the Treaty of Amiens to create the machinery of government. A Senate, Legislative Body and short-lived Tribunate provided the shadow of representative institutions, while the substance of power centred upon the Council of State and the reorganised ministries. Authority within regional *départements* was exercised by centrally appointed prefects, heirs of the royal *intendants*. There was a new legal system, the *Code Napoléon*; the nation's financial and educational systems were transformed; and the Concordat achieved compromise between the papacy and state control of religion.

Renewal of war could not be long delayed. French diplomatic activity in the Near East and the virtual assimilation of her Continental satellites into France alarmed her former enemies, and war broke out again in May 1803. The contest between France and Britain, which was to last until 1814, was in essence that between an elephant and a whale: Britain alone could not inflict a decisive defeat on France by land, but her naval mastery prevented French military power from striking at Britain herself. Such was the lesson of 1804–5. Napoleon concentrated his army in a huge camp at Boulogne and gathered barges ready for the invasion of England. But the destruction of the French and Spanish fleets at Trafalgar confirmed British control of the sea, and made the invasion impossible. Napoleon then struck inland and, in a masterly campaign, defeated Austria and Russia, continental members of the Third Coalition, at Austerlitz, forcing Austria out of the war.

Prussia, alarmed at the spread of French power, entered
the conflict, only to have her army shattered at Jena in
November 1806. Of Britain's allies Russia alone remained in
the field, but her defeat in the hard-fought battles of Eylau
and Friedland persuaded her to make peace at Tilsit in June
1807. The treaty left France dominant in Europe, her
frontiers pushed out to include not only the Rhine but also
northern Italy, her imperial family established on the
thrones of Holland, Westphalia and Naples, and with a
constellation of satellite states running to the borders of
Russia and Austria.

This unprecedented extension of French authority was a
tribute to the military genius of Napoleon and the temper of
the weapon he had forged. Building on foundations laid by
Lazare Carnot, 'organiser of victory' under the Republic, and
making full use of the experience and enthusiasm generated
in the wars of the First and Second Coalitions, Napoleon
created a *Grande Armée* recruited from French volunteers
and conscripts, and, increasingly, from foreigners, who
formed fully half of it by 1812. The Imperial Guard was the
new army's *corps d'élite*, though it grew from a small body
of veterans in 1804 to a larger and less selective formation by
1814. Regiments replaced the Revolutionary *demi-brigades*
in 1803, and the *Grande Armée*'s regiments were grouped
into brigades (two or more regiments) and divisions (two or
more brigades). Divisions were combined into the *corps
d'armée*, an all-arms formation, commanded by a marshal or
senior general.

The tactics of the *Grande Armée* were firmly rooted in the
eighteenth century. For the infantry, Napoleon favoured
l'ordre mixte, with some battalions in line and others in
column so as to produce both fire and shock, with a cloud of
skirmishers screening the unit's front and preventing the
enemy from concentrating his fire. Cavalry, which
improved rapidly in the early years of the Empire, only to
diminish in quality owing to heavy losses of horses in and
after 1812, had numerous tasks, from the massed charge on
the battlefield to scouting and outpost work. It was the
artillery, the Emperor's own arm of service, that typified his

approach to tactics. He favoured massing his guns as close to the enemy as possible, and capitalising on their fire effect with infantry assault and cavalry exploitation.

If there was little in his approach to war that was definitively new, the way that Napoleon buckled his tactics together to form cohesive operational plans was indeed remarkable. Impelled by the Emperor's directing brain from his *Maison Militaire*, nerve centre of Imperial headquarters, corps moved on separate routes but within supporting distance, able to react flexibly when the enemy was discovered. Napoleon favoured the ploy of *manoeuvre sur les derrières*, pinning the enemy to his position by a feint attack, then marching, by a screened route, to his flank or rear. There, ideally behind the 'strategic curtain' of river or mountain range, Napoleon would force his opponent to fight or surrender. When faced, as he often was, by a number of enemy armies, Napoleon usually attempted to bring his concentrated strength to bear on his opponents' divided forces, defeating them in detail. He had a sharp eye for the ground and was merciless in unravelling weaknesses in enemy deployment or 'hinges' where Allied armies joined. Speed, concentration and surprise were the essence of Napoleon's art of war: when he neglected these principles he won costly victories or suffered defeat.

After Tilsit, still grappling with an implacable Britain, Napoleon applied the Continental System in an attempt to exclude British trade from the Continent in order to undermine the British economy, which not only maintained the Royal Navy but also subsidised France's enemies in Europe. However, the British answer to the Continental System was a crippling blockade of French ports and Napoleon's desire to tighten the system's application was one of the motives behind an aggressive foreign policy, which caused more problems than it solved. Worse still, the economic crisis of 1811–13, which swept the whole of Europe, was widely blamed on the Continental System and it encouraged a flood of smuggling the French government was incapable of stemming.

Intervention in Spain set Napoleon on the downward

slope. Spain was already a French satellite when, in 1808, Napoleon put his brother Joseph on its throne, and tried to keep him there in the face of popular insurrection and a British army based in Portugal. The 'Spanish ulcer' was a steady drain on the French army and was all the more enervating because war broke out again in central Europe. Although the Austrians were beaten at Wagram in 1809 and forced to make a peace reinforced by the marriage of Princess Marie-Louise to Napoleon (who conveniently divorced his childless wife Josephine), Russia responded to the economic crisis by imposing duties on French imports and opening its ports to neutral shipping. Napoleon prepared the *Grande Armée* for a campaign in Russia and launched it on 24 June 1812. He won the bitterly contested battle of Borodino and entered Moscow, but the Russians declined to negotiate and the *Grande Armée* lurched back across western Russia in worsening weather, suffering appalling casualties in the process. Prussia entered the war, and Napoleon fought drawn battles at Lutzen and Bautzen in 1813. He spurned compromises that might still have kept Austria out of the war and broken the Alliance, and though he beat the Allies at Dresden in August 1813 he was thoroughly defeated at Leipzig, 'The battle of the Nations', in October.

It was a measure of Napoleon's success that until 1814 his battles had been fought on foreign soil, and the fatal avenue had housed only garrisons, depots, armies on the march and convoys of increasingly discouraged conscripts bound for the front. All this changed in early 1814 as the Allies approached the natural frontiers of France. The 1814 campaign lasted barely two months, but shone with flashes of real Napoleonic brilliance, with the Emperor darting across Champagne to strike at each adversary in turn and winning a string of glittering little victories. However, he failed to capitalise on the tactical successes of February 1814 by accepting an Allied offer of peace on the basis of the frontiers of 1791, and when he tried to reopen negotiations on that basis in March the moment had passed. He was still in the field when he heard, on 31 March, that Paris had surrendered.

Napoleon's resolve to continue the war collapsed when he heard that his marshals would no longer follow him and he abdicated, hoping to ensure the succession of his young son. The Allies insisted on complete abdication but granted him sovereignty over the Island of Elba and a substantial pension. Louis XVIII entered Paris, France was deprived of territory acquired since November 1792 and the future of Europe was to be decided by a congress, meeting that September in Vienna.

The debates of the Congress of Vienna were to be sadly interrupted. The Bourbons were maladroit in re-establishing themselves and, despite the sufferings caused by his protracted wars, Napoleon retained an amazing appeal for many Frenchmen. In February 1815 he sailed from Elba, landing near Cannes on 1 March. He won over soldiers sent to arrest him and the army deserted to him in droves. On 19 March Louis departed for the Belgian frontier and Napoleon entered Paris the following day. He had little hope that the great powers would assent to his reappearance and he was perfectly right. On 25 March a formal alliance was concluded and Allied mobilisation began. France responded by putting as many men as possible under arms. It was obvious that she could not match the Allied armies in manpower and that her enemies would strike at a number of points along the frontier. Such indeed was the Allied plan: Wellington and Blücher would attack through Belgium, Schwarzenberg's Austrians would advance on the upper Rhine, and an Austro-Italian force would move into the Riviera. Finally, a Russian army would tramp up into the central Rhine.

Yet time was not on the Allies' side. It would take weeks for the Austrians to arrive and the Russians would be later still. Napoleon considered a cautious solution, raising and training more troops, and meeting the Allied offensive between Seine and Marne, but rejected it in favour of a bolder plan. He would strike hard at the British and Prussians in the north, aiming at the hinge between the two armies so as to force each to fall back down its line of communications. Napoleon succeeded brilliantly in the first phase of his offensive and the Allies were badly wrong-footed. But although he had the best of the earlier encounters, on 18 June Napoleon was decisively defeated at Waterloo. He abdicated again on 22 June and this time was sent, not to the relative comfort of Elba, but to the rocky island of St Helena in the South Atlantic. The eagle was caged for ever.

From Restoration to Third Republic

The restored Louis XVIII, hauled back to Paris, it was unkindly suggested, in the baggage train of the Allies, governed according to a Charter that established a constitutional monarchy with a Chamber of Deputies and a house of peers. There were wide concessions over liberty and equality, religious toleration and land tenure – confiscated church and émigré lands were not restored. Yet there were inevitably tensions between the Ultra-Royalists, led by the Count of Artois, the King's brother, and the liberals, while on the extreme left there were not a few who sought the total overthrow of the monarchy. The Restoration army, with its mixture of returned émigrés and Napoleonic officers, was inevitably a melting pot and an early indication of its political uncertainty came in 1822 when four sergeants in the garrison of La Rochelle tried to subvert their regiment.

The conservative Artois succeeded his childless brother as Charles X in 1824 and his Ultra policies aroused increasing unrest. In 1823 a French expeditionary force had intervened in Spain to restore the Bourbon monarchy, but success led to a split within the right and made the monarch's position no more secure. A successful expedition to Algiers in 1830 similarly produced no wave of support for the government. Indeed, within days of the arrival of news of victory insurrection broke out in Paris and the government, its best troops in North Africa, could do little to quell it. After fighting in the capital in July, a Provisional Government offered the throne to the Duke of Orléans, in whose favour Charles promptly abdicated.

King Louis-Philippe's attempt to set a new style in monarchy was demonstrated by the reappearance of the revolutionary tricolour in place of the white flag and the fleur de lys. The army, too, changed in content and appearance. There was a purge of the officer corps – not least because ambitious sergeants saw that denouncing Ultra officers enhanced their own prospects. The Restoration army had reverted to the old white uniforms and adopted the unhappy compromise of calling infantry regiments 'departmental legions'. The blue infantry jacket and numbered regiments came back in 1820, and in 1830 red trousers – hallmark of the Line till 1915 – were introduced.

The army of the July Monarchy had its baptism of fire in the fatal avenue. The defeat of Napoleon had left Belgium and Holland in a forced union, much to the dissatisfaction of the largely Catholic Belgians, who felt that the relationship was anything but an equal partnership. Within weeks of the July revolt in Paris, rioting broke out in Brussels and Dutch troops were expelled from most of Belgium. After much debate it was decided that Belgium should be an independent kingdom. The possibility of one of Louis-Philippe's sons obtaining its throne was alarming to Britain, which feared an uneven tilt in the balance of power, and Leopold of Saxe-Coburg was eventually elected King. However, it took French intervention to remove the Dutch from their remaining strongholds in Belgium and in 1832 Marshal Gérard's *Armée du Nord* marched north. The expedition produced, in the siege of Antwerp, one of the last sieges of the age of classical fortification.

It was not until 1839 that Belgium and Holland came to terms, with the Dutch at last agreeing to the Twenty-Four Articles which had been drawn up at the London peace conference in 1831. Article VII declared Belgium 'independent and perpetually neutral' with its neutrality guaranteed by the great powers. This 'scrap of paper' was to have awesome consequences in 1914.

Louis-Philippe, the 'Citizen King', governed under a constitution that ended hereditary entry to the upper chamber and extended voting rights. But the 'political nation' remained tiny, and there was friction between the well-to-do bourgeoisie who formed its core and the politically powerless majority: Lyons, in particular, was the scene of serious unrest. The legitimists – adherents of the Bourbons – were also active and there was a minor legitimist flurry in the Vendée in 1832. Finally, although Napoleon's son the Duke of Reichstadt died in 1832, his nephew Louis-Napoleon kept Bonapartism alive, with unsuccessful coups at Strasbourg in 1836 and Boulogne in 1840.

Despite its conquest of Algeria, the July Monarchy's failure to produce a 'glorious' foreign policy, and to live up to the Romanticism that courses through the art and literature of the period, did it as much damage as political failure. Damaging, too, was the regime's refusal to do much for the lot of the growing urban population, which flocked to the towns from a saturated countryside – the *Misérables* of Victor Hugo's novel. Nor did French industry expand at the speed of that in Britain, and in railway construction, too, France lagged behind both Britain and Prussia.

Yet the July Monarchy left its mark. It was less obvious than the massive neoclassical architecture of the Empire, but every bit as pervasive. Romantic medievalism encouraged enthusiastic architects to tear down the genuinely medieval and rebuild spiky pastiche: Notre-Dame-de-Paris and the cathedrals at Amiens and Laon all show their scars. Louis-Philippe's military engineers also left their signature for posterity. In 1841–5, largely as a result of the initiative of Adolphe Thiers, President of the Council and Foreign Minister, a wall was built round Paris, enclosing outlying villages, which had in effect become suburbs with the growth of the capital's population. Thirty-three feet high, it was defended by ninety-four bastions and protected by a moat. Fifteen detached forts, like Issy, Montrouge and Charenton, were built far enough away to keep a besieger out of artillery range of the city itself.

Paris pulled down Louis-Philippe just as it had Charles X. In February 1848 riots directed at an unpopular minister got out of hand and there was no serious attempt to use the army to put them down. The King abdicated in favour of his grandson, the Count of Paris, but a Republican Provisional Government was proclaimed at the Hôtel de Ville in Paris. Elections produced a generally conservative Chamber and in the 'June Days' the Paris mob rose in the fury of desperation against a government that seemed to be doing nothing to alleviate its misery. The street fighting that followed burnt itself deep into French national consciousness. The insurgents threw up barricades in the narrow, winding streets of central Paris, and the army and National Guard suffered heavily as they subdued the capital.

After much debate it was decided to choose a president by universal suffrage and the ensuing election, in December 1848, was won by Louis-Napoleon. The new President had escaped from the fortress of Ham, where he had been imprisoned after the failure of the Boulogne adventure in 1840, and he was seen by many Orleanist politicians as a mediocrity whom they could safely manage. He soon proved them wrong. Although he appointed a conservative ministry and outlawed the extreme republicans, Louis-Napoleon had no intention of securing the longevity of the Second Republic and in December 1851 he staged a coup, with lavish military support, establishing what was to become, in late 1852, the Second Empire. Its constitution was unashamedly monarchical: the Emperor nominated all officials and the Legislative Body, elected by universal suffrage, was so much a creature of government patronage that it reflected official views.

The gaslight Empire of Napoleon III has had a worse press than it deserves. It ushered in a wave of economic development, most notably a surge of railway building, which in turn stimulated iron and steel production. A programme of public works helped shape the great cities of France. Its effects are most marked in Paris, where Georges Haussmann, appointed Prefect of the Seine in 1853, demolished street and tenement alike to make way for the wide boulevards and imposing vistas that still give Paris much of its distinctive character. True, it was not merely cultural enthusiasm that inspired the government to back Haussmann: the Second Empire lived in fear of urban unrest and long, straight streets helped give the government's troops the edge over insurgents.

Napoleon III announced that 'The Empire is Peace', but his foreign policy revealed the need to give Frenchmen the glory his uncle had sought. In 1853 French support of Roman Catholic interests in the Near East brought involvement in the Crimean War. The French army's achievement in that conflict has received scant justice at the hands of English-speaking historians. The experience of Algeria meant that the French were generally better at living in inhospitable surroundings than the British and for most of the war the French were the senior partners in the alliance. As the army of the Second Empire lost its way in the 1860s, some French officers were to look back with affection to the Crimean army. Based upon the conscription law of 1832, which produced an army of long-service conscripts with excellent *esprit de corps*, it was officered by a mixture of com-missioned rankers and graduates of the military academies – St-Cyr for the infantry and cavalry, and the Ecole Polytechnique in Paris for the gunners and sappers.

If France's involvement in the Crimea was something of an accident, her excursion to Italy in 1859 was altogether more deliberate. Napoleon had been involved with the Italian revolutionary society, the *Carbonari*, in his youth and genuinely hoped to do something for Italy. In 1858 he met the Piedmontese premier Cavour at Plombières and agreed to support Piedmont against Austria, then in occu-pation of much of northern Italy. The following year the Austrians obligingly furnished a *casus belli* and a French army, under the Emperor's personal command, advanced into Italy and beat the Austrians in two soldiers' battles, Magenta and Solferino. After the latter Napoleon hastily concluded the Peace of Villafranca, stopping well short of his war aims but making what was nonetheless the crucial move towards Italian unification. As a price of her assistance France received Nice and Savoy, and Napoleon emerged with gains from a war that reflected little credit on his diplomatic skill and far less upon his military ability.

Napoleon found it impossible to pursue a consistent

colonial policy, but during his reign France established herself firmly in West Africa and Indo-China. In 1862 she sent an army to Mexico as part of an international force, though it stayed on after other participants withdrew. The Mexican adventure deserves better than to be the butt of film-makers who delight in setting bearded and confused French soldiers against US cavalry or Texan freebooters. France subdued most of Mexico and established an Austrian archduke on its throne as the Emperor Maximilian I. But the adventure ended in tears. While much of the French army was campaigning in the New World, the balance of the Old had changed with Prussian victory over Austria in 1866. Moreover, the victorious Union had no intention of tolerating European involvement in Mexico, and the last French troops were withdrawn in 1867. The unlucky Maximilian stayed on, but was captured by his former subjects and farce turned to tragedy as a firing party delivered its ill-aimed volley on a sunny hillside.

Nemesis was approaching Napoleon himself. In the 1860s he felt secure enough to permit the growth of something approaching, albeit from a distance, parliamentary democracy. The early death of the capable Duke of Morny, the Emperor's half-brother, set the process back and the Liberal Empire limped uneasily on to the stage in 1868. So too did a new military service law, prompted by the Emperor's alarm at the effectiveness of Prussia's conscript army in 1866. It failed to make military service genuinely universal and although it would, in theory, increase the size of the regular army and create a *Garde Nationale Mobile*, it was hedged about by the hesitations and compromises that marked the Empire in its twilight, and in any event had little time in which to take effect.

The ostensible cause of war with Germany in 1870 was the Hohenzollern candidature – the offer of the vacant throne of Spain to a German prince. In fact the episode testified to the diplomatic skill of the Minister-President of Prussia, Otto von Bismarck, who saw war with France as the final step towards the unification of Germany, and to the maladroit handling of the crisis by the French government.

Although the French army had been substantially reformed since 1866, and had received both a new breech-loading rifle, the *Chassepot*, and a primitive machine-gun, the *Mitrailleuse*, it was certainly not the nation in arms in the Prussian sense. Its officers, though personally brave and often well-seasoned in hack-and-gallop affairs in Algeria or Mexico, were not prepared for modern European war. Its staff was little more than a collection of commissioned secretaries, for Napoleon III had made too much of the old Imperial connection to create a command structure that could function without the Emperor's guiding hand. And in 1870 Napoleon was tired and ill, reluctant to relinquish command to his marshals but well aware that the stakes were too high for him.

A chaotic mobilisation and concentration saw the French shuffle into position along the frontier. The Germans formed three armies, operating under the firm hand of Helmuth von Moltke, Chief of the Prussian General Staff, and though the French managed a half-hearted excursion into Germany, taking the heights above Saarbrücken on 2 August, on the 4th the Germans mauled an isolated French division at Wissembourg, and on the 6th they defeated the northern group of French corps at Spicheren and the southern group at Froeschwiller. Marshal MacMahon got the remnants of the southern group back to Châlons, where they formed part of the extemporised Army of Châlons. The northern group, eventually under the command of Marshal Bazaine, retreated through Metz towards Verdun, but on 16 August was checked at the Battle of Rezonville–Mars-la-Tour, and two days later fought the battle of Gravelotte–St-Privat before returning to stand siege in Metz. The Army of Châlons, accompanied by the Emperor, staggered forward in an attempt to relieve Metz, but was elbowed up against the Belgian border and destroyed at Sedan on 1 September. The Empress Eugènie, left in Paris as regent, fled to England with the Prince Imperial and a Government of National Defence was proclaimed, headed by General Trochu, military governor of the capital.

Thanks largely to the energy of Léon Gambetta, Minister

of War and the Interior, and his deputy at the ministry of war Charles de Freycinet, France managed to conjure up armies that kept in the field through a bad autumn and worse winter. The old Imperial armies melted away. Strasbourg, bombarded and besieged, capitulated in late September. A month later Bazaine surrendered at Metz after attempts to break out, which were so spiritless as to encourage his countrymen to court-martial him after the war. The Armies of National Defence, a heterogeneous collection of surviving regulars, *mobiles*, marines, sailors, *francs-tireurs* and foreign volunteers, were to subject the energies of the German armies and the skill of their commanders to no mean test.

In the west, the Army of the Loire made brave but fruitless efforts to relieve Paris. To the east, in the Saône and Doubs valleys, a mixture of newly raised troops and *francs-tireurs* struggled on, while in the north a growing army was based in Lille. Given time to organise and train, these forces might have achieved some lasting success against German forces at the end of long and vulnerable lines of communication: but as it was, the pressing need to relieve Paris led to hasty action.

The Army of the Loire won a crisp little victory at Coulmiers near Orléans on 9 November, but failed in its assault on Beaune-la-Rolande later that month and was badly beaten at Loigny on 2 December. This defeat coincided with the failure of a large-scale sortie from Paris, which had seized a mere toehold in German lines at Champigny, and the Germans, themselves prone to political bickering as final victory seemed so elusive, began to bombard the city on 5 January. The forts built in the 1840s to guard against just such an eventuality were now too close to protect Paris from modern rifled artillery. The bombardment did remarkably little damage, but it goaded the garrison into another hopeless sortie against the German lines in front of Mont Valérien.

This sortie coincided with bad news from the provinces. The Army of the Loire had been split in two after its loss of Orléans, and one of its portions staggered back to Le Mans and the other to Bourges. The Army of the North, in the steady hands of Faidherbe, briefly shone the light of success through the gloom of defeat, capturing Ham and defeating a German counter-thrust, in a grim foreshadowing of things a generation ahead, on the chalk slopes near Corbie on the Somme. Faidherbe judged that his army would not stand another day's pounding – or another night on freezing hillsides – and retired on Arras. Early in January he felt obliged to move to the relief of Péronne, and though he had the best of the fighting around Bapaume on the 3rd, he was ever-sensitive to his army's shaky condition and ignorant of the damage he had done the Germans, and did not press the matter to a conclusion. The Germans kept Bapaume and Péronne at last capitulated.

The rump of the Army of the Loire was in even worse straits: beaten at Le Mans on 11 January, its remnants shuffled off westwards, held together only by the indomitable personality of its commander, Chanzy, who deserves better of his country than many of the better-known commanders whose names litter these pages.

In the north, Faidherbe had at last lost his winner's touch. Urged on by Freycinet, he mounted a thrust towards St-Quentin in an effort to take the strain off the capital and help its garrison to mount a sortie. His move intercepted, Faidherbe stood his ground at St-Quentin, where he was decisively beaten by General von Goeben: the residue of the Army of the North spent the rest of the war ensconced in fortresses where the Germans were perfectly content to leave it. The fighting in the east showed the brief glint of false dawn before the fog of disaster and in mid-January the French were defeated on the Lisaine near Belfort; the shreds of the Army of the East made their way to internment in Switzerland.

An armistice was signed on 28 January and peace terms were agreed after painful negotiations. For France they were nothing less than shattering and rank not least amongst the causes of the First World War. France lost most of Alsace and Lorraine, though Belfort, which had held out until the armistice, was to remain French. The Germans were allowed a triumphal entry into Paris and were to receive a war indemnity of two milliard francs. German departure from Paris in early March was followed by an outburst which left its own livid scar on French history. The Bordeaux-based National Assembly, which had agreed the peace terms, moved to Versailles, with the veteran politician Thiers as head of its Executive. An attempt to seize cannon from the Paris National Guard sparked off riots and led to the withdrawal of legal authorities from the capital. A new municipal government, with the time-honoured revolutionary title of Commune, was elected and the Versailles army laid siege to the city. It forced an entry on 21 May and the ferocious fighting that followed saw a return to the old days of the barricades. Both sides committed atrocities, but the executions that followed the victory of the Versaillais made the wall of the cemetery of Père-Lachaise, where some of the shootings occurred, a place of pilgrimage for the Left.

The end of the war and the suppression the Commune left France with the familiar task of constitution-making. The Assembly contained a monarchist majority, but disputes between Legitimists and Orleanists, and the arrival of more republicans at successive by-elections, helped ensure the survival of a monarchist republic, dubbed, because of the three dukes in its cabinet, the Republic of Dukes. Marshal MacMahon became head of state in 1873, and two years later the Third Republic slid into existence with a flurry of constitutional laws setting up a bicameral legislature and a presidency with limited but significant powers. The Assembly was duly dissolved and elections produced a largely republican chamber, which MacMahon promptly dissolved. For all his attempts to use traditional electoral machinery to swing the 1877 election towards the right, the result was a marked victory for the Left and the reduction of the presidency to an almost entirely honorific appointment: the Third Republic was republican in form as well as name.

The war had also had a marked effect upon Germany. On 18 January 1871 William I of Prussia had been proclaimed Emperor of Germany in the Hall of Mirrors at Versailles, and although he was less than enthusiastic about the ceremony, it emphasised Prussian hegemony in a united Germany. This was mirrored in the post-war organisation of the German army: although the national components of the German army retained some individualistic trappings, the Great General Staff reigned supreme. The status of the military in German society had been enhanced by victory in 1866 and 1870–1, and the Wilhelmine era was marked by the increased impact of the army on foreign policy, domestic affairs and social attitudes. Kaiser William II tended to take the military line on many key issues, and his failure to apply a brake to the blinkered logic of his military planners helped accelerate Europe to the catastrophe of 1914.

From Longchamps to the Marne

On 29 July 1871, only two months after the Treaty of Frankfurt had formally ended the Franco-Prussian War, MacMahon had led 120,000 men past Thiers on the plain of Longchamps. The huge crowd applauded wildly, and the parade was symbolic of a union of army and nation, which would last for the next quarter-century. Left and Right alike agreed on universal conscription, a general staff was established and there was a veritable passion for military studies.

Plans for war with Germany were the abiding military preoccupation and were at first defensive. Following proposals by General Seré de Rivière, work began on new constructions whose turrets and ditches were scarcely less significant than the redans and ravelins of Vauban. A number of 'entrenched camps' were planned. Each consisted of a town with a ring of outlying forts, designed to act as a base for a field army. On the German border, forward entrenched camps were built at Verdun, Toul, Epinal and Belfort, with camps at Langres, Dijon and Besançon to give depth to the defence. It was hoped to cover the Belgian border with camps at Valenciennes and Maubeuge with a defensive curtain between them, but in the event the work at Valenciennes was left unfinished and it was replaced as an entrenched camp by Lille.

Defensive curtains, their forts five to ten kilometres apart, were used to link entrenched camps and isolated forts were built to command key routes. The line between Toul and Verdun ran along the Meuse Heights while the Epinal–Belfort line strengthened the already difficult country of the Vosges. A gap – the *Trouée de Charmes* – was deliberately left between Toul and Epinal, to encourage an attacker to put his head into the guillotine formed by the two entrenched camps.

The exits from the Ardennes were covered by works at Hirson, Charlemont, Les Ayvelles, Montmédy and Longwy. Only the forts at Hirson and Les Ayvelles were new: the others were old works with some improvements. A curtain ran from Maubeuge, through Valenciennes, to Lille and a thin continuation to the sea at Dunkirk. Some work was done on a second line running from Péronne, through La Fère and Laon, to Rheims, but was unfinished. The Swiss, Italian and Spanish borders were not neglected, although in most cases old forts simply required modernisation. Expense prevented the construction of a new ring of forts round Paris, but three segments of defensive curtain and three detached forts were constructed.

We shall have the opportunity of studying some of Seré de Rivière's work in more detail later on, but the general principles behind it merit early mention. Originally a Seré de Rivière fort was a fortified gun position, its guns – sixty for a large fort – mounted on a flat *terreplein* atop the barracks or around the fort's perimeter, firing over a parapet. Forts were polygonal, surrounded by a deep, dry ditch with its own defences, and contained barracks for up to 1,200 men. They were built of stone or brick, with earth-covered roofs, but the improvement of explosives in the 1880s led to the addition of a thick skin of concrete with a 'burster layer' of sand between it and the stonework. Some guns had been put into armoured cupolas even before the new explosives arrived on the scene, and the improvements in artillery around the turn of the century led to the removal of *terreplein*-mounted pieces and their replacement by fewer turreted guns. A remarkable total of 166 forts, forty-three *ouvrages*, or small works, and about 250 batteries was built between 1873 and 1885, and only about one-third were modernised before the outbreak of the First World War. Most modernisation was carried out on forts in the Belfort, Toul and Verdun groups, leaving the Belgian border more thinly defended than had originally been the case.

The Germans, too, were building forts – on what had once been French territory. During the Second Empire forts had crowned the high ground just west of Metz, with the familiar object of keeping an attacker out of bombardment range of the main body of the fortress. These were not fully completed when war broke out in 1870, and after the war the Germans modernised them and added new works, creating a formidable fortress system known as the *Moselstellung*. The fact that German engineers tucked so much steel and concrete into the steep slopes above the Moselle reflected a marked change in German military policy. In 1892 France and Russia had signed a military accord, presenting the Germans with the dreadful spectre of a war on two fronts. The response of Count Alfred von Schlieffen, who became Chief of the Great General Staff in 1891, was to develop a war plan which envisaged winning a swift and complete victory over France in the early stages of a future war: this could be followed by more leisurely operations against the Russians.

The Schlieffen Plan – a rough piece of historical shorthand for a lengthy series of memoranda – was audacious. The French fortress system – at its best along the Franco-German border – compelled Schlieffen to look to the north, where he could slam his mighty army into the pit of the French stomach, albeit at the cost of violating Belgian neutrality. Schlieffen's concept of a wheeling attack through Belgium and northern France, pivoting on the *Moselstellung*, taking Paris within its sweep and swinging round to encircle the French armies somewhere in Champagne was, as its author recognised, a hazardous venture. But the work of the diplomats who negotiated the Franco-Russian entente, no less than the efforts of engineers who made the Franco-German border such a forbidding obstacle, pushed the

Germans into a solution which offered at least the illusion of quick victory and the avoidance of a head-on assault into Seré de Rivière's handiwork.

While Schlieffen's projects evolved, the French army, too, was changing. The 'Golden Age' that followed the Franco-Prussian War, when the nobility had flocked back into the officer corps, conscription had been widely accepted and armaments were comprehensively modernised, ended in tension and scandal. The use of soldiers against strikes and riots revived the Left's old suspicions of the army. The conviction of a Jewish officer, Captain Alfred Dreyfus, on charges of treason, caused a lengthy and damaging political dispute. Slow promotion and low pay provoked frequent resignations: in 1904 another scandal broke when it transpired that the War Minister had been assisted by Masonic officials in maintaining records on officers in an effort to make the officer corps genuinely republican.

From 1904 the General Staff grew increasingly concerned at the prospect of war with Germany, and the fact that the demographic balance favoured the Germans encouraged discussions with Russia and Britain alike. There were difficulties in both cases: the Russian army was recovering from defeat in the Russo-Japanese War and although Britain's newly formed General Staff cheerfully participated in military conversations, it was emphasised that these were not politically binding. Although France had developed some first-rate equipment, she was perilously short of heavy guns. The famous 75mm quick-firer field gun was regarded as 'God the Father, God the Son, and God the Holy Ghost'. One cynic remarked that it would have been nice to see it surrounded by a few saints of heavier metal.

There was some reason for the lightness of French artillery. A combination of popular philosophy and a military theory, which reacted against the passiveness of 1870–1 by emphasising the importance of 'a conquering state of mind', led to the emergence of the doctrine of the offensive. This was not as rash in principle as it was to be in practice: after all, attacking robbed the enemy of initiative and offered the possibility of recovering the lost provinces.

And although the French had some evidence of Schlieffen's projects, they failed to recognise that the Germans would make their right wing so enormously strong and argued that if their enemies did extend their flank into Flanders, it could only be at the expense of their centre – target for the French thrust. The final French war plan, Plan XVII, consisted of an offensive by four French armies, attacking in two groups on either side of Metz. It would have been hard to design a scheme that played more completely into the German hands: Schlieffen had described this sort of attack as 'a kindly favour', since it would inevitably denude the northern flank, where the full weight of the German blow was to fall.

The First World War

There had been so many potentially serious crises in the years leading up to 1914 that the assassination of the Archduke Franz Ferdinand, heir to the throne of Austria-Hungary, did not seem more alarming than the rest. But Austrian pressure on Serbia – believed to be behind the assassination – was answered by Russian mobilisation. France could not assure Germany that she would remain neutral in a Russo-German conflict, and Germany duly declared war on Russia on 1 August and France on 3 August. Mobilisation transfigured Europe. Reservists reported for duty, often amid scenes of wild enthusiasm, and rattled off to teeming concentration areas in a meticulously organised railway system.

Britain's position was ambivalent. Although the military conversations with the French General Staff had produced plans to send an Expeditionary Force to northern France, they had never been politically binding and Asquith's Liberal government hesitated. But Belgium had announced that she would defend her neutrality and Germany duly declared war on her. Britain was a guarantor of the Belgian neutrality and German violation gave Britain just cause – or reasonable pretext – for war.

Although Belgium's army was tiny, it was to inflict

serious delay on the Germans. Moltke had rejected the possibility of invading Holland, which meant that much of the German right wing had to pass through the narrow gap between the 'Maastricht appendix' of Dutch territory and the hilly Ardennes south of Liège. The Belgians had fortified the Liège gap with underground forts, their guns mounted in armoured cupolas. An attempt to rush them failed and it was not until heavy siege howitzers were brought up that the forts were battered into submission; the last fell on 16 August. The Germans swept on through Belgium, their tired and nervous troops sometimes committing atrocities that Allied propaganda lost no time in inflating.

The French Commander-in-Chief, the laconic General Joseph Joffre, had established *Grand Quartier Général* (GQG) at Vitry-le-François on the Marne, roughly equidistant between the headquarters of his armies. He initially paid little attention to reports of the German advance in Belgium and duly launched the planned offensive into the lost provinces. It went well enough at first: Mulhouse was briefly recaptured and Sarrebourg fell on 18 August. The German armies in this sector were commanded by Crown Prince Rupprecht of Bavaria, who resented fighting a defensive battle and asked Moltke for permission to counter-attack. Moltke weakly gave way and on 20–24 August the battle raged. It ended with the 1st, 2nd, 3rd and 4th French armies bloodily repulsed in Alsace and Lorraine, while the 5th Army, on Joffre's left, was beaten at Charleroi and fell back south-westwards. Still further to the left was the British Expeditionary Force under Field Marshal Sir John French. This had concentrated around Maubeuge, then conformed to the French plan by advancing into Belgium. On 23 August it clashed with the Germans at Mons, after which it fell back in a long retreat, through Le Cateau, where its II Corps paused to administer a sharp rebuff to its pursuers.

Joffre patched together a new 6th Army to shore up his left flank and dismissed hesitant or incompetent commanders as his whole line jolted backwards. On 1 September he ordered withdrawal to the line Verdun–Bar-le-Duc–Vitry-

le-François–Arcis-sur-Aube and the Seine at Nogent, with
the BEF at Melun. The government left Paris, leaving
General Galliéni, its governor, with instructions to defend
the capital to the last. If Joffre's nerve remained equal to the
strain, Moltke's did not. The German 1st Army, on the outer
flank, edged eastwards and swung in front of Paris, giving
Joffre the opportunity for a counter-attack. The Battle of the
Marne, which began on 6 September, was a confused affair,
but ended with the demoralised Moltke agreeing to a
withdrawal. German troops fell back on to the Aisne, where
the war quickly went to earth. Over the following weeks the
'Race to the Sea' ensued as each side sought to outflank the
other by edging further to the north. The BEF moved from
the Aisne to the area around the little Belgian town of Ypres,
where it fought the First Battle of Ypres, initially attacking,
then defending as the Germans sought to break through to
the Channel ports. The autumn ended with a line running
from the Swiss border to Nieuport on the Channel coast,
signalling the end of open warfare and setting the pattern of
what was to follow for four long years.

    The sheer scale of suffering in what became known as 'The
Great War' was to stun the generation that endured it and to
encourage subsequent commentators to ask if there was any
alternative to the bloodletting on the Western Front. Two
points merit emphasis here. The first is that the status quo in
early 1915 was clearly unacceptable to the French. The
Germans held a huge tract of French territory – including
much of France's coal and steel production. Noyon, where the
tip of the German salient bulged out, was as close to Paris as
Canterbury is to London. Standing on the defensive was,
therefore, not an option that appealed to the French. Belief
that the war could be won elsewhere – the solution
championed by the 'Easterners' – did not commend itself
either to the bulk of French military and political opinion, or
to British commanders in France: French, or his successor
Haig. Secondly, there was no easy solution to the problem
posed by advances in weaponry which had given the
entrenched defender the advantage over the exposed attacker.
Better communications would have helped, but battlefield

radios were not available and even when the defensive crust was broken it was usually impossible for the attacker to exploit success before the defender could seal the gap. Most of the generals of 1914–18 may have few claims to military genius, but we must recognise that they were grappling with difficulties of genuinely staggering proportions.

In March 1915 the British achieved a limited success at Neuve Chapelle, attacking on a narrow front behind the heaviest bombardment yet fired by British gunners. In April a larger attack, delivered by the British at Aubers Ridge and the French further south in Artois, proved a costly failure. That autumn an ambitious plan for a two-pronged assault on both flanks of the German salient, with the British and French attacking in Artois and the French in Champagne, also failed, and the carnage at Loos cost Sir John French his job. The one German offensive that year, the Second Battle of Ypres, in April, was an ugly foretaste of things to come, for it involved the large-scale use of poison gas and might have achieved greater results had the Germans been able to take advantage of initial success.

Both sides produced offensive plans for 1916. At a conference at Chantilly in December 1915 the Allies agreed upon a concerted offensive on the Western, Russian and Italian fronts to prevent the Germans from switching reserves to meet each emergency in turn. The British and French were to mount a major attack astride the River Somme. German strategy was now in the hands of Falkenhayn, who had replaced the exhausted Moltke. He too argued for an offensive, selecting an objective in whose defence 'the forces of France will bleed to death – there can be no question of voluntary withdrawal . . .' Accordingly, on 21 February the Germans launched their assault on Verdun, making excellent progress in the first few days but gradually grinding to a halt in fierce fighting, which imposed the most intense strain on French and Germans alike.

The need to reinforce Verdun compelled the French to reduce their contribution to the Somme battle and to press the British to attack as soon as possible. Although con-scription was introduced in Britain early in 1916, the

Somme was substantially the battle of the New Armies –
troops raised in response to Lord Kitchener's appeal for men
in 1914. The plan was for a heavy bombardment to break the
German defences on Pozières Ridge and for the infantry of
the British 4th Army to follow up. When the assault was
delivered on 1 July it was discovered that the artillery had
done less damage than expected, and the lines of advancing
infantry suffered cruelly from artillery and machine-gun
fire. The French, attacking south of the Somme – albeit,
because of Verdun, on a much smaller scale than originally
planned – did rather better. The battle went on till
November, ending with much of the German defensive
system in British hands, though at a frightful price.

Failure at Verdun finished Falkenhayn, who was replaced
by the formidable combination of Hindenburg and
Ludendorff. Conversely, success made the reputations of
Philippe Pétain, army commander at Verdun during the
most desperate days, and Robert Nivelle, whose methodical
attacks had enabled the French to recover most of the lost
ground. Such was Nivelle's almost magical status – 'I have
the secret,' he announced – that he replaced Joffre as
commander-in-chief of French armies on the Western Front
in December 1916. He immediately planned to replicate the
methods that had brought success at Verdun, but on a much
greater scale. Nivelle's 'vehement attack' in Champagne
would be co-ordinated with a British effort in Artois,
attracting German reserves just before the French blow fell.

The Germans moved first. In March 1917 they gave up the
nose of their salient, pulling back to the Hindenburg Line, a
new fortified position behind the old Somme battlefield,
shortening their line and making nonsense of Nivelle's
scheme. Such was the pressure upon him that he pressed
ahead. The Germans had constructed deep defences on the
Chemin des Dames and had captured a copy of the French
operations order. The French assaulted in appalling weather
and suffered heavy casualties for no real gain. The strain and
disappointment almost broke the French army: mutinies
began in late April and there were serious outbreaks in no
less than fifty-four divisions. Pétain replaced Nivelle as

commander-in-chief and set about restoring morale.

Information on the state of the French army combined with the urgent need to take action against German submarine bases on the Flanders coast to encourage Haig to mount an offensive of his own around Ypres. In June his 2nd Army took Messines Ridge, south of the town, and the following month the 5th Army began its long ascent of calvary, fighting up to reach the village that has earned the battle, properly called Third Ypres, the name of Passchendaele. Haig's men were stopped well short of their final objective, but he was to argue, not without some reason, that the terrible fighting had done lasting damage to the German army and had prevented it from concentrating on the French, giving Pétain the chance to restore its morale and confidence. During the Somme battle the British had made limited use of tracked, armoured vehicles known, in an effort to preserve security, as 'tanks'. These produced a small-scale success and by 1917 many more were available. They were almost helpless in the mud of Flanders, so the British used them on the downland west of Cambrai in November 1917, achieving impressive results early in the battle, but losing most of their gains to a well-staged counter-attack.

German satisfaction at the results of Cambrai and the fact that Russia had effectively been driven out of the war were more than balanced by grimmer news. America had entered the war and American troops were already trickling into France. Now it was the Germans who could not afford to stand on the defensive. With the Americans on their way and the British blockade causing growing dissatisfaction on the Home Front, Germany had to force a conclusion in the west in 1918 or lose the war.

The Germans came close to winning. They developed tactics that combined a lightning barrage with the infiltration attacks of lightly equipped, fast-moving infantry and in March they sent the British 5th Army reeling back across the old Somme battlefield. In April they launched another offensive around Armentières on the Lys and this too made useful gains. Finally, in May they smashed the

French 6th Army on the Chemin des Dames. Although the Ludendorff offensive captured large numbers of men and guns, and overran huge tracts of territory, it was indecisive. The Allies did not disintegrate and the appointment of Ferdinand Foch as Commander-in-Chief of the Allied Armies in France helped co-ordinate their efforts.

Ludendorff found himself with a salient to hold and an exhausted army with which to do so. Allied counter-attacks began with the French 10th Army's advance at Villers-Cotterêts in July. On 8 August – 'the black day of the German army,' as Ludendorff called it – the British attacked at Amiens and by September the Germans were back behind the Hindenburg Line. Although some American troops had helped stem the German tide, General John J. Pershing, commanding the 1,500,000 troops in France, was anxious that his army should be used as a cohesive whole. On 12 September he nipped out the St-Mihiel salient in Lorraine and in September formed the southern prong of the general Allied advance, attacking in the difficult country of the Argonne.

Ludendorff's resolve, already damaged by the failure of his offensive, was further bruised by bad news from other theatres of war and in late September he advised making peace on the basis of the Fourteen Points put forward by the American President Wilson in January. It soon became clear that the removal of Germany's monarchy and military leaders were prerequisites for an armistice. Ludendorff resigned and the Kaiser abdicated, and an armistice duly came into force on 11 November.

The Peace that Failed

On 13 November President Poincaré entered Metz with Clemenceau, the premier whose resolve had done so much to maintain French will, and Pétain, who rebuilt the army after the near collapse of 1917. 'A day of sovereign beauty,' Poincaré wrote. 'Now I can die.' The Treaty of Versailles, which ended the First World War, confirmed what Allied armies had won. Alsace and Lorraine became French again, and further north Germany lost territory around Malmédy and Eupen to Belgium. The German army was restricted to a mere 100,000 men, without tanks, heavy guns or aircraft, and the General Staff was abolished. The Rhineland was demilitarised and France was to administer the coal mines in the Saar for fifteen years to compensate for the mischief done in France by the Germans. Germany accepted guilt for causing and prosecuting the war, and was to pay huge reparations. The peacemakers went much further, redrawing the map of eastern Europe, setting up the new state of Czechoslovakia and re-establishing Poland. The latter caused particular problems, for the 'Polish corridor' separated East Prussia from the remainder of Germany and the German city of Danzig was isolated in Polish territory.

The peacemakers unwittingly helped create the conditions that would permit the rise of extremism in Germany and brought into being states whose security depended on external support. France entered the peace haunted by the past. Not only had she suffered terribly – one-third of young Frenchmen were dead or crippled – but the fighting had left a belt of murdered nature across the north. The war memorials which sprang up in towns and villages emphasised the moral credit that France believed she had earned by shouldering such a burden, but it was depressing to discover that the sacrifice they commemorated counted for little in the new world. Rampant inflation hit peasant and bourgeois alike, and wartime allies seemed reluctant to help force Germany to pay its reparations. When Germany defaulted in 1923 the French and Belgians occupied the Ruhr, and the episode only worsened Germany's already desperate plight. The Wall Street crash of 1929 plunged the entire Western world into a financial crisis, encouraging extremist politics on both sides of the Rhine. The French army found its self-confidence and prestige eroded by the lack of a sense of mission, fall in the value of pay, slow promotion and growth of political divisions, which mirrored the tensions in a divided and disillusioned France.

Security had been France's goal at Versailles and it loomed large in her post-war policy. She had a permanent seat on the Council of the League of Nations and tried, by alliances with Czechoslovakia and Poland, to maintain a network of support within Europe. Militarily, the need for security found expression in a fortress system known – from the wounded veteran who was Minister of War when credits for the work were voted – as the Maginot Line. In the early 1920s military commissions had considered the problem of frontier defence and the relatively small amount of internal damage suffered by the Verdun forts helped shape their decisions. It was decided to build a few field fortifications along the Belgian border, for Belgium, at the time, was a French ally and, as a French minister remarked: 'We cannot build up impregnable defences in the face of our friends the Belgians.' The main effort of the *Commission d'Organisation de Régions Fortifiées* was concentrated on the eastern frontier. A double line of infantry casemates covered the Rhine frontier from Basle to Fort Louis. The Lauter Fortified Region covered the area between Bitche and the Rhine, and the Metz Fortified Region from the Moselle north of Thionville to Longuyon. Work was also carried out on the defences of Alpine passes, and defences blocked the gap between Mount Mounier and the sea.

Although the Maginot Line has gone down as one of the bad jokes of military history, it was not without merits. When the Germans marched into the Rhineland in 1936 and garrisons moved into the Line for the first time some design faults were detected, but many of these were rectified. Belgium's declaration of neutrality in the same year caused a more serious problem, because by then funds for defence works were almost exhausted and the generally low-lying Franco-Belgian border was less well suited to underground forts than the hillier terrain on the Franco-German frontier. Some defences were constructed along the Belgian border, but these were on nothing like the same scale as the Line proper.

Perhaps more dangerous than the Line's physical limitations were its psychological and diplomatic consequences. Frenchmen were encouraged to believe that it would prevent German aggression and thus reduce the chance of war. Its cost helped inhibit production of other war material and when the left-wing Popular Front came to power in 1936, with the avowed aim of improving France's military preparedness, it was clear that arms production would fail to match that of German industry. Finally, France looked towards Eastern Europe to counterbalance the disparity between French and German manpower, but it was impossible to link fixed defences with distant alliances.

In September 1938 France's premier Daladier and the British Prime Minister Chamberlain met Hitler at Munich and agreed to German occupation of part of Czechoslovakia. It was at best an exercise in buying time, for when Hitler invaded Poland a year later France and Britain had no alternative but to fight. The Line was put on a war footing and the interval troops deployed. However, the large-scale mobilisation of reservists emptied armaments factories and many men had to be released to civilian life.

The strength of the Line was no secret to the German General Staff, which had already made tentative enquiries to the armaments manufacturer Krupp for guns capable of dealing with the forts. However, the rapid collapse of Poland and the ensuing redeployment of the German army to the west caught her planners flat-footed.

Success in Poland had hinged upon fast-moving armour, intimately supported by ground-attack aircraft. Karl-Heinz Freiser's important work *The Blitzkrieg Legend* suggests that the tactics soon christened *blitzkrieg*, lightning war, were actually less cohesive than was once suggested, and that German successes at the beginning of the Second World War owed much to preparedness to react to fluid situations and accept as inevitable the chaos of battle. The initial German plan for the invasion of France and the Low Countries, *Case Yellow*, looked very similar to the Schlieffen Plan, prescribing a wheeling attack through Belgium and down into northern France. As the Allies planned to move forward into Belgium – taking up a line on the River Dyle, covering Brussels – if the Germans violated its neutrality, *Case Yellow* would probably have produced an indecisive clash somewhere in central Belgium. It was a clash which never came. A number of factors – amongst them the compromising of *Case Yellow* when a German courier accidentally landed in Belgian territory – persuaded Hitler to back a new plan, in essence the work of Erich von Manstein, one of the sharpest brains in the German army. The Manstein plan left Army Group B, with three panzer and twenty-six other divisions, to attack in Holland and northern Belgium, while Army Group C simply masked the Maginot Line. In between, Army Group A, with forty-five divisions, seven of them panzer, was to attack through the Ardennes, crossing the Meuse between Sedan and Dinant, and driving hard for the coast. The Allied plan remained unchanged. In the north, the British Expeditionary Force and the 1st, 7th and 9th French Armies were to swing up into Belgium, pivoting on Sedan. This crucial area was held by the French 2nd Army, while two whole army groups – some fifty divisions – garrisoned the Maginot Line.

The offensive began on 10 May and achieved almost total surprise. The Luftwaffe hammered Allied airfields and the striking force of Army Group A entered the Ardennes. The Allied left wing, obedient to Plan D, moved forward into Belgium, fixing the attention of General Gamelin, Allied Commander-in-Chief. On 13 May German armoured spearheads crossed the Meuse, then broke out to make rapid progress across northern France. Six days later Charles de Gaulle, then a colonel commanding a part-formed French armoured division, put in a counter-attack at Montcornet, but could not deflect the Germans. On 21 May the British launched a more serious counter-stroke at Arras, but neither Allied efforts nor the hesitations of a German high command part-bemused by its own success could prevent the Germans from reaching the Channel coast, cutting the Allied armies in half.

Operation *Dynamo*, the evacuation of Allied troops from Dunkirk, began on 26 May. It was made possible partly by the resolute defence of the Channel ports and the retention of Lille by part of the 1st French Army, as well as by Hitler's decision to order his tanks to halt. By 4 June, nearly 340,000 personnel, almost one-third of them French, had been evacuated from Dunkirk.

The alliance was under great strain even before the evacuation and a dispute over the RAF's commitment to the battle of France put extra weight on it. Gamelin had been replaced by General Weygand on 20 May and although he failed to buckle together a plan to produce co-ordinated attacks on the panzer corridor, his men fought for the lines of the Seine and Aisne with a vigour not always acknowledged by English-speaking historians. But Paris was untenable: the government left for Tours on 10 June and the Germans entered the capital four days later. As the Germans made fresh progress in eastern France the Italians entered the war and the government moved to Bordeaux. Paul Reynaud, the Premier, resigned and the aged Marshal Pétain, brought back from his post as ambassador to Madrid, took over. An armistice was concluded, leaving a huge tract of France under German occupation, with the French government, at the spa town of Vichy, in control of a mere rump of the country. There were some who accepted neither the military verdict of 1940, nor the establishment of Pétain's *Etat français*, coloured as it was by the darker tinge of the hard, anti-Semitic right. De Gaulle broadcast from London that although France had lost the battle she had not lost the war, and a number of French soldiers, sailors and airmen joined him in his struggle for Free France.

Yet there were many more who did not. Pétain enjoyed great personal prestige and it was tempting – more tempting at the time than it seems with the clarity of hindsight – for Frenchmen to remain loyal to Vichy, despite tussles within its cabinet between some, like Pierre Laval, who believed in the inevitability of German victory, and others who sought to avoid total collaboration. A wave of anglophobia swept the country, partly the result of the misunderstandings of the 1940 campaign, worsened by a maladroit British attempt to secure the neutrality of the French fleet at Mers-el-Kebir. However, German invasion of Russia in 1941 not only inspired increasing doubt about ultimate German victory: it brought the Communist Party, with its experience of clandestine organisation, into the Resistance against the Germans.

In November 1942 an Anglo-American force landed in North Africa, hoping to be welcomed by the French authorities. Admiral Darlan, head of the French armed forces, was there at the time and eventually decided not to oppose the Allies. After some confusion a ceasefire was agreed, and the Allies duly occupied Morocco and Algeria. The German response was abrupt. In late November they invaded the Unoccupied Zone of France, with the seizure of the French fleet at Toulon as one of their objectives. The fleet had orders to scuttle itself rather than fall into German hands: three battleships and seven cruisers were amongst the vessels sent to the bottom.

After November 1942 Vichy became, in Alfred Cobban's words, 'little more than an agency for the Nazi exploitation of France'. Occupation costs enabled the Germans to amass credits, which they used to secure food and raw material from France, and growing numbers of French workers were sent off to feed the insatiable appetite of German war industry.

The conscription of labour helped drive young men into the arms of the Resistance. This had made a slow start and its activities were impeded by disagreements, often politically inspired, between various groups. Arms and operatives were parachuted in from England, with much valuable work being done by the Special Operations Executive. The establishment of a National Resistance Council in 1943 helped improve co-ordination, but the Resistance was never able to free itself from the constraints of politics. The Communists grew increasingly strong, especially in the 'red belt' around Paris and the great industrial cities of the north. De Gaulle had tightened his hold on the Free French movement, soon to be called 'Fighting France', despite a challenge briefly posed by General Giraud in 1942. He established a *Délégation Générale*, intended to represent the interests of the state and to be above political parties, and this was eventually to assist the takeover of power from the Vichy provincial authorities in 1944.

De Gaulle enjoyed an uneasy relationship with the British

and Americans, who generally underestimated his growing authority over the Resistance and, indeed, that body's increasing military potential. He was not informed of the Allied invasion of Europe until just before it took place and partly in consequence the Resistance was less effective than it might have been. Worse, there were some premature risings in areas like the Vercors. Nevertheless, the Resistance made a useful contribution to the Allied invasion, carrying out selective acts of sabotage and providing useful intelligence. A heavy price was paid. The Germans had long responded to Resistance attacks by seizing and executing hostages, and in July 1944 they carried out one of their most spectacular atrocities when the village of Oradour-sur-Glane was destroyed and its inhabitants massacred.

The German soldiers who riposted so harshly against Resistance attacks were facing a military balance that had tilted sharply against Germany. The invasion of Russia had proved a fatal mistake and the German army spent its strength on the Eastern Front, leaving large but generally low-quality forces to garrison France and the Low Countries. Even so, the problems confronting the Allies were enormous. In August 1942 the 2nd Canadian Division and two British Commandos had mounted a large-scale raid on Dieppe. The operation was a costly failure, but drew attention to the difficulties facing an invader, enabling specialist equipment and techniques to be developed. Allied planners eventually decided to mount their attack not in the Pas de Calais, but in Normandy. An elaborate deception plan helped conceal the invasion's objective from the Germans, many of whose forces remained ready to meet an assault across the Channel at its narrowest point.

Much work had been done on coastal defences across the whole of northern France. Rommel, who had last served there as a divisional commander in 1940 and had made his reputation at the head of *Panzerarmee Afrika*, and now, as a field marshal, commanded Army Group B in the invasion sector: his personal energy helped improve defence works all along the coast. Rommel believed that the classical solution of letting the invaders land, containing them and then counter-attacking, would not work: the battle had to be won on 'the Longest Day' of the invasion itself. His superior, Field Marshal von Rundstedt, Commander-in-Chief West, disagreed, and favoured the conventional approach of identifying the enemy's real thrust and concentrating to deal with it. Neither was helped by the fact that the armoured reserves, which would prove crucial to the battle, could only be moved with Hitler's permission.

Operation *Overlord*, launched on 6 June 1944 under the supreme command of the US General Dwight D. Eisenhower, was the largest amphibious operation in history. With airborne troops dropped to secure the shoulders of the invasion sector, over 150,000 troops and 1,500 tanks were landed on five assault beaches in the first forty-eight hours. The results of D-Day itself were mixed. The British seized secure footholds at Gold, Juno and Sword beaches north of Caen and Bayeux, although they failed to push as far inland as had been expected. Omar Bradley's Americans were less fortunate. They were successful at their westernmost beach, Utah, but at Omaha they took heavy casualties on the beach and fought their way inland with difficulty.

Over the next weeks the fighting ground on with the Allies making slow progress amongst the orchards and hedgerows, with the Americans pushing through the *bocage* towards St-Lô and the British failing in several attempts to take Caen. General Sir Bernard Montgomery, overall land force commander till 1 September, was under growing pressure (not least from the air forces, who wanted space to build temporary airfields inland from the invasion beaches) to break the deadlock. It is likely that his own accounts of operations give them more coherence than they possessed at the time, and we cannot be sure that he always thought in terms of the British and Canadians pinning down the Germans around Caen while the Americans broke out further westward. His offensive, code-named Operation *Goodwood*, mounted on 18 July, was oversold to senior Allied commanders as an attempt to break out on to the

Caen-Falaise plain. It made disappointing progress, but on 25 July the Americans launched Operation *Cobra*, breaking out and swinging deep into France. Field Marshal von Kluge, who had replaced the dismissed Rundstedt, was pressed by Hitler to counter-attack, but when he advanced against the Americans at Mortain on 7 August he was firmly rebuffed.

The British, Canadians and Poles fought their way down from Caen, and the Americans of Patton's US 3rd Army swung round to all but encircle the remnants of the German armies in Normandy west of Falaise. Bradley, commander of the US 12th Army Group, halted Patton's leading corps at Argentan ('Better a hard shoulder at Argentan than a broken neck at Falaise'), a decision that helped the Germans to keep the neck of the pocket open. However, most of the Germans who escaped did so on foot and the litter of burnt-out vehicles in lanes and fields around Falaise bore dreadful testimony to the efficacy of Allied air power.

The Resistance rose in Paris even before the city's liberators arrived and it was fitting that the German garrison commander surrendered, on 25 August, to the Frenchmen of General Leclerc's French 2nd Armoured Division. The Allied armies hurtled on across First World War battlefields, which had already become part of British and American folk memory, liberating Brussels on 4 September. And there the advance faltered. This was in no small measure a tribute to the remarkable recuperative powers of the German army. Field Marshal Walter Model had taken over from Kluge and the day after Brussels fell Rundstedt emerged as Commander-in-Chief West once again, with the tough Model commanding Army Group B. The halt also reflected Allied logistic difficulties. Most of their supplies were still being landed over open beaches in Normandy, and despite the heroic efforts of their logisticians – the American 'Red Ball Express' used nearly 6,000 vehicles to shuttle supplies down a one-way loop of road from St-Lô to Versailles – it proved difficult to sustain an advance on all fronts.

There was also a disagreement between Montgomery, now commanding 21st Army Group on the Allied left, and Eisenhower, who had assumed direct command of Allied

land forces. Montgomery favoured advancing on a narrow
front into the very heart of Germany. Such an operation
would require both British and American troops, under a
single commander – and Montgomery was the obvious
candidate. Eisenhower, however, recognised that American
public opinion would not tolerate such an arrangement and
was in any case under pressure from Bradley, whose 12th
Army Group was on the Allied right, to allow Patton's 3rd
Army to continue its advance in the area of Metz.

Montgomery nonetheless obtained Eisenhower's agreement to lay an 'airborne carpet' across the water obstacles
ahead of the British 30 Corps in northern Belgium and
Holland. The US 82nd and 101st Divisions were successful,
but the British 1st Airborne Division was unable to secure
the bridge at Arnhem and its survivors were withdrawn.
With the failure of Montgomery's plan, painfully slow
progress made by Patton's offensive in Lorraine and heavy
fighting as the US 9th Army pushed forward east of Aachen,
the Allied advance had definitely lost its momentum. And
although the vital port of Antwerp had been captured on 4
September, it was not until late November, after the seizure
of the island of Walcheren, which commands the approaches
to the Scheldt, that it was at last open to traffic.

Although the German military situation on the Eastern
Front was profoundly gloomy, Hitler decided to use
some recently refurbished units alongside newly raised
Volksgrenadier divisions to mount a surprise attack in the
Ardennes, with the aim of reaching Antwerp and cutting the
Allied armies in half. Hitler hoped, unrealistically, that this
would shake the Anglo-American alliance to its very
foundation. The offensive, which began on 16 December,
started well, aided by the atrocious weather that kept Allied
aircraft out of the skies. On 19 December Eisenhower
adjusted command boundaries so as to give Montgomery
command over all forces north of the breakthrough and this,
together with the resolute American defence of the road
hub of Bastogne, helped check the German drive. Patton
counter-attacked from the south, and as the weather cleared
Allied aircraft intervened to hammer German columns and

railheads. By early January the attack had evidently failed and a subsidiary thrust into the Saar met with even less success.

The Allies resumed their own offensives in the aftermath of the Ardennes battle, and there was bitter fighting in a wet and dismal February. The US 1st Army crossed the Rhine at Remagen in early March, Patton forced a crossing south of Mainz and on 24 March Montgomery unleashed a lavishly staged masterpiece to achieve a mighty crossing at Wesel. With the Rhine behind them, the British, Americans and Canadians fanned out to complete the defeat of Germany, linking up with the Russians on the Elbe on 25 April.

German withdrawal faced Frenchmen with the familiar task of constitution-making. Old scores were settled, some judicially, others privately. In some areas the local Resistance took over as the Germans disappeared and many collaborators were killed out of hand. Where Gaullist authorities took over such excesses were quickly brought under control. Thousands of collaborators were dealt with in the courts and the more conspicuous appeared before a specially established High Court. Laval was shot after a botched suicide attempt and Pétain was sentenced to life imprisonment. A Constituent Assembly established the Fourth Republic, which staggered on with a succession of weak and divided governments. It was not without achievements. France was well to the fore in forging economic and political links with her European neighbours, while her economy, nourished by Marshall Aid, rose swiftly from the ruins. Rampant inflation caused difficulties, but it was the continuing strain of colonial conflicts that eventually brought the republic down.

France's colonial power was broken in the 1950s. In 1954 the battle of Dien Bien Phu marked her defeat in the long war for Indo-China, and Tunisia and Morocco subsequently gained their independence. Algeria presented a more serious problem, for it contained a large and vociferous European community. A lengthy guerrilla war threw unbearable strain on the tottering Fourth Republic. In 1958 a military revolt in Algeria swept de Gaulle, who had withdrawn from political life to await the call to return, back into power. After ruling by decree for six months, he submitted a new constitution for approval by popular referendum and the Fifth Republic, with its strongly presidential constitution, came into being.

The history of France under the Fifth Republic lies outside the confines of this book. Yet the new Republic was to leave its own mark on the fatal avenue. The regeneration of French industry splashed new factories across the north and east in particular, turning the Moselle valley around Metz and Thionville into a swamp of dust and chimneys, and poisoning sections of the Channel coast with chemical works. As some industries rose, others fell. The decline of iron and steel production around Longwy is marked by disused works and blighted towns. The expanding web of autoroutes criss-crosses the north, making it easier to drive south to the sun without battling through Paris or round its *péripherique*, but inevitably changing the character of the countryside. The beautifully named Pigeon Ravine Cemetery, on the back edge of the Somme battlefield, thrums to the traffic of the A26 and at St-Privat the A4 follows the attack axis of the Prussian Guard on 18 August 1870.

Armed forces garrison the fatal avenue as they have for centuries. The French army was shorn of an element of its panache when Algeria became independent: the *Armée d'Afrique* marched into that frozen world of costume prints and military museums. Thousands of its soldiers still lie on French soil, beneath headstones topped with an arabic arch, an ironic testimony to the vagaries of politics that took young men from bled or casbah to the Chemin des Dames or the Meuse Heights. With its flexible attitude to the transfer of military traditions, though, the French army has managed to retain a little of the *chic exquis* of its African army, and the *Tirailleurs d'Epinal* now wear the crescent moon badge and their band turns out in the bright glory of *turco* full dress.

Most of the old garrisons towns of the eastern frontier house soldiers no longer, though some French regiments still live in the high-eaved barrack blocks built in Metz to house the *Kaiserheer*, and the old artillery and engineer *école d'application* above the Moselle, topped by a curious gazebo built to accommodate young officers practising field-sketching, is now a joint mess for officers and NCOs. Châlons-sur-Marne (now rechristened Châlons-en-Champagne) remains the Salisbury Plain of France and the army retains a good deal of other land. Often old forts, peeling quietly amongst the undergrowth, keep their *Terrain Militaire* warning notices.

The oddest irony of all is at Verdun. On an early visit to Fort Douaumont I heard the thump of rifle fire echoing over the pitted concrete. It was neither a ghostly manifestation of the Brandenburgers taking the fort, nor evidence of an overactive imagination, but came from a French army rifle range running out from the fort towards the start line of the German assault in 1916. All around the scarred landscape was masked by pine trees, but the range was open and bare, laid out almost like a glacis intended to give good fields of fire to the fort's guns. Nearby, young conscripts carried out section attacks amongst the shell holes.

I found it unspeakably ironic, but the poignancy was lost on the sentry manning a barrier. Yes, of course he knew about the battle, but his preoccupations were those of live soldiers not dead heroes. He was bored and hungry, and hoped the firing would soon be over so that he could get back to barracks. In any case he came from further south and the bleak uplands of Lorraine with their searching wind did not suit his temperament. Neither did conscription, whose logic eluded him. An invasion? Again? By whom? Surely the Germans, who flocked to the fort in their well-appointed coaches, were no danger. He snorted into the blue smoke of his cigarette. 'That, that's all past.' Let us hope that he is right.

II

Flanders, Artois and Picardy – I

THE ANCIENT PROVINCES of Flanders, Artois and Picardy, so familiar to Chaucer's squire and to generations of British soldiers, now constitute the French *départements* of Pas de Calais, Somme and Nord, together with the Belgian provinces of East and West Flanders. Picardy, with its capital at Amiens, lies on a chalky plateau, which stretches from the Paris basin to the Artois escarpment. Its rolling uplands are slashed by the valleys of the Somme, the Authie and the Canche, which meander through ponds and marshes rich with fish and waterfowl. Most major towns lie in the valleys, Amiens and Abbeville on the Somme, Hesdin and Montreuil on the Canche, and Doullens on the Authie. Its once busy ports, St-Valery and Le Crotoy at the mouth of the Somme, and Etaples on the Canche, now shelter only fishing boats and pleasure craft.

Artois, too, stands on chalk, which ends abruptly at the escarpment of Vimy and Notre-Dame-de-Lorette, where the ground drops down to the Flanders plain. Its main river, the Scarpe, does not flow north-west to the Channel like the rivers of Picardy, but curls north-eastwards from the high ground behind the Lorette spur, through the provincial centre of Arras, to join the Escaut south of Tournai. On the eastern borders of Artois, across the Scarpe, lie the old county of Hainault (the modern Belgian province of the same name is to its north), with its capital at Valenciennes, and the Cambrésis, with its capital at Cambrai. Boulogne, the principal port of Artois, perches on a knuckle of limestone, with Capes Griz-Nez and Blanc-Nez to its north

FLANDERS, ARTOIS & PICARDY

© Hugh Bicheno

looking out across the Channel. It remains France's main fishing port and retains a charm now stripped from Calais, on the Flanders coast.

Flanders rolls away from the foot of Vimy Ridge. Behind the coastal dunes, dykes and canals criss-cross an open vista of reclaimed land with only church towers, civic belfries and a few surviving windmills to break the monotony. Further inland, woods dot the landscape and big Flemish farms, with a square of white-walled, red-tiled buildings around a courtyard, sit squatly amidst dark brown furrows. The little hills of Flemish Switzerland, the Mont des Recollets at Cassel, the Mont des Cats, Mont Noir, Rodeburg and Kemmel Hill, pimple the landscape between St-Omer and Ypres. Two of the rivers of western Flanders play their part in the story which follows: the Yser or Izer, which rises near St-Omer and curls round to enter the sea at Nieuport, and the Lys, flowing north-east from Armentières to Ghent.

Lille, the capital of French Flanders, now merges untidily with Roubaix and Tourcoing to its north, while to its south-west lies *le bassin houiller*, the mining belt, which stretches from Auchel, through Lens and Douai, away across the Belgian border to Mons and beyond. The towns of the French mining belt bear the imprint of the rapid urbanisation that took their population from 45,000 in 1851 to 267,000 in 1911 in response to a hunger for coal, which drove its price up and sucked workers in. Depressing rows of miners' cottages – *corons* – first began to give way to better-planned *cités* of workers' houses in the 1870s. Three-quarters of miners' houses were destroyed in the First World War, and more *cités* were built in the 1920s, their semi-detached houses standing on decent plots, but with a soul-destroying uniformity.

Most of the pitheads (*fosses*) with their towers for winding gear have now gone, but the *crassiers*, slag heaps, are more obdurate, although usually vegetation shrouds their slopes. Indeed, the whole area is vegetating. In 1947 there were 109 working pits in the area; by 1986 this had fallen to six and now there are none left. Unemployment is high, emigration rife and local politics is frequently old left laced with a hard edge of racism of the hard right. Factories and houses stand empty, property is often a drug on the market.

The region is no stranger to the ebb and flow of economic fortune. On the eve of the Hundred Years War it was remarkably prosperous. During the twelfth century it had come increasingly under the control of the Counts of Flanders, but during the next century French authority edged steadily northwards, until by 1300 the boundary between the counties of Artois and Flanders ran just south of the modern Franco-Belgian border. Towns flourished on both sides of the frontier, and the thriving ports of Calais and Boulogne imported smoked herrings and coal from England, and wine from Gascony, while the cloth products of Artois and Flanders left by sea for England, Spain and Italy. The names of the cloth towns insinuated their way into the English language: lisle came from Lille, chambray and cambric from Cambrai, and Shakespeare's Polonius hid 'behind the arras', a tapestry from the bustling capital of Artois.

The slide began shortly before the Hundred Years War. The cloth towns of Artois were hard hit by competition from Flanders, unemployment grew and workers left the towns for the illusory security of the countryside. A series of riots greeted rising corn prices and food shortages, and bourgeois dignitaries were easy targets for the mob: in 1356 seventeen leading citizens of Arras were flung from the windows of the council chamber. The Black Death took a heavy toll. In 1348 St-Omer lost a quarter of its 10,000 citizens, and in 1400 half the population of Lens perished.

To the miseries of depression and pestilence were added those of war. There was desultory fighting in the region in 1337–40, but in 1346 war came to Picardy and Artois in earnest. Edward III landed in Normandy on 12 July and marched for Rouen with the intention of joining his Flemish allies. He soon found that the Flemings had marched slowly, while his adversary Philip VI had moved far more swiftly, concentrating his army at Rouen. Early August saw both armies marching on opposite sides of the Seine. Edward found the bridges demolished or held against him, but on the

13th he jinked back to Poissy, pushed a detachment across the damaged bridge and set fire to several villages under the walls of Paris to keep Philip guessing. Then, with the bridge repaired, he crossed it with his whole army and marched for the Somme, hoping to cross it, too, and join the Flemings, who had now laid siege to Béthune. Philip beat him to the river. Moving with unusual speed, he reached Rouen, cutting Edward off from the Flemings and then following him westwards along the widening Somme.

Edward's position was grim. The marshy valley of the Somme lay between him and the Flemings, and the French held all the bridges. His fleet, which had kept pace with his march through Normandy, had now returned home, and his men were tired and hungry. Yet the King's nerve held. He addressed his prisoners, offering a reward to any who would tell him of a crossing place: one duly revealed a ford at Blanchetaque or Blanquetaque, near the village of Saigneville, which was passable at low tide. The English arrived at the southern entrance to the ford at about five on the morning of 24 August and had to wait three hours for the tide to fall. The ford probably consisted of a narrow causeway of raised bedrock, running some 2,000 metres across the tidal valley. Although the canalisation of the Somme and peat-gathering in its valley have long since swept the ford away, the probable line of Edward's advance can still be seen, where a little road winds its way due north from the centre of Saigneville to end by the quiet banks of the Somme.

The banks were much less quiet in August 1336, for Godemars du Fay, Bailiff of the Vermandois, had arrived on the northern bank on the 23rd with a small detachment of mounted men and a larger force of peasant militia. At about 8 a.m. the water was low enough for a man to cross with difficulty, and Edward sent Hugh Despencer with a mixed force of horsemen and archers to clear Godemars from the crossing. The archers struggled through the water, holding their bows above their heads to keep them dry, while the horsemen splashed along behind. As soon as they were within range, the archers, standing ten deep on the narrow

causeway, unleashed a storm of arrows. The French horsemen had probably advanced into the shallows to contest the
English advance and by doing so they lost much of the
advantage of the ground. The foot soldiers, to the rear, did
not stand the iron hail for long and ran, leaving Godemars to
his fate. The French knights made a brief resistance against
the cavalry, then seem to have done their best to rally
between Sailly-Bray and Nouvion. By now they were hopelessly outnumbered, for the main body of Edward's army
was across the river. Godemars, wounded and his banner
lost, eventually halted at St-Riquier, where he received the
unwelcome news that Philip had declared his intention of
hanging him for his failure to hold the ford.

Philip had every reason to be vexed. He had caught the
tail-end of the English rearguard on the southern bank,
capturing some baggage wagons and killing a few men, but
the tide was now too high for him to cross. He rested briefly,
then swung back to Abbeville, where he spent the 25th,
while new contingents arrived to strengthen his army. That
evening the King addressed his leading noblemen, urging
them to be 'one to the other friendly, courteous, without
envy, hatred or pride', and no doubt doing his best to lay
down an order of march for his cumbersome host. At dawn
on the 26th the French heard Mass and moved off to seek
their enemies.

Edward had made good use of the respite. He was now in
his mother's fief of Ponthieu, but its inhabitants had no
great love for the English, known – from their common oath
'God damn' – as *Goddons*. Edward took the little town and
castle at Noyelles, while Despencer spurred on to Le Crotoy,
capturing a number of ships laden with wine from Poitou
and Gascony, before going on to seize more provisions up
the coast at Rue. The English spent the night of the
25th–26th north of Noyelles. The following morning
Edward changed direction. Instead of heading north for
Montreuil, along the marshy shores of the Channel, where
the going was poor, he turned inland, probably skirting the
southern fringe of the Forest of Crécy. This beautiful and
peaceful place is still very much the size and shape that it

was in medieval times and a drive down the little D111, which bisects its eastern end, reveals why Edward avoided pushing directly through the forest. A dozen men could delay a hundred moving down its paths, and its wooded canopy would reduce the firepower of the archers. It was probably through the little villages of Nouvion, Forest-L'Abbaye and Canchy that the English marched in three great bodies, banners unfurled.

Edward's army had moved at a cracking pace, covering some 230 kilometres in the nine days since crossing the Seine. Although Despencer's captures on the 25th had helped, it was hungry as well as tired and there were rumblings of discontent. Edward concluded that he could not count on outrunning his pursuers and looked for a good position in which to offer battle. He made a virtue of necessity by announcing that he would defend his mother's fief and sent his most experienced commanders off to look for suitable ground. They found it between the villages of Crécy and Wadicourt, where a low ridge running north-eastwards falls away into the Vallée aux Clercs. To its south the village of Crécy, hub of a network of small roads, lay on the edge of the forest and the little River Maye followed the line of the forest edge. Behind the position stood Crécy Grange Wood. The French, moving from Abbeville up the main road through Canchy, the modern D928, could not outflank the position from the south because of the twin obstacles of forest and village. Swinging round to outflank it from the north would leave the French army strung out in line of march, offering its flank to the English, and was in any case the sort of enterprise which no fiery feudal army could be relied upon to perform. Edward could therefore be confident that Philip would deploy from the main road, probably through Fontaine-Maye, and launch a frontal assault across the valley, and it was to meet just this threat that he drew up his army on the morning of 26 August.

Edward himself took post atop a windmill just north-east of Crécy, a spot now marked by a galleried tower, which provides visitors with a useful vantage point. To his right front, his son the Prince of Wales commanded the right 'battle', with 800 dismounted men-at-arms in its centre, flanked on either side by 2,000 archers and 1,000 Welsh spearmen. The Prince's right flank probably ran as far as the Crécy–Estrées road (D938), well within bowshot of the Maye. The second battle, under the Earls of Northampton and Arundel, lay further along the ridge, to the King's left front. It numbered perhaps 500 dismounted men-at-arms, with 1,200 archers on the flanks, and ran to within bowshot of Wadicourt. The King's own battle, with 700 men-at-arms, 2,000 archers and 1,000 spearmen, stood in the centre of the ridge, north-east of the windmill, somewhat drawn back from the two forward divisions but covering the gap between them. The baggage was left in Crécy Grange Wood with a small escort. In all, the English position was some 1.5 kilometres wide. Precise numbers must remain a matter of some conjecture, but it is fair to assume that the chronicler Froissart overestimated the total size of the army at 15,000 men, while Sir James Ramsey's more scholarly 1914 estimate of over 8,200 men is a little too low. Sir Charles Oman put the English strength at a little over 9,000, and C. H. Rothero, writing in 1981, at something over 10,700. I have accepted Oman's figures for the army's three divisions, and adding a baggage guard and allowing for sick and stragglers, would put the total figure in or near the field at around 10,000.

When the army was drawn up, with supplies of spare arrows dropped amongst the archers by baggage wagons, sharpened stakes were hammered in to protect the archers, and holes about thirty centimetres deep were dug across the whole of the front. At about mid-morning Edward mounted a small white palfrey and rode along each battle in turn, urging his men to stand steady and not to rush forward to strip the dead, and stressing that his claims against Philip were just. He then rode back to the windmill, dismounted and summoned his son, knighting him in full view of the army to the accompaniment of loud cheering. To this sound grasp of psychology Edward added good practical leadership, for he then ordered his men to fall out to eat and drink, leaving their positions marked by helmets for the men-at-arms and bows for the archers. A sudden rainstorm brought the archers scampering back to protect their bowstrings from the wet and by early afternoon the whole army was sitting in position, waiting for the French.

The French were coming. They had left Abbeville early on the 26th – Froissart reports that it took the army half a day to leave the town – and swung north-westwards, lured on by the smoke of the villages fired by the English the previous day. Having failed to find the English south of Crécy Forest, the French then marched around its eastern end. Four seasoned knights rode well ahead and in early afternoon they sighted the English, still sitting down, on the Crécy–Wadicourt ridge. As they rode back to report to Philip they passed scattered bodies of French soldiers, rushing on in enthusiastic disorder. When they found Philip, probably near the village of Lamotte-Buleaux, one of them compared the good order of the English with the confusion of the French and recommended that the army should be properly formed up to give battle the next day. Philip consulted his brother, the Count of Alençon and other senior noblemen, and tried to halt his army. He might as well have tried to stem a torrent. The heterogeneous mass poured forwards, confident of victory and anxious not to miss the battle, and soon the head of the column reached Estrées, with the remainder strung out behind through Froyelles.

On the Estrées ridge, facing the English army across the open Vallée aux Clercs, the French formed up into some sort of order. A screen of Genoese crossbowmen led the way, though their effectiveness was limited by the fact that the heavy wooden shields – *pavises* – behind whose cover they were accustomed to load and fire were miles to the rear with the baggage. Then came a body of horsemen under Alençon and another under Philip himself. Further to the rear came more horsemen, including a body of northerners under John of Hainaut, and huge numbers of the communes, peasant infantry snatched from byre and plough. The lowest contemporary estimate puts the French at 40,000, although only the knights, some 12,000, and the Genoese, another 6–8,000, had real fighting value. Philip, riding beneath the oriflamme, the sacred red silk banner of St Denis, flown when no quarter was to be offered to the enemies of France, lost his temper when he saw the English calmly waiting for him less than two kilometres away. He gave the order to advance, only to receive a request for delay from the commander of the Genoese, a plea that Alençon regarded as evidence of cowardice. At this juncture a heavy rainstorm swept the field and when it had passed the Genoese trudged down into the valley to open the battle.

The odds were not in their favour. The sun, setting behind the English, was in their eyes and they had marched all day carrying their heavy crossbows. Nevertheless, they were tough professionals, and as they came into effective range they gave their customary three shouts and leaps to alarm their enemies, and began to wing crossbow bolts up the ridge in front of them. The response was awesome. Froissart tells us how 'the English archers stept forth one pace and let fly their arrows so wholly together and so thick that it seemed like snow'. The Genoese could not compete with the sheer volume of fire and fell back in disorder. Their retirement infuriated Philip and Alençon, who ordered them to be ridden down, and the first French cavalry casualties came as desperate Genoese fought for their lives with crossbow and dagger, ducking low to hamstring steeds and knife unhorsed men-at-arms.

The first French cavalry charge proper went forward through the debris of the Genoese at about 7 p.m. The valley was good going for horses. Ann Hyland's perceptive work on the medieval warhorse and the practical experience of re-enactors suggest that chargers were actually a good deal smaller than they were once believed to be and we must take stories about knights being hoisted into the saddles of huge horses with more than a pinch of salt. A knight needed to be able to mount, dismount and remount, and a man with well-fitting armour, who had spent part of his life training for just such an eventuality, would have been able, visor lowered and lance levelled, to raise a gallop if the moment demanded it. Whatever the pace of the French advance, its momentum must have seemed irresistible to those in its

path. Again the archers stepped forward and again the arrow
storm drenched the slopes. Knights and men-at-arms were
transfixed, and horses maddened by the arrows. Some
gallant souls wallowed on through the carnage to reach the
Prince of Wales's division: Jacques d'Estracelles led a body of
horsemen to handstrokes with the Prince's men, causing
one knight to seek Lord Arundel's support, while another
asked the King for help. Edward asked how his son was, and
when he heard that he was unhurt and fighting well, replied,
'Tell him this is the time to win his spurs.'

The French attacks went on till well into the night. There
were perhaps fifteen or sixteen in all, each a disorderly
scramble up the trampled slope and into the arrows. The
blind King John of Bohemia, hearing that the attacks were
all failing, begged to be led into the battle, and went forward
with a small party of knights and squires, their reins tied
together. Their bodies were found amongst the wreckage of
the French defeat the following day. As the assaults died
away, the English remained on the ridge, lighting large fires.
Philip stopped briefly at the little castle of Labroye, where
the Abbeville–Hesdin road crosses the Authie, then rode
northwards for the Somme.

Sunday, 27 August dawned foggy – indeed, thick mist still
occurs in the area on even the brightest of summer days.
Edward sent a mixed force of archers and men-at-arms to see
what had happened to the French, and this detachment met
droves of unlucky foot soldiers, who had spent the night
under hedges and bushes, and were milling about in search
of their lords, and fell on them 'as wolves among sheep and
killed them at will'. Stray French units were snapped up,
although Abbeville resisted an attempt to take it by *coup
de main*. Edward, having attended Mass, detailed Lord
Cobham, with a party of knights skilled in heraldry and a
substantial work detail, to count the French bodies and
identify noblemen so that they could be buried in
accordance with their rank. Froissart gives the losses at
eleven princes, two prelates, 1,286 knights and 15–16,000
foot soldiers, while Ferdinand Lot puts infantry casualties
at a staggering 30,000. It seems reasonable to accept

contemporary estimates of 1,200–1,500 knightly dead, with at least another 4,000 fatalities amongst the foot, and probably very many more. Few French noble families had not suffered: the Count of Alençon, the Duke of Lorraine, and the Counts of Blois and Flanders were amongst the dead. The English had lost 300 men at the very most.

There were certainly few enough English dead for them to be buried, on 28 August, some distance from the battlefield, in what one chronicler calls 'the abbey in the forest on the edge of the woods', possibly Forest-L'Abbaye, eight kilometres from the battlefield via a track through the forest. French heralds had arrived to negotiate a truce to bury their dead. Some French noblemen were taken home for burial or interred in Crécy itself, but most of the French dead were stripped of their armour and thrown into communal pits, one of them traditionally where the Vallée aux Clercs meets the Maye valley in what is now a builder's yard. Cistercian monks from Valloires Abbey, north-west of the battlefield, had come down and tended some of the French wounded, and those who died were buried in the monastery's estate at Crécy Grange.

Valloires Abbey itself, now a children's home, but well worth a visit as much for its measured eighteenth-century architecture – it was rebuilt after a series of fires – as for its sylvan setting, was the last resting place of many noble French dead, and there are reliable accounts that John of Luxembourg was buried there, although all traces of his tomb are now lost. He is commemorated on the battlefield itself: the Luxembourg Cross stands beside the French approach route, long known as *le chemin de l'armée*. It was built not long after the battle, but its present base dates from only 1906. Its location is something of a puzzle, for it stands well short of the English battle line, so the *Guide Michelin*'s suggestion that the cross stands on the spot where the king fell seems unwise. Even if the King was not killed outright, but died in Edward's tent, as some sources suggest, he would have breathed his last at least two kilometres north-west of his memorial. However, recent failure to confirm the generally accepted site of the battle with any archaeological evidence must throw at least the slightest shadow of suspicion, so there may be something more to the Luxembourg story than meets the eye. Another memorial to the King is more familiar to Englishmen – and even more so to Welshmen. Edward kept the ostrich feathers from the dead King's helmet, with the German motto *Ich dien* (I serve) engraved below them. He presented them to his son as a reward for his valour in the battle, and successive Princes of Wales have retained the plumes and the motto in their coat of arms. This is a landscape full of poignancy. In the little communal cemetery at Marcheville, due south of the Luxembourg cross, lies Private Herbert Ashworth of 11th Battalion The Royal Fusiliers, who died on 26 December 1916 at the age of thirty-four, and is the only British soldier buried there.

A contemporary account of Crécy, by the Italian Giovanni Villani, speaks of cannon being used by the English: this, if true, would mark the debut of artillery on the battlefields of the West. Unfortunately Villani was not present at the battle – he was in prison in Florence at the time – so his evidence is inconclusive. We know that the English used cannon at the siege of Calais shortly afterwards, so it is certainly not inherently impossible that cannon were present at Crécy. A few cast-iron cannon balls have been found on the battlefield – three are at present in a small museum in Crécy itself. Henri de Wailly, author of the most recent account of Crécy (1985), doubts whether these date from the battle, for cast-iron cannon balls were not to be in use for another century. Historians are divided on the issue and the best that we can say is that it is not proven one way or the other. Had there been cannon at the battle, they would have been short, stubby bombards, carried on carts, and their effect – except, perhaps, for the alarm caused to horses and superstitious peasants – would have been decidedly limited.

With Philip's army no longer a threat, Edward marched northwards at his leisure. He arrived at Montreuil on 29 August, but found the lofty castle held against him by the Count of Savoy, whose force of 500 lances had arrived too

late to fight at Crécy and slipped behind the English army to
reach Montreuil. Edward could make no impression on the
castle and contented himself with burning the surrounding
town. This was not to be Montreuil's last stroke of bad luck:
in 1537 the troops of the Holy Roman Emperor Charles V
stormed and sacked the place. The present citadel, partly the
work of the great engineers Errard and Vauban, encloses the
remnants of the castle and is open to the public, although its
custodians take a liberal Gallic lunch break. A walk round
the ramparts, which follow the line of the sixteenth-century
defences and include five thirteenth-century round towers,
is a good antidote to a heavy lunch, or a picturesque way of
killing time until the guardians of the citadel return from
theirs.

Montreuil housed Napoleon's headquarters when his
army was in camp at Boulogne in 1804, preparing to invade
England and by a curious irony British GHQ was there in
1916–18. Both headquarters were in the former Caserne
Duval, on the Rue de la Porte Becquerelle at the Place des
Carmes, which housed the *Ecole Militaire Preparatoire de
Montreuil* between 1884 and 1924. The site is now taken up
by two lycées, but two of the old barrack buildings remain.
Brisk walks along the breezy ramparts were the only
recreation for staff officers living a more pressurised life
than popular mythology gives them credit for. In 1918 the
town was regularly bombed and GHQ staff not on duty
moved out at nights, much to the derision of the inhabitants, who stayed behind. Field Marshal Haig was quartered
in the modest Château de Beaurepaire, four kilometres
south-west of the town on the D138, which is once more a
private house. In Montreuil itself a statue of the Field
Marshal mounted on his charger 'Miss Ypres' stands in the
Place du Théâtre, effectively part of the market place, the
Place Général de Gaulle. The abbey church of St-Saulve
provides another link with the town's military past: the
chapel on its northern side, now dedicated to the Virgin
Mary, was once the crossbowmen's chapel.

Edward marched on northwards, firing Etaples, Waben
and St-Josse, before laying siege to Calais. The town was

defended by Jean de Vienne and a determined garrison, and
the siege went slowly. However, French attempts to relieve
Calais failed, and the English land and sea blockade slowly
starved the defenders. In July 1347 six leading citizens gave
themselves up as hostages. Eustache de St Pierre and his five
colleagues approached the King barefoot, stripped to their
shirts, with ropes round their necks, carrying them the keys
to the city and castle. Edward, enraged by the long resistance, was inclined to execute them, but his wife, Philippa
of Hainault, interceded on their behalf and the King spared
them. Rodin's fine bronze statue commemorating the
burghers of Calais stands between the imposing town hall,
built in the steep-roofed Flemish style, and the Parc St
Pierre.

Calais, with an enclave called the pale beyond its walls,
was to be held by the English until the Duke of Guise
recaptured it in 1558: its loss was deeply felt by Mary Tudor,
who lamented that when she died the word 'Calais' would
be found engraved on her heart. It lay behind the Allied lines
during the First World War and in 1940 it was stoutly
defended by the Rifle Regiments of Brigadier Nicholson's
30th Brigade, whose heroism tied up German forces that
might otherwise have increased the pressure on the Dunkirk
perimeter. A few hundred metres away from Rodin's statue,
in the Parc St Pierre, is a very large bunker, which housed
the military telephone exchange during the German
occupation. It now contains a museum of exhibits relating
to the occupation of Calais, the local Resistance and the
battle of Britain.

Despite the damage it suffered during the Second World
War, Calais still retains other items of military interest. The
thirteenth-century Tour du Guet stands in the Place
d'Armes, the centre of medieval Calais, and a few hundred
metres to its south-east is the church of Notre-Dame,
finished under English occupation in the fourteenth century
and, with its almost Perpendicular style, looking more
English than French. It was there that Charles de Gaulle
married a local girl in 1921. Calais also bears the unmistakable stamp of Vauban. His citadel, on a site first fortified

in 1560, stands just west of the old town centre and a little park, aptly named the Square Vauban, contains a ravelin covering the Neptune Gate, its eastern entrance. Since 1965 the citadel has housed a memorial stadium, with three football pitches, an athletics track and a concrete grandstand. Just west of the town, on the D243, almost opposite the Channel Tunnel terminus, is Fort Nieulay, a fine example of a bastioned fort. The first fort on the site was English, built in 1525 and taken by the Duke of Guise in January 1558. The present fort was built by Vauban from 1677, has four bastions and the remnants of two ravelins, and is open on summer afternoons.

Edward III's capture of Calais ended the Crécy campaign very much in the English favour, but the war rumbled on. In 1415, however, it came back to Artois with a vengeance. Henry V landed near Harfleur in Normandy on 14 August, taking the town on 22 September. He had originally intended to make for Paris, but the siege had consumed time and resources. His council of war recommended garrisoning Harfleur and returning to England with the main body of the expeditionary force, but Henry determined on a *chevauchée* across Normandy and up to Calais. The French army, much larger than the English but weakened by factional disputes, concentrated at Vernon, with its advance guard at Rouen. It was probably on 8 October that Henry set off from Harfleur, his army marching light to make best speed. He crossed the Béthune at Arques on the 11th and the Eu at Bresle on the following day, and was making for the ford at Blanchetaque when he heard that Marshal Boucicaut was holding the northern end of the crossing in strength. Henry then had to feel his way eastwards along the Somme, with his army on short rations and the French holding the bridges and fords against him.

Near Corbie, just east of Amiens, Henry left the river and struck inland, cutting across the loop of the Somme and gaining ground on the French who had to march further round the northern bank. On the 18th he reached Nesle, between Roye and Ham, and managed to reach the fords at Voyennes and Bethencourt the following day. The river is now bridged in both villages, but the woods and marshy pools between Voyennes and Offoy on the far bank give a good idea of the difficulties facing the English, who slipped a force of 200 archers across the river to hold a small bridge-head while the damaged causeways were repaired with timber torn from houses in the two villages. The last man was not across until after dark, and that night the army bivouacked at Athy and Monchy-Lagache, up towards Péronne. The French had reached Péronne the same day, and on the 20th their heralds visited the English camp to bring a formal challenge to Henry from the Dukes of Orléans and Bourbon, commanding the army in the absence of King John, who had been persuaded to stay in Rouen by the Duke of Berry's assertion that it was 'better to lose a battle and save the king than lose a battle and lose the king too'. Henry announced his intention of marching straight to Calais and fighting the French if they opposed him, but took the precaution of taking up a defensive position, probably on the ridge that runs east-west across the Péronne road and looks down on the valley of the Cologne.

Far from attacking, the French fell back on Bapaume, probably to join another column under the Constable of France, Charles d'Albret. This detour took the French off Henry's direct route to Calais, and the King threw out a flank guard to cover his right as he marched to Albert. Thereafter the two armies moved on parallel routes, the English through Lucheux and Frévent to Blagny on the Ternoise, the French through Coullemont, on the eastern edge of the Forest of Lucheux, and St-Pol, their advance guard approaching Blagny, but on the northern bank of the river, just as the English had crossed on 24 October. The French had made better speed than the English, probably because the latter had strayed from the direct route from time to time, which was scarcely difficult as there were no accurate maps available and English knowledge of the ground was sketchy. Henry's route had taken him through what were to be the British rear areas in the Somme fighting of 1916, while the men of his flank guard, covering Bapaume, had marched through Combles, Ginchy and

Martinpuich, which were to become depressingly familiar to their successors of 1916.

On the night of the 24th, the vigil of the feast of Saints Crispin and Crispian, the two armies camped within earshot of one another. The English were undoubtedly tired – Lieutenant-Colonel Burne calculated that they had covered 260 miles in seventeen days with only one day's rest. The French, too, had made good speed – 180 miles in ten days – but many of them were mounted. There is general agreement that Henry's army was reduced to around 9,000 men, the majority of them archers, by the losses of Harfleur, but we can be far less confident about the precise size of the French army. The distinguished French historian Ferdinand Lot, perhaps impelled more by patriotic fervour than military logic, declared that since the French deployed three ranks of horsemen on a front of 800 metres, they could fit only 1,800 men-at-arms into the available ground. He adds 3,600 infantry to give the French an overall strength of 5,400 – rather less than the English. This total tallies with neither the known casualties nor with the fact that the sources make it clear that the French were very tightly packed into the available space. The French chronicler, the monk of St Denys, reckoned that the French outnumbered the English by four times, and it once seemed safe to assume a French army of 20,000 at the very least. Placing its total much higher than this, as does Dr J. H. Wyllie, who assumes a ten-to-one superiority, must give us pause for thought. How were the men and horses to be fed? Even an army of 20,000 men, most of them mounted, moving as a concentrated body would have imposed an immense strain on the local resources. Feeding upwards of 50,000 would have required in the order of a million pounds of fodder a day for the horses alone, a task beyond the means of wagon and packhorse logistics, and the resources even of fruitful Picardy.

In 2005 both Juliet Barker and Anne Curry produced scholarly accounts of the battle. The former thought that the Burgundian Jehan Waurin's contemporary estimate of 36,000 was most likely, if only because he listed the number of men assigned to each position in the French host.

Professor Curry, with careful analysis of its pay records, suggests that the French army was very much smaller, with 'a total around 12,000, with at least two-thirds of its strength being men-at-arms'.

It was not a comfortable night. Rain fell steadily and from the noisy French camp could be heard shouts as lords called for straw to cushion them from the soggy ground, sown with autumn wheat and already churned up by men and horses. The English, much better disciplined (noisy gentlemen were to lose horse and harness, and their social inferiors were to have an ear cropped), passed a quiet night, though they can scarcely have awaited the drizzly dawn with much enthusiasm.

Henry was billeted in the village of Maisoncelle, with his army in the village and surrounding fields. He sent out a small scouting party, which reported that the French appeared to be far more numerous. The Calais road (now the little D104) crossed a rectangle of open ground about 1,000 metres wide, with a slight dip in its centre, falling away more sharply to woods on either side. To the English left stood the village of Agincourt (now Azincourt) and to the right, Tramecourt. At dawn the two armies deployed about 1,000 metres apart, the English just in front of the lane that joined Maisoncelle to the Calais road: the centre rear of this position is now marked by a monument and an orientation table, though sadly the main battlefield cannot be seen from the spot. Henry, dressed in armour covered by a surcoat with the leopards of England quartered with the fleur-de-lys of France, rode a little grey horse along his line to address his men, then took post at the head of his centre division. A golden crown, also provocatively adorned with fleurs-de-lys, encircled his helmet.

There is much disagreement about the precise English deployment. Historians agree that the English formed up in the customary three divisions, though shortage of man-power and the size of the frontage between the woods meant that these were deployed side by side, not one behind the other. The Duke of York commanded the vanguard, now the right division, replacing Sir John Cornewaille and Sir Gilbert

Umfraville, who had led it up from Harfleur: it is said that
York begged the honour on bended knee. The veteran
Thomas, Lord Camoys replaced York in command of the
rearguard, now on the left, and the King himself was the
centre. Each division probably had dismounted men-at-arms
in its centre, with archers echeloned forward on each flank.
One English source speaks of 'wedges' of archers and it may
be that there were four such wedges, two on the left and
right flanks respectively, skirting the woods, and two
between the divisions. It has to be said, though, that contemporary accounts are obscure and often contradictory. On
the King's order the archers protected themselves with
sharpened stakes, driven diagonally into the ground, making
a chequerboard of obstacles rather than a solid fence, a
reflection of Henry's concern that the French might try to
ride down his bowmen.

The French formed up less tidily. Although the Constable
was in nominal command, the presence of so many royal
dukes made his task more than usually difficult.
Nevertheless, it is clear from an unusual extant document
that there was indeed a plan of deployment and attack,
drawn up by Marshal Boucicaut. It was not followed in
detail, but it did include provision for flank and rear attacks
by mounted troops, and sought to combine these with
advance of the dismounted men-at-arms in the centre. In the
event, there were probably two 'battles', a vanguard
followed by the main body, with two smaller wings on
either side of the vanguard, all consisting of dismounted
men-at-arms: the French were to repeat the policy, which
had failed so dismally at Poitiers, of sending their knights
forward on foot to avoid the risk of arrow-maddened horses.
The available archers and crossbowmen were divided
between the two wings. There were two bodies of mounted
men in the rear, one tasked (as Henry had indeed suspected)
with riding down the archers on the flanks, and the other
smaller part tasked with hooking round to the English camp
to seize the horses of Henry's dismounted men. The deployment was complicated by the fact that, in sharp contrast to
their opponents, the French enjoyed no clearly defined chain

of command, and there were many 'independent and independently minded' noblemen in their ranks that day.

The armies faced one another throughout the morning. Henry's only chance of victory lay in provoking an attack, for his archers were essentially a defensive, not an offensive asset, but the French, the evidence of Crécy and Poitiers before them, were reluctant to assault into arrows and, in any event, had much to gain from simply blocking the road and letting fear and hunger work in their favour. At about 10 a.m. the King decided to move forward. Every man dropped to his knees, kissed the earth and took a morsel of it into his mouth, in what Juliet Barker calls an 'extraordinary ritual . . . conducted with all the solemnity of a genuine Church sacrament', echoing both the Eucharist and the words of the burial service. Then he ordered: 'In the name of Almighty God, and of Saint George, Advance banner!' This was more easily ordered than executed, for the archers had first to pull out their stakes before the army advanced in line, slowly, and with frequent halts to preserve alignment and to allow dismounted knights to catch their breath, until it reached the centre of the field, possibly around the Azincourt–Tramecourt road that now crosses it and the copse marking one of the grave pits. At this point the line halted, the archers hammered in their stakes: it is a measure of poor French co-ordination that the archers were allowed to place their stakes without interference.

Provocation had the desired effect. The cavalry designated to attack the archers streamed off at once, but in swirls of horsemen rather than formed ranks, and made slow going across the muddy ground, with the first division of dismounted men-at-arms squelching along behind them. The experienced Sir Thomas Erpingham, in overall command of the archers, threw his baton of office into the air and shouted, 'Now strike!' The French horsemen were pelted by the arrow storm: some were killed by arrows or amongst the stakes and more streamed back in disorder. The vanguard, meanwhile, made heavy going across the mud, pelted by arrows, buffeted by the survivors of the cavalry attack coming off at the gallop and compressed as the space between the woods narrowed. The chaos worsened as the knights lumbered on, visors down and heads lowered, like men walking into driving rain, towards their natural rivals – the English men-at-arms – and shied away from the projecting wings of archers. Nevertheless, the sheer momentum of the attack and the bravery of the men who made it eventually brought the survivors up to handstrokes with the English line and the weight of their attack drove it back several feet. At this juncture the archers abandoned their bows and fell to with swords, axes and the heavy mallets they had used to drive in their stakes. The French were no match for the combination of men-at-arms and nimble archers, and the men of the vanguard died in their hundreds, stabbed and suffocated, with their comrades main body, trundling on in sweaty blindness, pressing on to add to the confusion. 'Great people of them were slain without any stroke,' records one chronicler.

Anne Curry is right to observe that the mêlée – 'bloody murder' at its worst – receives less attention than it deserves and certainly the English did not have things their own way. A French knight, probably one of a small group that had sworn to kill Henry or perish in the attempt, lopped a fleuret from the King's crown before he died. Humphrey, Duke of Gloucester took a sword thrust in the groin and the Duke of York was killed with almost a quarter of his original retinue, suggesting that the fighting might have been heaviest on the English right. After something between two and four hours of fighting, with successive waves of Frenchmen eddying amongst the heaps of dead and wounded, the fighting had abated sufficiently for Englishmen to seize prisoners for ransom and sometimes great men were captured by comparatively humble ones: the Duke of Bourbon was taken by Ralph Fowne and Marshal Boucicaut by William Wolf Esquire.

This went on for some time, but the combination of a raid on the English baggage and threatening moves by uncommitted parts of the French host, some elements of which straggled in as the day went on, presented Henry with a cruel dilemma. Most of the prisoners were still in armour

and discarded weapons lay about. His men could not guard
the prisoners and deal with a new attack. He ordered that
the prisoners should be killed, warning that any Englishman
who disobeyed would be hanged. The order was rescinded as
soon as it became clear that the threats to front and rear
were illusory, but in the interim hundreds, perhaps
thousands, of Frenchmen had been butchered. Henry has
been harshly judged for this decision, but he was acting in
the spirit of a cruel age. No French chronicler condemned
him for it: one indeed suggested that it was the natural
consequence of his own countrymen attempting to renew a
lost battle. Moreover, the fact that about 2,000 prisoners
later accompanied the English to Calais indicates that the
slaughter was by no means complete. Even if humanitarian
considerations did not weigh heavily upon them, the
English cannot have welcomed the order to kill men whose
ransom could make their fortunes and the order to kill the
prisoners cannot have been easy to transmit or supervise.

French casualties, in the battle itself and the throat-
cutting that followed it, were appallingly heavy. Constable
d'Albret fell and with him the Dukes of Alençon, Brabant
and Bar, at least eight counts and perhaps ninety other
noblemen and the Archbishop of Sens, as well as over 1,500
knights, though, just as we cannot be sure of the strength of
the French army, it is hard to be sure of the scale of its losses.
Local men featured prominently in the lists of those killed
or taken, depriving the region of leadership it needed during
these turbulent years. The French King's household had
suffered severely: Guichard Dauphin, its grand master, and
Guillaume de Martel, who bore the oriflamme (though the
banner itself was not present that day as the king was not in
the field) were killed. So too was the Ponthieu nobleman
David de Rambures, grand master of the crossbows, with
three of his five sons. Some were buried in nearby church-
yards, with many being carried to Hesdin, seven miles away,
or embalmed for burial at home; 5,000 or more were interred
in pits on the battlefield itself and it is said that one such pit,
marked by a nineteenth-century cross and surrounded by
trees, is on the north edge of the Calais road, just in front of

the second position of the English line, though without archaeological evidence it is impossible to be sure that this does indeed mark a common grave.

English losses, both killed and wounded, numbered 500 at the very most. The story that the Duke of York had been suffocated in the press is probably a Tudor fancy, but both he and the young Earl of Suffolk, whose father had died before Harfleur, were among the slain. Sir Richard Kyghley and Daffyd ap Llewellyn, known as Davy Gam, were also killed, and of the 112 named English dead, two-thirds were archers. Most of the dead English were burnt in a barn at Maisoncelles, but the bodies of York and Suffolk were boiled to remove the flesh, and their bones were sent back for burial at home.

There is a good visitor centre in Maisoncelles and the local roads are edged with life-sized cartoon medieval soldiers: a row of them marks the English first position. The woods on both flanks are still there, though it is little more than an educated guess to say that they retain their medieval extent. Azincourt castle, once visible from the field, has long since disappeared, but the visitor centre has some floor tiles discovered in excavations in 1976. The Château of Tramecourt post-dates the battle by three centuries, but stands quietly in soft red brick at the end of a wide avenue just off the battlefield's northern edge. A nearby monument pays tribute to three members of the noble family of de Chabot Tramecourt who perished in Second World War concentration camps.

With the French cleared from his path, Henry marched on to Calais, where he was welcomed by its captain, the Earl of Warwick. He sailed for England shortly afterwards, at the close of a campaign that had again demonstrated the superiority of a small, cohesive force over a larger, ill-disciplined rabble of gentility. Nor should the role of the longbow be underestimated. It was the dominant defensive weapon of its age. The fact that it was used by men who definitely did not come from the knightly class astounded contemporaries. 'The might of the realm', wrote the chronicler Froissart, 'standeth upon archers which are not rich men.'

But archers were no panacea. They could not be improvised, for it took a man years of practice to attain the skills – and develop the muscles – required for effective use of the longbow. French and Scottish kings failed in their attempts to transplant the weapons of their enemies into their own kingdoms, and by the end of the fifteenth century the social fabric that had sustained the archer was changing even in England. There were increasing complaints that young men no longer practised at the butts after church on Sunday, and the expertise which had littered the slopes of the Vallée aux Clercs with French dead at last vanished towards the end of the sixteenth century. Moreover, the tactics which had won Crécy and Agincourt worked less well against French forces exhibiting a discipline and cohesion absent from the armies of Philip VI and Charles VI. Standing rigidly on the defensive in the face of a more mobile enemy brought Sir Thomas Kyriel defeat at Formigny in 1450 and perhaps Talbot's headstrong assault at Castillon, three years later, reflected recognition of the fact that the days were long gone when French noblemen would squander their lives in a welter of bodkin points and grey goose feathers.

The archers of Crécy and Agincourt have left their mark upon us in the strangest ways. Many English towns have an open space called the butts, on which our forefathers would have practised their archery. And the less than polite gesture of extending two fingers in a derisive version of the Churchillian V sign dates back to the Hundred Years War. Captured archers sometimes had two fingers cut from their right hands to prevent them drawing a bow again – cruel, perhaps, but less cruel than execution and cheaper than perpetual imprisonment. An archer, temporarily at a dis-advantage and scuttling for cover, might disdainfully flaunt the appropriate digits as he reached refuge to show that he was still intact and would be back – with a bow in his hand and arrows in his belt.

Commercial life in the towns of the north blossomed as the risk of English *chevauchées* receded. A new sense of confidence and civic pride was embodied in the public

buildings of the late Middle Ages, like the town hall and
clock tower of Arras, sadly destroyed in the First World War
but rebuilt on the same grand scale, and the superb clock
tower in Béthune. Guildhalls and cloth halls reflected the
power of the guilds, and bear testimony to the volume of
trade that flowed through the towns of Flanders, Artois and
Picardy alike.

The waning of the English menace did not, however, mean
lasting safety for the comfortable bourgeoisie of the bustling
towns of the north. The area north of the Somme had been
under the control of the Dukes of Burgundy. Duke Charles
the Bold died fighting the Swiss at Nancy in 1477, and his
daughter and heiress Mary married the Emperor Maximilian
of Austria. Their grandson Charles V was elected Holy
Roman Emperor in 1519, and his election ushered in a long
conflict between the houses of Hapsburg and Valois.

The period began with an attempted rapprochement
between France and England, when Henry VIII and Francis I
met between Guines, English because it was within the
'pale' surrounding Calais, and the French town of Ardres.
Each monarch tried to outdo the other in magnificence and
the meeting place became known as the Field of the Cloth of
Gold. A signboard on the A26 just outside Calais now marks
its site, although it would be hard to imagine anything less
like cloth of gold than Calais's dreary industrial fringe. The
meeting failed to achieve its desired result, for Henry,
instead of concluding an alliance with Francis, supported
Charles instead. This might have been an omen for French
fortunes, for the war went badly. The French took Hesdin,
and the rival armies scoured Artois and Picardy. Pestilence
followed in the wake of war, and both Arras and St-Omer
were ravaged. Disastrous defeat at Pavia forced Francis to
come to terms, first with the Treaty of Madrid in 1526, then,
on 3 August 1529, with the Treaty of Cambrai, which ceded
the sovereignty of Artois to the Emperor.

Charles established a Council of Artois in an effort to
mark out the province's judicial independence from Paris
and his sister, Mary of Hungary, governed the block of

Hapsburg territories in the north, which now included it. When war broke out once more in 1536, Spanish troops failed before Péronne, and Francis took Hesdin, St-Pol, Lillers and St-Venant. The Imperialists counter-attacked, but were unable to take Thérouanne, which remained a French island in a Spanish sea.

A ten-year truce, signed in 1537, broke down in 1542 and the French pressed deep into Artois, sacking Tournehem and Lillers. This encouraged the Emperor to persuade his English ally to attack the tiny French enclave around Boulogne, joined to France by a thin coastal strip defended by Montreuil. Boulogne was defended by Jacques de Coucy with a French garrison of 1,800 men, 500 Italian mercenaries and the town's militia under its mayor, Antoine Eurvin. It was besieged by Henry VIII in person with an army of 30,000 men and powerful siege artillery. The walls of Boulogne had been built in the thirteenth century by Philip Hurepel, Count of Boulogne, son of the French King Philip Augustus. They covered a rectangle 325 by 400 metres, pierced by four gates, each defended by two flanking towers.

On the eastern corner of the town's defences and joined to them by a bridge across its moat, stood the castle, a striking example of the polygonal work without a central keep, which Philip Augustus's engineers had developed. Boulogne castle had nine sides, with sharply roofed round towers at their angles. The governor maintained a stout resistance till mid-September when, against the advice of the leading citizens, who no doubt scented what was to come, he agreed to surrender. The garrison and some of the townspeople were permitted to march off for Etaples and Abbeville, but were harried as they went by pillagers. The town was sacked and the miraculous statue of the Virgin, left in Boulogne by the Merovingian King Dagobert in 636, was carried off to England.

The Treaty of Crépy-en-Valois, signed a few days after the surrender of Boulogne, ended the Franco-Spanish war, but it took tortuous negotiations and the payment of 400,000 gold *écus*, to persuade the English to relinquish Boulogne. The town's return to French rule was marked by Henry II's gift of a silver statue of the Virgin, to replace that stolen by the English. Fourteen years later the boot was on the other foot. The Duke of Guise took Calais, the last English possession in France, depriving the English of a base from which they could raid the towns of the north.

The busy fishing port of Boulogne is still dominated by the high town, with Philip Hurepel's walls around it and his castle looking down on the Boulevard Eurvin, named after the town's valiant mayor. The ramparts offer stunning views and the Gayette Tower, at their western corner, directly opposite the castle, marks the spot where, in 1785, the balloonist Pilatre de Rosier took off in his unsuccessful attempt to cross the Channel.

Boulogne is rich in Napoleonic associations. In 1803 Napoleon established a huge camp around the town, with the ostensible purpose of invading England, but the defeat of his fleet at Trafalgar meant that he was unable to secure safe sea passage for his troops and they eventually set off westwards instead, to march deep into what is now Czechoslovakia and beat the Austrian and Russian armies at Austerlitz. The site of the camp is marked by the *Grande Armée* Column, three kilometres north of the town. The imposing marble tower can be ascended by 263 steps and the grey cliffs of Dover can be seen from its top on fine days. The Legion of Honour monument, just off the D940, marks the spot where Napoleon held the second investiture into the Legion of Honour, in August 1804. It was at Boulogne that the Emperor's nephew Louis-Napoleon attempted an abortive coup in 1840. The venture failed and he was imprisoned in the fortress of Ham, whence he escaped to England to re-emerge as president of the Second Republic and, later, as the ill-starred Emperor Napoleon III.

Fighting broke out again in 1551. This time the emperor concentrated on eliminating Thérouanne and Hesdin, the French outposts in Artois: in 1553 both were razed to the ground. A few fragments of Thérouanne cathedral, once a jewel of Gothic architecture, remain in the new town. Hesdin was rebuilt on its present site and the village of Viel-Hesdin, five kilometres to the south-east, off the D340, marks where the old town stood, but there is no trace of the palace of the Counts of Artois with its garden filled with mechanical practical jokes – the 'wonders of Hesdin'. Although Charles V abdicated in favour of Philip II, the war went on. The French were defeated at St-Quentin in 1557, but Guise's capture of Calais the following year was a more than adequate compensation. The Treaty of Cateau-Cambrésis, signed in 1559, left Calais and parts of Lorraine in French hands in return for renunciation of French claims in Italy.

The growth of Lutheranism in the Spanish Netherlands was followed by its spread into Artois. Coastal areas were most seriously affected because of their commercial contact with England, but by the 1560s, despite repeated royal decrees against the reformed religion, it was firmly established across much of the north. Although Margaret of Parma, Philip II's governor of the Netherlands, warned that repression would provoke an uprising, the King persisted in abolishing fiscal and political privileges in the Netherlands. In 1565 there was a popular outburst of iconoclasm, which was especially strong in the valley of the Lys. Order was restored with the assistance of the Prince of Orange and the counts of Egmont and Horn, but Spanish heavy-handedness was soon to make a difficult situation infinitely worse.

In 1567 Philip sent the Duke of Alva with an army of Italian veterans up to the troublesome provinces. Margaret resigned as regent and Alva set about drowning resistance in blood. He could not see that there were complex political as well as religious elements in the dispute. Egmont, for example, was a staunch Catholic, but valued the traditional rights and privileges of the Netherlands. Egmont and Horn were beheaded, and perhaps 6,000 other citizens were executed. Alva's heavy hand reached down to the southern borders of Flanders and the iconoclasts in the Lys valley suffered severely.

French territory was also rocked by disturbances: the citizens of Boulogne murdered or expelled its priests in 1567, but Catholicism was re-established there the following year. Initial Catholic successes failed to stamp out

Protestantism and Alva was recalled in disgrace in 1573. His successors faced a deteriorating situation, not only in Holland, where the Prince of Orange's influence grew steadily, but also in Artois, where there was a serious Protestant revolt at Arras in 1577. Two years later the Duke of Parma, who had taken over as regent of the Netherlands in 1578, reached an accommodation with the Catholic nobility of Flanders and Artois, recognising some of their demands for traditional liberties and in so doing detaching them from the hardline Protestants to the north. Parma then turned north and achieved much military success, taking Antwerp in 1585. But Philip's insistence that he should aid the Catholics in the French wars of religion and prepare for the invasion of England into the bargain meant that his strength was dispersed and while he was busy in Artois, Maurice of Nassau, William of Orange's successor, consolidated the Protestant position, laying the foundations of an independent Netherlands. Parma was slightly wounded in a skirmish at Caudebec, just down the Seine from Rouen, but he sickened and died at Arras in 1592. Fighting went on, with Frenchmen of the Catholic League being supported by the Spaniards, until the Peace of Vervins at last brought the war to an end in 1598.

The balance of power in Flanders, Artois and Picardy was transformed over the next few years. Spanish recognition of the independence of the United Provinces – the modern Holland – left Flanders and Artois in Spanish hands, and in the early 1600s the Catholic revival saw the foundation of Jesuit colleges at Arras, Aire-sur-la-Lys, Hesdin and St-Omer. With the outbreak of the Thirty Years War the prosperity of the first three decades of the century evaporated as the French and Spanish contended for the possession of Artois. The Spaniards achieved early successes and in 1636 reached the Somme, taking Corbie. As the Dutch renewed the pressure on Spanish forces in the north, so the French counter-attacked, capturing Arras in 1640, and going on to take Béthune and Lens in 1645. Condé won a brilliant victory at Lens on 19 August 1648, but France's weakness during her internal troubles enabled the Spanish to push back into Artois. In 1654 Condé, now fighting for the Spaniards, besieged Arras, but on 24 August Turenne led a French army to the town's relief, inflicting a sharp defeat on Condé, who lost 3,000 men. It was a measure of Condé's skill and determination that he held his army together and made a fighting retreat to the safety of Cambrai.

The same commanders clashed again near Dunkirk on 14 June 1658. Turenne, operating on the Flanders coast with a French army, which included an 8,000-strong English contingent, assisted by the English fleet, took Mardyk and moved along the coast to besiege Dunkirk. Condé and Don John of Austria approached with a relieving army: Condé recognised that he was at a disadvantage, for although numbers were roughly equal, most of the Spanish guns had not yet come up. His advice was overruled and the Spanish drew up with their line stretching from the shore to the area where the Fort des Dunes now stands, just north of the N1, six kilometres from the centre of Dunkirk. Turenne saw that the infantry on the Spanish right, down on the beach, were in difficulties because of the rising tide and heavy fire from the English fleet, and threw the weight of his attack against the Spanish left. After four hours of fighting the Spanish were defeated with the loss of 6,000 men for only 400 French and English casualties. The victory enabled Turenne to take Dunkirk, then to move into Flanders to threaten Brussels and Ghent.

The Treaty of the Pyrenees, signed in November 1659, set the seal on Turenne's gains. Artois, with the exception of Aire and St-Omer – 'reserved Artois' – became French. The inhabitants were less than enthusiastic. They feared that their rights and privileges would be swallowed up, and that French taxation – particularly the hated *gabelle* salt tax – would bear heavily upon them. Despite the fact that the Treaty spoke of the observance of the 'Catholic, Apostolic and Roman religion' many Catholics were profoundly suspicious of a French government that had allied itself to English and Dutch Protestants, and chose to leave Artois for Spanish Flanders. A sharp revolt flared up in the Boulonnais in 1662 and it was heavily put down by a government, which feared that unrest might spread across the whole of Artois.

During the War of Devolution the French strengthened their grip on the north. Dunkirk, ceded to the English, was bought back and the French moved up into Flanders, taking Douai and Lille. In the Dutch War Aire and St-Omer were both captured, and the Treaty of Nijmegen (1678) put an end to 'reserved Artois' and gave the French a firm hold on Flanders. This grip was strengthened by the great French military engineer Vauban, who laid out what he called the *pré carré*, a term best translated idiomatically if inelegantly as 'defensible frontier zone'. While his distinguished predecessor Jean Errard (1554–1610) had concentrated on defending the Picardy-Artois border, with works at Montreuil, Doullens, Amiens and Ham, the new Flanders frontier was Vauban's chief concern. An outer line ran from Dunkirk on the coast, to Bergues, Furnes, Knocke, Ypres, Menin, Lille, Tournai, Mortagne, Condé, Valenciennes, Le Quesnoy, Maubeuge, Philippeville and Dinant on the Meuse. The inner belt stretched from Gravelines to St-Omer, Aire-sur-la-Lys, Béthune, Arras, Douai, Bouchain, Cambrai, Landrecies, Avesnes, Mariembourg, Rocroi and Charleville.

Much of Vauban's work is still extant – indeed, few of these towns do not retain at least a ravelin or section of ramparts. Some have much more and almost all have a Rue Vauban or a bar that bears the genial Marshal's name: Maubeuge even has three tower blocks, Vauban I, II and III, which would have given the great man pause for thought. But the defences of Le Quesnoy remain in such excellent preservation that it is best to let one speak for all and to let this sleepy little town tell us something about the principles of artillery fortification.

Le Quesnoy was part of the Spanish Netherlands until it became French in 1659. It had already been fortified, but the thoughtful Spaniards demolished part of the works before abandoning the place to Turenne in 1654. Vauban found it bounded by an old wall dating from the time of Charles V, strengthened by later bastions. He set about thorough rebuilding, basing the defence on eight powerful works – the

Imperial, Royal, Caesar, Soyez, St-Martin, Guard, Castle and Forest (or Green) Bastions.

The bastion is the quintessential ingredient of artillery fortification. It is a four-sided work, projecting from the main rampart, with two faces jutting out to the country beyond, and two flanks looking along the line of ditch and rampart. Guns stood atop its *terreplein* and fired over its broad parapet. Those on the bastion's faces reached out across the glacis beyond the ditch. Those on the flanks swept the ditch that attackers would have to cross, with the guns in the next bastion providing mutual support. These flanking pieces were often sheltered by an *orillon*, a projecting shoulder at each end of the bastion's faces that makes it hard for the attacker's cannon to hit the bastion's flanks. Several bastions at Quesnoy have *orillons*. The Caesar Bastion, just north of the causeway carrying the Valenciennes road over the ditch, has well-defined redbrick *orillons* covering its gently curving flank.

A bastion might be given extra firepower by casemates, masonry chambers housing guns, which fired through embrasures. These could be added to strengthen an existing bastion or simply – as improvements in artillery techniques made the attacker's fire more damaging – to get guns off the vulnerable *terreplein*. Sometimes casemates are subterranean, forming in effect the bastion's lower gundeck: the embrasures grinning out from the Green Bastion have bricklined casemates behind. Sometimes, however, they are up on the *terreplein*, like the casemates on the Guard Bastion. These date from the 1870s and make the point that fortification was in a constant state of evolution, with additions and modifications meeting changed threats or new fashions. Engineers of Vauban's generation would increase the firepower of a bastion by siting a *cavalier*, a raised interior battery, in the middle of it.

Between the bastions ran the curtain, the main wall of the fortification. It formed part of a carefully planned profile designed to ensure that an attacker had no vulnerable masonry to engage and that his advance towards the body of the fortress took him across a series of features, which made the defender's life easier and his own shorter. Seen from the enemy side, the ramparts barely showed above the gentle slope of the glacis. This open area, swept by the defenders' guns, was the subject of peacetime regulations which prevented permanent structures from being built upon it, so that when a fortress was put into a 'state of siege', a phrase that had legal effect, clapboard summerhouses and garden sheds could be easily knocked down. Even today a change in the character of architecture marks the glacis of many towns of the *pré carré*, and the glacis at Quesnoy is easy to identify: to the south a convenient campsite gives a good feel for its original expanse.

The fortress side of the glacis ended with a palisade and a drop down into the covered way, so called because it was just deep enough to permit infantrymen on it to be covered from fire and view. There was a fire-step running just below the palisade, so that the covered way could provide a line of defence as well as a sheltered communication route.

To its rear stood the counterscarp, the wall which retained the outer side of the ditch. Galleries within the counterscarp gave access to ditch defences – an increasingly common feature of later fortification – or were sited to help the defender disrupt an attacker's attempt to dig his way beneath key points to establish powder-filled mines. The ditch might be wet or dry: wet ditches were usually wide and shallow, dry ditches narrow and deep. The hands of the engineer were tied when it came to ditches, for their character was largely determined by the amount of spoil extracted from them to build the ramparts, the level of the water-table or the presence of a reliable water supply. A wet ditch was smelly and unhealthy in summer, and could lose its value altogether if the garrison did not break the ice regularly during a sharp winter: it also meant that the defenders could not form up in the safety of the ditch before launching a sortie. It did, however, give the besiegers a trying time if they attempted to carry out a 'passage of the ditch' by sapping their way forward across it.

Even if its constructors had decided against a wet ditch, no fortress could survive without water for drinking and for

operating the mills that ground its corn. Windmills were all
very well in the piping times of peace, but in war they
disintegrated in a pleasing rain of brick dust and splintered
sails. Quesnoy had a number of water mills, and until the
eighteenth century a windmill stood on what is now the
Green Bastion and was once known as the Forest or
Windmill Bastion. Vauban sought to put Quesnoy safely
behind a very wide wet ditch, but the town did not lie on a
river and ensuring an adequate supply of water was always
a problem. In the 1660s and 1670s an elaborate system of
dams converted a marshy stream, which ran north to join
the Rhonelle at Villers-Pont into the Horse-Shoe Lake
(*Etang du Fer à cheval*), which still covers the defences
between the Valenciennes and St-Martin Gates. Subsequent
work produced the Red Bridge Lake (*Etang du Pont Rouge*)
protecting the south-west front of the fortress and a canal
was eventually dug connecting Quesnoy with watercourses
in the Forest of Mormal.

Flooding its ditches put the final gloss on a refurbished
fortress and in the summer of 1673, with Spain about to
enter the war against France, governors were ordered to
place their commands 'beyond insult', that is, to ensure that
they could not be taken by a sudden *coup de main*. The
governor of Quesnoy warned that the redoubt covering the
sluices on the Horse-Shoe Lake was not yet finished and
attackers might easily drain the lake. In September the
sluices were opened and water poured into the ditches.
Although there were some problems with the leaking dams
(*batardeaux*), the results were generally successful, and
Vauban's earthworks and masonry seemed to sit securely
behind a sheet of water. The governor had little time for self-
congratulation, for on 11 October part of the curtain
between the Royal and Caesar Bastions slid into the ditch. It
had been built higher than originally intended to bring it
level with the flanking bastions and the scarp revetment,
the masonry supporting the earth above the fortress side of
the ditch, proved unequal to the added strain. The hard-
headed Louvois ordered the contractors to repair the damage
at their own expense.

The scarp revetment at Quesnoy is of brick, the usual material in the fortresses of the *pré carré*. Further south, where good local stone is readily available, revetments tend to be of ashlar. Along the revetment's upper edge ran the *cordon*, a coping stone, with the *tablette*, or parapet retaining wall, on top of it. The parapet was made of well-packed earth, usually at least eighteen feet thick and proof against all the weapons of the age. Guns could either be mounted on high fortress carriages to fire over the parapet – the cheapest solution – or could stare out through embrasures, slits in the parapet lined with brick or stone and splayed so as to combine a good field of fire with maximum protection. On the town side of the parapet the *banquette*, an infantry fire step, rose from the *terreplein*. The ground then sloped down to the town at an angle of some forty-five degrees, and occasional inclined ramps enabled the garrison to haul guns and march men up to the ramparts.

Two centuries of urban development mean that it is seldom easy to see ramparts from a besieger's point of view, though at Quesnoy the visitor is more fortunate. It was theoretically possible to carry a fortress by escalade, putting scaling ladders in the ditch and climbing over the parapet, but revetments were usually so high as to make this difficult even if the defenders did not complicate matters with gusts of cannister shot, musketry and grenades. Quesnoy was taken by storm only once, in November 1918 when the New Zealanders fought their way in between the St-Martin and Guard Bastions. The New Zealand Memorial, a handsome relief monument set in the scarp revetment, shows the scene, with steel-helmeted figures scrambling up the wall and a winged victory crowning their efforts with laurel. Breaching the curtain by battering the revetment to bring the ramparts tumbling into the ditch could not be accomplished by guns on the glacis, for they could only see and engage the parapet. Mining was a possibility, although rocky subsoil, a high water-table, like that at Quesnoy, and the defenders' counter-mines made it a risky business.

If an attacker was unable to enter a fortress by a ruse, shrank from the perils of immediate escalade and could not afford the time to starve the defenders out, he had no alternative to a formal siege. He would begin by opening a trench line parallel to the defences, just out of effective range of artillery (the first parallel), then sapping forward in zigzag lines, open a second parallel, and later a third, building batteries whose guns concentrated their fire on the parapets opposite, while mortars sent bombs curling into the body of the fortress. The defenders might mount a sortie and could do much harm if they managed to get amongst the attacking artillery, spiking pieces by driving nails or files into their touch-holes, or inflicting more permanent damage by stuffing a cannon's muzzle with earth and turf, and firing it.

It was rare for a defender to be able to do more than buy time by such means and the siege would grind inexorably on. Eventually the defending guns in the works under attack would be dismounted by direct hits. Two Russian bronze *licornes* in the Royal Armouries' artillery collection in Fort Nelson, above Portsmouth, and French guns in the forecourt of the *Invalides* in Paris show the effects of hits square on their muzzles, blows that would produce a characteristic bell-like ring and send the piece bounding from its carriage. With the cannon opposite dismounted and the parapet ploughed by shot, the attacker could assault a section of the covered way. This is where grenadiers came into their own, as the song reminds us:

> When e're we are commanded to storm the palisades
> Our leaders march with fusees, and we with hand grenades:
> We throw them from the glacis, about our enemies' ears . . .

With the covered way duly stormed, the attacker would entrench himself there and establish breaching batteries. Heavy guns – 24-pounders were the workhorses of siege artillery – would then engage the revetment, their fire carefully controlled so as to knock away a line of masonry at the foot of the wall. Knowing engineers might have

constructed a *tenaille*, a low wall at the foot of the revetment, which prevented the attacker from hitting its very
base: the viewpoint for the New Zealand Memorial at
Quesnoy stands on just such a *tenaille*. But cunning
attackers tried to select sections of the front that lacked this
extra layer of protection. After a day or two of hammering,
the wall would come down, and a wise governor might
decide to beat the *chamade* and negotiate for terms, for if the
besieger's infantry came roaring through the breach, across
the mutilated bodies of their comrades, they could be
expected to respect neither the lives of the garrison nor the
property and chastity of the citizens.

This unpleasant outcome could be delayed by the addition of defences covering the bastions, forcing the attacker
to chew his way into the fortress bite by bite. The ravelin,
sometimes known by its descriptive French name *demi-
lune*, was the favourite outwork. It was triangular, with a
scarp revetment and parapet facing the enemy, and as time
went on it became bigger than Vauban's ideal sixty-five
metres wide by sixty deep, and was embellished by added
defences like squared flanks and interior redoubts. The
ravelin between the Guard and St-Martin Bastions is a good
example of its tribe. Just behind it, where the path comes
out of the ditch, is part of a twelfth-century tower, more
evidence that ground judged important by one engineer is
often as attractive to his successors. This ravelin's well-
defined flanks each provide a good thirty metres of rampart
to mount cannon covering the ditch to either side.

Between the ravelins and the bastions themselves can be
found counter-guards, detached bastions that stand just in
front of the bastions proper, shielding the masonry of the
bastions and the curtain from enemy fire. Sometimes they
might be linked together to form a complete outer skin of
fortification or *couvreface générale*. At Quesnoy there is a
plain counter-guard in front of the Castle Bastion, while the
counter-guards stretching round the western side of the
fortress do indeed form a *couvreface générale*. These are
eighteenth-century works, begun under the direction of the
Count of Valory, governor of Quesnoy and director of

fortifications in Hainault in the early years of the century.

The hornwork is rightly described by Christopher Duffy as 'a most spectacular piece of architecture'. It consisted of two lines of rampart running out into the country, ending in a 'head front' with half-bastions at its corners – the 'horns' – and a length of curtain in between, usually with a ravelin in front. Its spiky outline gives it the alternative name of crownwork. Vauban was fond of hornworks and we shall shortly see how two such works at Lille made Prince Eugène's life difficult in 1708. The huge hornwork at Quesnoy, however, was the work of Valory. His predecessors, from Vauban onwards, had mulled over the problem posed by the suburb of Fauroeulx, just to the south-west. It stands on higher ground than the area of town behind the Green Bastion and its capture would provide an enemy with a useful point of vantage.

Work on the Fauroeulx Hornwork eventually began in 1738 and most of it was completed in five years, although as late as 1777 it was noted that more remained to be done. Each wing, running some 400 metres to the salient of its half-bastion, was composed of three lengths of curtain with short flanks linking them. A ravelin (now demolished) protected the head of the work and counter-guards shielded the two half-bastions. The road joining the suburb to the town was defended by a parapet on either side, and the whole south-west front was improved by further work on the Green Bastion and the Fauroeulx Gate.

The Tourist Office, just inside the Fauroeulx Gate, gives visitors an excellent tourist guide, which includes a map of the fortifications, and informative signboards point out the various features of the defences. The visitor can spend anything from half an hour to a whole day exploring the town, and those who can cope with the language will find Bernard Debrabant's scholarly paperbacks, available at the Tourist Office, an invaluable guide. In addition to the features mentioned above, Quesnoy also contains an eighteenth-century barracks, now sensitively converted into flats (just off the southern edge of the square), and a large bombproof magazine on the St-Martin Bastion. The word 'bombproof' pre-dates aerial bombardment and originally referred to casemates or magazines with curved masonry ceilings covered with earth. They were intended to keep out the mortar bombs a besieger would lob into the body of a fortress, lowering civilian morale and blowing unprotected magazines sky-high, sometimes ending a siege with a single dramatic blast.

The defences of Quesnoy were put to the test several times. In June 1712 the Allies approached from the south, attacking the Guard, Castle and Forest Bastions. By cutting off the water supplied from the Forest of Mormal they deprived the town and its ditches of water, and the place duly surrendered. In September–October the French sapped their way forward from the north-west. The Allied governor, a Huguenot in Dutch service, had drained the Red Bridge Lake in order to profit from the sale of its fish, and when the French reached the covered way and took soundings in the ditch they were delighted to discover that the water was only three feet deep. The embarrassed governor surrendered the next day.

Le Quesnoy was unlucky again in 1793, when it surrendered to the Austrians after a bombardment that threw 57,000 projectiles into the town, and did fearful damage to private houses and public buildings alike. In 1794 victories at Tourcoing and Fleurus enabled the French to re-establish themselves, and Quesnoy was recaptured after another siege. The French fired red-hot shot into the town, igniting most of the buildings which had survived the Austrian attack. News of the town's fall was flashed to Paris on the Chappe optical telegraph, reaching the capital only an hour after the French garrison had marched in. In both these sieges the bombardment had proved decisive, for the attackers had failed to breach the curtain and in 1793 the garrison still held the entire length of the covered way and several outworks. Nevertheless, the fortifications were badly damaged in 1794 and were not immediately repaired, and Quesnoy, in no state to stand siege, fell easily on 24 June 1815.

With Quesnoy standing as our prime example of artillery

fortification, let us return to military developments in eighteenth-century Flanders. There was fighting in Hainault during the War of the League of Augsburg, and the War of Spanish Succession brought conflict back to Flanders and Artois. For the first years of the struggle operations in the north centred on Brabant, where Marlborough won his great victory at Ramillies in 1706.

In 1708 the French began by seeking to draw Marlborough away from French Flanders, where he might be expected to move following successes in Brabant. The two French commanders, the young Duke of Burgundy, ultra-conscious of his royal blood, and the blunt and scruffy Marshal Vendôme, got on badly, and much marching and counter-marching between Brussels and Louvain in June produced no useful result. The French then went into camp at Gembloux, between Brussels and Namur, and discussed future strategy. Vendôme wanted to push south and besiege Huy, while Burgundy hoped to find an opportunity in Flanders. Events proved Burgundy right. The French heard that the citizens of Bruges and Ghent were dissatisfied with Anglo-Dutch occupation, and planned to send flying columns into the towns, covering the operations with their main field army, which would remain between Marlborough and the target towns.

The first phase of the plan went well. Bruges and Ghent fell on 5 July, and Vendôme and Burgundy followed up, crossing the Dender at Ninove and breaking down its bridges behind them. Although Marlborough had his suspicions of Flemish loyalty and had sent a force to watch the citizens of Ghent, the French move took him by surprise and, at a stroke, he found himself cut off from his base at Ostend, with his garrisons in Menin and Courtrai isolated behind the French army.

Marlborough followed the French columns at a cracking pace and nearly caught them at Gooik, just north of the N28 between Halle and Ninove. He was not at his best: his recurrent migraine was worsened by fever and his trusted lieutenant Cadogan was temporarily away with Prince Eugène. Fortunately for him, Burgundy and Vendôme fell out again, the former wishing to besiege Menin, the latter preferring Oudenarde. Orders from Versailles supported Burgundy: Vendôme was authorised to blockade but not besiege Oudenarde. In the meantime the two generals had decided that Marlborough would probably make for Charleroi or Namur, and prepared to march to Lessines, where the Dender was bridged. They moved too slowly. Marlborough pushed reinforcements into Oudenarde and sent Cadogan, back from his mission to Eugène, to spearhead the army's advance on Lessines. On 10 July he secured the place and the Allies crossed the Dender in strength as the French fell back to the north, anxious to put the Escaut between themselves and Marlborough.

The French spent the night of the 10th on the east bank of the Escaut at Gavre, just north of Oudenarde, and began to cross the river the following morning. At the same time Cadogan, with a powerful advance guard and the bridging train, reached the river between Gavre and Oudenarde itself, and soon had five pontoon bridges operating and a detachment forward on the line of the Diepenbeek. Behind him came Marlborough with the Allied cavalry and the infantry followed, stepping out briskly in the July heat. In the early afternoon, elements of the French advance guard, making for the village of Heurne, bumped Allied cavalry on the Diepenbeek line and galloped back eastwards. The Duke of Biron, commanding the advance guard, quickly deployed his horsemen to check the pursuit and climbed the tower of Eine church, whence he could see clouds of dust to the south-west. He sent word back to his commanders at Gavre, giving them the almost incredible news that Marlborough's army had caught up with them.

The town of Oudenarde lies astride the Escaut. It had been taken by the French in 1674 and, with Lille, Courtrai, Armentières, Ath and Tournai, formed part of French gains at the Treaty of Nijmegen. In 1706 it fell easily to the Allies, and 1708 found it with a good garrison and modern fortifications – parts of them extant today. The banks of the Escaut were marshy in 1708, but have long since been drained. The Courtrai–Ghent railway line brushes the river

just north of the town and the N60 bypasses Oudenarde to
the west before joining the line of the old Ghent road near
Eine. For the three kilometres between Oudenarde and the
Diepenbeek the ground is flat and was well suited to cavalry
action in 1708, but is now cloaked by busy road junctions
and an industrial suburb.

The Diepenbeek joins the Escaut near Eine and its
tributary the Marollebeek runs north from their confluence
before curling off westwards. Further north the Leedschebeek
flows westwards along the foot of the heights of Huise to
enter the Escaut near Gavere. The area between the
Diepenbeek and the Leedschebeek is, now as then, spattered
with small villages, though in 1708 it was considerably
more broken than it is today, with hedges and gullies
making it difficult going for large bodies of troops. The main
road stretching north-west from Oudenarde (the modern
N59) runs past the village of Ooike and across the hilly,
wooded feature of the Boser Couter, which was to establish
the left flank of the battlefield as the Escaut did the right.

The French enjoyed a slight numerical superiority, with
85,000 men formed in ninety battalions of infantry and 170
squadrons of cavalry, while Marlborough commanded some
80,000 in eighty-five battalions and 150 squadrons. Neither
side could make much use of artillery, for their heavy pieces
were rumbling along well behind in the line of march, but
Cadogan was accompanied by thirty-two 'regimental pieces',
light guns that accompanied the infantry and deployed in
their firing line.

It was in the French interest to fight on the open ground
immediately north of Oudenarde, which would restrict the
space in which Marlborough could deploy and give full
opportunity to the numerically superior French cavalry.
Accordingly, Vendôme ordered Biron to use his seven Swiss
battalions to push Cadogan back so as to enable the French
to form their line of battle just south of the Diepenbeek. By
the time Biron had raked together just four battalions and
his twelve squadrons, he found that he was facing the whole
of Cadogan's force. Worse, Vendôme's chief of staff heard
that the area around the Diepenbeek was unsuitable for

cavalry. The reinforcements destined for Biron were accordingly called off and a new position was sought just below the heights of Huise. Unfortunately, news of the changed plan was too late to help Biron. At about 3 p.m. Cadogan fell upon him, swamping three of his Swiss battalions near Eine and catching the fourth at Heurne. Biron's survivors fell back across the Leedschebeek and Cadogan's pursuit was checked by the fire of a French battery on the high ground at Mullem.

With the French advance guard brushed away, Cadogan formed up facing west parallel with the Ghent road, with Natzmer's Prussian cavalry covering his right rear and Argyll's infantry coming up literally hotfoot to extend his southern flank to the confluence of the Diepenbeeke and Marollebeek. While the Allies were forming their line, the French launched a series of attacks. First, Grimaldi's cavalry were ordered across the Leedschebeek against the Allied left, but found the going too soft and wheeled about to join Burgundy at the windmill near Roijgem. Burgundy then ordered an attack on Groenewald. This, too, failed, but Vendôme arrived to rally the attackers and, without reference to Burgundy, fed more and more fresh troops into the battle.

Fortunately for Marlborough, a series of errors prevented the French from gaining full advantage from Vendôme's assault. Burgundy seems to have believed that Vendôme wanted the army deployed on the heights rather than in the valley at their foot and busied himself keeping the French left wing out of the battle. Then, when Vendôme – in the thick of the fray, redolent of snuff, sweat and black powder – called for Burgundy to throw his weight against Cadogan's right flank, Burgundy allowed himself to be convinced that he could not get cavalry across the Leedschebeek and sent a message lamely announcing that he was unable to support Vendôme.

Vendôme never received the message and launched his main attack towards Schaerken, on Cadogan's left, confident that Burgundy would roll up his right. The French were making good progress when fresh Allied infantry pounded up from the bridges. Twenty battalions of Lottum's

Dutchmen wheeled into line on Cadogan's left, another eighteen Hanoverian and Hessian battalions were hard on their heels, and seventeen British squadrons under Lumley were further back, just ahead of the Dutch General Overkirk and his wing of the army, moving up through Oudenarde.

At this crucial juncture Marlborough handed over command of his centre to Eugène and rode down to his left, where he performed a remarkably slick piece of tactical juggling. Lottum's men were pulled out of the line, though their colour parties were left behind to make it look as if they were still there. Then, along with Lumley's horse, they were fed behind Argyll and Cadogan, to strengthen the threatened right flank. Lumley's arrival enabled Eugène to launch Natzmer's cavalry north of Herlegem in a charge that bit deep into the French line before it was vigorously counter-attacked by the *Maison du Roi*. This costly venture had two useful functions: it took much of the pressure off the hard-pressed infantry of Cadogan and Argyll, and helped attract French attention to the eastern flank of the battlefield.

The real danger to the French now came from the west. At about 7.00 p.m. Marlborough sent the Hanoverians and some Dutch battalions against Diepenbecke village to threaten the French right. But at the same time Overkirk, with twelve Danish squadrons under Tilly and Oxenstiern's sixteen Dutch battalions, was moving up the Ooike road on to the Boser Couter. Overkirk swung right in Ooike and at about 7.30 p.m. pushed hard into the French right rear, making for Roijgem. The French line was already bent like a great U, and the combined attacks of Overkirk from the west and Eugène from the east threatened to encircle the bulk of the French infantry, still fighting hard on the Marollebeek south of the windmill. Rain and darkness saved the French from total disaster, but even so their losses were heavy. To the more than 3,000 Frenchmen killed or wounded in the battle must be added perhaps 7,000 prisoners. The Allies lost 3,000 men in all.

Marlborough was disappointed not to have accomplished more, but the victory was remarkable by any standards. The

Duke had snatched the initiative and held it, and his ability to turn a fluid situation to his advantage marks him as a commander of the highest ability. In the weeks that followed the Allies debated how best to exploit their success and eventually decided to lay siege to Lille. The capital of the old county of Flanders had been taken by Louis XIV in person after a short siege in 1667 and fortified by Vauban in 1668–74. It now sprawls far beyond its former boundaries, but the old town, around the Palais Rihour and the Place du Général de Gaulle, contains some well-restored examples of the Flemish baroque style, an attractive mixture of stone and red brick garnished with cherubs and cornucopias. Most of the ramparts surrounding the town have long since been elbowed aside, but the Ghent Gate in the old quarter sits in a section of curtain wall between two bastions. Even more remarkable is the striking Paris Gate, in a large grassy roundabout at the foot of the Rue de Paris. Built by Simon Vollant in 1682–95, in honour of Louis XIV, it is both city gate and triumphal arch, its shining ashlar embellished with the arms of France and Lille, and crowned by Victory.

The citadel of Lille, which was to cause the Allies such difficulties in 1708, remains intact. Rightly known as 'Vauban's masterpiece', it covers an area of over thirty-six hectares (ninety acres) and the 2,000 labourers who built it used sixty million bricks, made in four specially built brick kilns. Its six mighty bastions are covered by ravelins and were once totally surrounded by a moat fed by the River Deule. The main entrance, at the end of the Boulevard de la Liberté, takes the visitor across the ditch, with its formidable set of defences, including a section of *tenaille*. The gateway, the Porte Royale, reached by a short drawbridge between *tenaille* and ramparts, is surmounted by a massive relief bearing the arms of France.

The visitor who walks through the Porte Royale to enter the citadel itself is fortunate, for it is still a military barracks, named Quartier Boufflers – Boufflers Barracks – after its governor in 1708, and currently houses the 43rd Infantry Regiment, once more rejoicing in its old regime title *Royal des Vaisseaux*. Guided tours, available only on one day a week, must be booked via the office of tourism in the Place Rihour. The enthusiast who is unable to grasp this fleeting opportunity can nevertheless wander along the ditch and gain a good impression of the enormous strength of the bastions and ravelins, and see part of the water defences on the citadel's western side. The ditch was used by German firing parties in both world wars: fortress ditches have a grim track record in this respect, for they provide safe rifle ranges conveniently close to secure prison accommodation and city centres. A noble monument in the nearby Square Daubenton was inspired by Lille patriots shot in 1915. The Porte Royale gives on to a vast pentagonal *place d'armes*, surrounded by the brick and stone façades of Simon Vollant's stylish barracks. The two-hour visit includes the chapel, armoury and ramparts.

It took all Marlborough's skill to bring his 'Great Convoy' of ninety siege guns and twenty heavy mortars to within battering range of Lille and in mid-August the siege began. Eugène took charge, with 95,000 men under his hand, while Marlborough's 55,000 covered operations from their camp at Helchin, just north of the city. The governor of Lille, that squat and energetic Picard, Marshal Boufflers, had 12,000 men at his disposal and enjoyed the advice of the engineer du Puy de Vauban, nephew of the great Vauban. The first parallel was opened on 22 August. These trenches, just north of the present line of the Boulevard Robert Schuman as it curls round the city from its intersection with the Avenue A. Max to that with the Rue du Général de Gaulle, faced the Water Gate, near the bridge now carrying the Avenue A. Max across the Deule, and the Magdalene Gate, which stood just along the Rue du Pont Neuf from the church of Saint Mary Magdalene. Bastions II and III, jutting out of the curtain wall in this sector, were covered by two formidable hornworks. One of these was in the area of the barracks north of the Rue St-Sebastien, the other between the Rue du Pont Neuf and the Rue de Gand.

Eugène was not well served by his Huguenot engineers du Muy and Le Vasseur des Rocques, and although the trenches were driven ever closer, and the second and third parallels established, artillery fire was poorly co-ordinated, the attack covered too wide an area, and the two flanking hornworks poured a merciless fire on troops making for the bastions and curtain between them. An assault by 15,000 Dutch infantry on 21 September cost 1,000 men and Eugène himself was slightly wounded. Another attack, on 7 September, left the attackers with four small handholds on the covered way, and 3,000 dead and wounded scattered across the glacis.

The French took care not to let the Allies pursue the siege at their leisure and Marlborough had to cope with a blockade, which cut him off from his base at Brussels. He reopened a line of communication through Ostend and when the French intercepted a huge convoy under Major-General James Webb at Wynendaele on 28 September they were thoroughly beaten. But the French, too, had their successes and on the same day the Chevalier de Luxembourg managed to bluff his way into Lille with 1,500 cavalry, carrying some 40,000 pounds of powder. A few of his men were blown up when the besiegers realised that they were French and opened fire, but the episode did much to raise the garrison's morale.

Nevertheless, Eugène's engineers were learning and now concentrated on a more systematic attack. They eventually took the ravelin between the two bastions and were at last able to establish batteries to pound the curtain. The breaching batteries thundered into life on the morning of 21 October and, after a day of this treatment, which left 110 metres of the curtain tumbled in the ditch, Boufflers's drummers beat the *chamade*. The French were given generous terms: Boufflers and his 4,500 able-bodied defenders were permitted to retire into the citadel, and the sick, wounded, cavalry and other worthies who would consume rations without assisting the defence were allowed free passage to Douai.

These gentlemanly exchanges completed, Eugène opened his first parallel before the citadel on 26 September, when the truce was still in force, and did so on the Esplanade, although he had agreed not to attack from the town side.

The next phase of the siege took place across the Champ de Mars, the open ground between the Avenue des Maronniers and the Citadel. This time the Allied engineers were supremely cautious and sapped forward laboriously, stitching five parallels across the Champ de Mars and the ravelins above it before throwing up breaching batteries to hammer the bastions and curtain on either side of the Porte Royale. On 9 December Boufflers capitulated, marching out with the honours of war two days later. The siege had cost the Allies 15,000 men – only 1,250 of these lost in the methodical attack on the citadel.

In 1709 Marlborough took Tournai and in September narrowly won the bloody battle of Malplaquet. In the following year he was again active in Flanders, taking Douai, Béthune, St-Venant and Aire-sur-la-Lys. The capture of most of these fortresses was a wearisome and costly business, for the Allies were sadly short of talented engineers, and resolute defenders were usually able to command a heavy price for their earth and masonry. For all Marlborough's genius and a string of major victories, the Allies had merely dented the surface of the French fortress belt. Louis XIV, not surprisingly, became increasingly enthusiastic about frontier defence lines and in 1711 Marlborough found himself facing Villars in what the confident marshal called the lines of *Non Plus Ultra.* The French had thrown a great fortified barrier from the River Canche on the Channel coast to Namur on the Sambre. It made good use of natural obstacles, running along the sharp divide between the uplands of Artois and low-lying Flanders for part of its length. The long curtain of fieldworks was shored up by fortresses – Montreuil, Arras, Bouchain, Cambrai, Valenciennes and Maubeuge – and the wily Villars, with some 100,000 men, lurked behind the Sensée marshes in his centre, with Arras covering his left and Cambrai his right.

Mid-June found Marlborough with 122,000 men on the plain of Lens, facing the central section of the lines. He recognised that he had little chance of forcing them by frontal assault and so embarked upon a plan intended to make Villars move quickly from flank to flank, attacking when the French were off balance. He first ordered the governor of Douai to send a small force to secure the little town of Arleux, just in front of the lines and commanding the approaches to Bouchain. The place was captured and fortified, and, as Marlborough had predicted, Villars lost no time in attacking it. Marlborough sent Cadogan with a flying column to its aid, but let him know privately that he did not want relief to arrive in time. Arleux fell and Marlborough gave an uncharacteristic display of fury – for the benefit of French agents. Villars demolished the defences of Arleux and felt confident that he had won the first round. Having attracted Villars to his centre-right, Marlborough then spun him back to the centre-left, by moving up to Villers Brulin, between Avesnes-le-Comte and Bruay-en-Artois, threatening the lines as if seeking revenge for the loss of Arleux. Villars concentrated to face the threat and on 4 August Marlborough ostentatiously reconnoitred the lines.

The sword was already glimmering behind the matador's cloak. Marlborough had sent his guns and bridging train off westwards, behind Vimy Ridge, and had ordered the bridging of the Scarpe at Vitry, between Arras and Douai. On 4 August Cadogan slipped away to Douai to join General Hompesch and a force of twenty battalions and seventeen squadrons drawn from the garrisons of Douai, Lille and St-Amand. Marlborough's army marched off from its campsite, leaving the fires still burning, on the night of the 4th. As soon as Villars learnt what had happened he set off in pursuit and in the small hours of the 5th the two armies were marching within sight of one another in the moonlight, with the Sensée and the Scarpe between them. Cadogan reported that Arleux and the lines beyond it were clear of the French, and Marlborough, now at the bridging site at Vitry, begged the foot to step out. By eight that morning the Duke was through the lines and it was only with difficulty that the exasperated Villars, well to the fore of his marching columns, escaped capture.

On the following day the armies faced one another just north of the whaleback mass of Bourlon Wood, but

Marlborough declined to risk another expensive assault like Malplaquet and sidestepped to besiege Bouchain, a fortress that blocked the line of the Escaut between Valenciennes and Cambrai. The fortifications have long since been demolished, and the combination of the N30 and the thundering A2 makes it difficult to follow the Duke's operations with any accuracy. However, Wavrechin Hill, across the river from Bouchain, is clearly visible, and this feature was occupied by a strong French force, which strove to disrupt the siege and maintained contact with the garrison across the valley of the Escaut, still wide and marshy, by a 'cow path' running south of the present D943 Bouchain–Auberchicourt road. Villars was close by with his field army, with his headquarters in the village of Etrun at the confluence of the Sensée and the Escaut just upstream from Bouchain. Marlborough cut the cow path with a detachment of picked grenadiers and fended Villars off while pressing on with the siege. Bouchain fell on 14 September, its garrison being marched off as prisoners of war.

It was a fitting close to Marlborough's last campaign. His duchess, long a confidante of Queen Anne's, had fallen from royal favour and the delicate political balance at home at last tipped against the Duke, who was dismissed in January 1712. His men knew the debt they owed him and sang:

> Grenadiers now change your song
> And talk no more of battles won
> No victory shall grace us now
> Since we have lost our Marlborough.

They were all too prescient, for in the campaigning that preceded the signing of the Treaty of Utrecht in 1713 the Allies lost several of the fortresses they had captured under Marlborough.

III

Flanders, Artois and Picardy – II

FLANDERS AND ARTOIS, scene of so much fighting in the seventeenth and eighteenth centuries, were largely spared during the nineteenth, despite some sporadic activity in 1814–15 and 1870–1, and it was not until the First World War that devastation again laid its heavy hand on the area. The war came to the north in 1914 largely because of the strength of French defences along the Franco-German border, which inspired Schlieffen's plan for a wheeling attack through Belgium and northern France. He had spoken of the last man on the right 'brushing the Channel with his sleeve', but in August 1914 the advance of the westernmost German army, Kluck's 1st, took it along the eastern borders of Flanders, Artois and Picardy, and it was not until the autumn of 1914 that war came to the provinces in earnest as each side grappled for an advantage, extending their lines across north-westwards in 'The Race to the Sea'.

Military operations in the north divide naturally into three sectors whose boundaries coincide more or less with those of the old provinces. In the north, Flanders saw some of the worst fighting of the war, with the Belgian town of Ypres forming a British-held salient in the German lines. In Artois, the Arras–Vimy sector was bitterly contested in 1915–17, while the valley of the Somme and the Picardy downland to its south was fought over in 1916–18. It is convenient to consider the war in these three areas, focusing on a pivotal town in each case: Ypres for Flanders, Arras for Artois and Albert for Picardy.

Ypres, like its northern neighbours Bruges and Ghent, owed its importance to the cloth trade. In the mid-

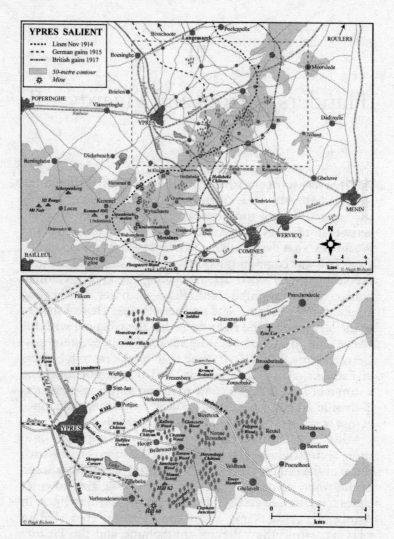

YPRES SALIENT

Lines Nov 1914
German gains 1915
British gains 1917
50-metre contour
Mine

thirteenth century work began on the cloth hall, a covered market in the spiky Flemish style. The huge vaulted ground floor could be approached from the cobbled square outside or from a quay on the Yperlee canal, there were storerooms upstairs and the imposing clock tower dominated the town. Ypres's great wealth and natural position – the ground to the north-west, between it and the coast, was low-lying, lacerated by small waterways and difficult for military movement – made it a frequent target for ambitious noblemen or quarrelsome dynasts. In 1383 it was besieged by an English force under the improbable command of the Bishop of Norwich. The Duke of Parma sacked the town in 1584 and over the next century Ypres was repeatedly in the firing line as French and Spaniards clashed in Flanders.

Ypres became French by the Treaty of Nijmegen and was fortified by Vauban. Captured by the Dutch in 1713, it was held by them until the French recovered it in 1744. It changed hands several times in 1792–4 and during the Waterloo campaign of 1815 its fortifications were hastily patched up by a British garrison. Ypres was Dutch after Waterloo and became Belgian when Belgium gained her independence in 1833. The town ceased to be a fortress in 1852: most of its defences were demolished and all the gates except one were widened so as not to obstruct the flow of traffic.

The Lille Gate retains two medieval round towers, which Vauban incorporated into his defences. Shelters beneath the towers and in the ramparts nearby proved useful during the First World War, although there is no truth in the legend that the rooms in the eastern tower were Plumer's head-quarters during the Messines Ridge battle: they were, more prosaically, a signal centre. A walk along the edge of the moat from the Lille Gate takes the visitor to a bastion forming the south-west shoulder of the defences. Further along the ramparts are two long shallow bastions with *orillons* and at the end of the second is the Menin Gate. This has changed name and appearance frequently over the years. It was once the Hangoart Gate, then the Antwerp Gate; between 1804 and 1814 it was the Napoleon Gate. Vauban built the substantial Antwerp Hornwork on the far side of the ditch, but nothing of it remains.

The Menin Gate now takes the form of an arch in white stone and red brick. Designed by Sir Reginald Blomfield, it was unveiled by Field Marshal Viscount Plumer on 24 July 1927, and commemorates the 54,896 British and Common-wealth soldiers who perished in the Ypres salient before 16 August 1917 and have no known graves. All British and Commonwealth soldiers killed in the world wars are com-memorated by either a known grave or an inscription on one of the memorials to the missing. Names are arranged by regiments in order of regimental seniority: the list on the Menin Gate is headed by 'Commands and Staff', and the first name is that of Brigadier-General Charles FitzClarence VC, killed in action commanding 1st Guards Brigade in November 1914. It is an appropriate setting, because it was through the Menin Gate – then a simple cutting with a stone lion to either flank – that soldiers passed along the Menin Road and climbed the calvary of the Ypres salient. The Last Post is sounded beneath the Menin Gate at eight o'clock every night. This has been done regularly since 11 November 1929, with a break during the German occu-pation in 1940–4. Such was the strength of the tradition that on 6 September 1944, the very day the Germans left, the call echoed out again beneath Blomfield's domed roof.

Modern Ypres has more features of military interest than I can hope to do justice to. The cloth hall, badly damaged by fire in October 1914 and subsequently pulverised by shelling, was rebuilt between 1920 and 1962, and now houses *In Flanders Fields*, one of the biggest museums on the whole of the Western Front. The clock tower can be climbed by those with the stomach for it and on the second Sunday in May model cats are thrown from its top as part of the local 'Cats' Festival'. St Martin's Cathedral, likewise rebuilt from ruins, lies behind the cloth hall and contains some British memorials. In nearby Elverdingestraat stands St George's Memorial Church, built in 1928–9 as the church for the town's then thriving British community. It is filled with memorials to units and individuals who served in the salient, and deserves to be better known. I am never sure where I stand with the Almighty and always find that some quiet minutes in this haunted spot helps put the world into perspective.

The war came to Ypres in October 1914. The British Expeditionary Force had fought at Mons and Le Cateau, then made 'the retreat from Mons' all the way back to the stately River Marne. After 'The Miracle of the Marne' it advanced to the Aisne, where the war of movement solidified into trench lock. Sir John French, the British Commander-in-Chief, who assured the King that the spade would now be as important as the rifle, asked Marshal Joffre, his French counterpart, for permission to withdraw from the Aisne to the Allied left. This was a logical scheme. It might permit the Allies to outflank the Germans, for at that stage both sides still thought in terms of finding an open flank, and would in any case give the British better access to the Channel ports and to their main line of communication running up from Le Havre. On 8–9 October the British II Corps under Sir Horace Smith-Dorrien detrained at Abbeville and advanced on Lille. Pulteney's III Corps followed, arriving at Hazebrouck and St-Omer on the 11th, and Douglas Haig's I Corps reached Hazebrouck on the 19th, and was ordered to advance on Bruges and Ghent by way of Ypres.

A strong British detachment was already in the vicinity, for an extemporised Naval Division had been sent to help the Belgians defend Antwerp, and Sir Henry Rawlinson's IV Corps had landed in Zeebrugge in early October. Antwerp fell on the 10th, but the remains of its garrison and the Belgian field army, together with the Naval Division's survivors, escaped down a corridor held open by Rawlinson. The Belgians went into line on the Yser from Nieuport to Dixmude and were to remain there for the rest of the war. Rawlinson was ordered to fall back on Ypres and duly entered the town on 13 October. A line of sorts solidified across Flanders as successive British corps arrived. Although it was GHQ's oft-repeated intention to attack, unexpected German strength – as the Germans themselves groped for an

open flank – soon fought the advance to a standstill. By 21
October II Corps held a line from Givenchy to Aubers, with
III Corps astride the Lys on its left. The Cavalry Corps,
fighting as infantry, held the Messines–Hollebeke sector,
including the high ground of Messines ridge. IV Corps, soon
to be merged with I Corps, stood between Hollebeke to
Zonnebeke, and I Corps between Zonnebeke to Bixschoote,
whence de Mitry's Frenchmen held the line as far as the
Belgian positions at Dixmude.

Even GHQ's optimism, sustained by the ebullient
Ferdinand Foch, commander of the French Northern Group
of Armies, could not ignore the German attacks, which
began on 20 October, and far from attacking, the BEF was
soon fighting for its very life. After four days of battle south
of the Lys, Aubers Ridge was lost and with it a battalion of
the Royal Irish, cut off in Le Pilly. At the end of the month
II Corps was briefly withdrawn from the line and replaced by
part of the newly arrived Indian Corps.

North of the Lys the situation was every bit as desperate.
IV Corps had advanced down the Menin Road on the 19th,
reaching the Kruiseecke crossroads east of Gheluvelt, and I
Corps took over its left-hand positions when it arrived on
the 21st. Over the next few days the Germans nudged away
at the salient that now curved round Ypres. Finally, on 29
October Army Group Fabeck, over three corps of fresh
troops, was launched against the face of the salient on the
heels of a bombardment fired by the then unprecedented
number of 262 heavy and 484 field guns.

Disaster soon loomed. Gheluvelt, on the Menin Road,
attacked on the 29th, fell on the 31st, opening a gap in the
British line. Immediately behind it, Hooge Château,
headquarters of the 1st and 2nd Divisions, was hit by shells,
which mortally wounded Major-General Lomax of 1st
Division, stunned Major-General Monro of the 2nd and
killed or wounded every staff officer but one. More shells hit
I Corps headquarters in the White Château, further
westwards, just before French himself arrived there. He had
made his way forward against a press of wounded, stragglers,
transport and – as he noted with alarm – heavy field guns

coming out of action at the trot. French left Haig to seek Foch, without whose help his position must shortly collapse. He had not gone far when one of Haig's aides rushed up with momentous news – Gheluvelt had been retaken and the hole was plugged.

The counter-attack was the work of three companies of 2nd Worcesters under the battalion's second-in-command, Major Hankey. Ordered to attack by Brigadier-General FitzClarence, Hankey took his companies from the south-west corner of Polygon Wood to the belt of trees at Polderhoek Château in column, then shook out into line. A long downward slope, swept by fire and strewn with fugitives, lay before them and Hankey decided that the danger would be minimised if it was crossed quickly, so he ordered: 'Advance at the double. Advance.' Although over a hundred men were hit over the next few minutes, the Worcesters caught the Germans off balance as they entered the grounds of Gheluvelt Château where some of the village's garrison of South Wales Borderers were still holding out. After a brief close-range battle the Germans fell back and the Worcesters had restored the line. They held it until they were withdrawn that night, by which time a new line had been established behind them, in front of Veldhoek.

The Worcesters' counter-attack, which took place on St Crispin's Day, the anniversary of Agincourt, deserves mention even in a book that cannot usually devote such space to a single battalion. It is a shining example of how a small number of brave men, well led, can influence major events, and the ground over which it took place, although traversed by the A19, is much the same now as it was then. Looking north-westwards from the edge of Gheluvelt (the village's tiny cemetery offers a good vantage point), up the gentle slope towards Polderhoek and Polygon Wood, only increases one's admiration for the Worcesters – and re-emphasises the difficulties facing troops who, like the Germans that afternoon, are slow in reorganising after a successful attack. Gheluvelt Château, knocked to bits by shelling as the war went on, has been rebuilt to its old plan and, though it is not open to the public, its front can be seen from the top of its long drive and an information board reproduces a painting of the Worcesters' relief of its defenders.

The village of Zandvoorde sits on a hillock, dominating the ground south of Gheluvelt and on 30 October it was attacked by Germans in such great numbers that a Royal Flying Corps pilot reported that 'the country is black with the sods'. 1st Royal Welch Fusiliers, under Lieutenant-Colonel Henry Cadogan, held the sector between the village and the Menin Road: it went into battle with thirty officers and 1,070 men, and in an action that gives eternal credit to this fine regiment, lost twenty-three officers and 864 men. Colonel Cadogan was killed. So too was the nineteen-year-old Second-Lieutenant Rowland Grey Egerton, eldest of the twin sons of Sir Bryan Grey Egerton of Oulton Hall, Tarporley, Cheshire. His brother Philip, then in the Cheshire Yeomanry, transferred to the 19th Hussars and was killed in a mounted action in October 1918, ending the family line. Their mother Mae had been courted by the novelist Anthony Hope and seems to have been the model for Queen Flavia in *The Prisoner of Zenda* and *Rupert of Hentzau*. Oulton Hall burnt down in 1926 and its park is now a race circuit.

More evidence of the fact that *noblesse oblige* meant something in 1914 stands in Zandvoorde itself. The 1st and 2nd Life Guards with the machine-gun section of the Blues, all fighting dismounted, held the village, having been warned: 'If we give way, the war's lost. Positions will be held to the last man and the last round.' On the 30th the Household Cavalry were overwhelmed by the German 39th Division, nine battalions strong, its attack lapping round Zandvoorde so that most of the defenders were cut off. Only seventy men escaped, and all fifty-four officers and men captured had been wounded. Lieutenant Charles Sackville Pelham, Lord Worsley, was reported missing, but later information was received, through diplomatic channels, that the Germans had found and buried his body. In December 1918, using the map provided by the Germans, a British officer found the base of the cross marking Lord

Worsley's grave and his widow bought the ground her husband rested in. In 1921, when the Imperial War Graves Commission was concentrating on isolated burials, Lord Worsley's body was exhumed for burial in Ypres Town Cemetery Extension and lies there in Plot II D 4. The Household Cavalry Memorial, built in 1924, now stands on the spot of Lord Worsley's orginal grave. So great was the damage done to Britain's peerage and baronetage in the first six months of the war that no edition of *Debrett* was published the following spring. Just south of Zandvoorde is one of the best German bunkers in the salient, made, as a sign erected by the pioneer company that built it informs us, in 1916. It served as a divisional headquarters and, though it is evidently damaged by shellfire, is still perfectly usable, albeit musty in summer and damp in winter.

At the same time that the German XV Corps and XXVIII Reserve Corps were gaining ground along the Menin Road, Prussian and Bavarian infantry were attacking the Cavalry Corps on Messines Ridge. On the 31st the London Scottish, the first Territorial battalion to fight, went into action in its hodden grey kilts and lost 321 of its 750 men: its action is commemorated by a fine cross, with saltire and lion rampant, on the N365 just north of Messines. By the time the battle died away the Cavalry Corps, despite French assistance, had been pushed off the ridge, but still held firm between Messines and the commanding height of Kemmel Hill.

Early on the foggy morning of 11 November the Germans made a last effort, attacking with twelve and a half divisions between Messines and Polygon Wood. They opened a gap south of Polygon Wood and elements of the Composite Guard Division entered Nonne Bosschen – Nun's Copse – level with the British gun line. They were evicted by a quick counter-attack by 2nd Oxford and Buckinghamshire Light Infantry. The fighting died away on 17 November. In two months the BEF had lost 58,155 officers and men: in the battalions that had been out since Mons and Le Cateau there remained on average only one officer and thirty men of those who had landed in August.

It was a dismal winter. The British commander-in-chief visited the lines and wrote that: 'The weather was terrible and the ground only a quagmire. The rain, the cold, the awful *holding* ground seemed to damp down my energy.' There was much debate over Allied plans for the coming year: Winston Churchill, First Lord of the Admiralty, favoured an amphibious attack up the Flanders coast, and French flirted with this idea while being encouraged to join Joffre in concentric attacks in Artois and Champagne. There were more reorganisations in the salient, with the British first giving up most of it to the French and then, once the original corps had been organised into two armies – Haig's 1st and Smith-Dorrien's 2nd – extending to their left once more.

In February GHQ and the French *Grand Quartier Général* (GQG) agreed on a joint offensive, but Joffre made it clear that his contribution depended on the arrival of another regular division, which would enable the British to extend their line further to the south, freeing French troops. When it was decided that this division – the excellent 29th – would be sent to the Dardanelles, Joffre withdrew from the project. French decided to press on regardless and 1st Army was ordered to attack the village of Neuve Chapelle at the foot of Aubers Ridge, with a view to taking the old Smith-Dorrien trench on its eastern edge – the British front line until the German attacks of October 1914. The gap smashed in the German line would be enlarged if possible. The plan emphasised careful artillery preparation followed by a rapid advance to 'carry the Germans off their feet'. For once there was sufficient artillery available, 276 field guns and howitzers and sixty-six heavy guns.

The Neuve Chapelle position was like a shallow V with its point towards the British, with the La Bassée road forming its southern and the Fleurbaix road its northern leg. The German line jutted out towards the junction of these roads west of Neuve Chapelle. The British planned to pinch out the salient, with the Meerut Division of the Indian Corps attacking from the south-west towards the Port Arthur sector and the 7th and 8th Divisions of IV Corps jabbing down from the north-west.

The brief but effective bombardment began at 7.30 on the morning of 10 March, and the assaulting infantry went forward thirty-five minutes later, making their best speed across muddy, low-lying ground. Where the shelling had been accurate they did well, although the right-hand Indian battalion, 1/39th Garhwalis, lost direction and suffered severely, losing all its British officers. On the left, in 8th Division's sector, two German machine-guns survived the shelling and ripped into 1st Middlesex and 2nd Scottish Rifles, the division's left assaulting battalions. A renewal of the bombardment, at about 11.30, dealt with the offending machine-guns and a further advance included the capture of the 'Moated Grange', a distinctive Flanders farm on the northern edge of the attack frontage. By nightfall the attackers had captured 4,000 metres of German front line to a maximum depth of 1,200 metres and taken 749 prisoners.

There was no further progress, by either renewed British attacks or German counter-attacks, over the next few days and on 11 March GHQ ordered the battle to be called off. It had cost the British and Indians 11,652 all ranks and provided both sides with invaluable lessons – if only they chose to learn them. For the Germans, the battle did serious damage to the notion that there should be one main defensive line and that building positions behind it encouraged faint-heartedness amongst its defenders. The front line had been taken relatively easily, but machine-guns further back, on the line of the Layes Brook and the edge of the Bois du Biez, had proved invaluable. The concept of defence in depth, which the Germans were to develop to a fine art, had its genesis at Neuve Chapelle.

The British learnt less soundly. The results of the bombardment suggested that there was both a close relationship between shells fired and ground gained, and particular merit in concentrating the shelling – in time and space – to maintain surprise and achieve moral paralysis. However, the British were often to attack behind a less concentrated bombardment than at Neuve Chapelle, and as often they forfeited surprise by extending it over several days. It was also clear that a plan based on the logic of the Aldershot

Tattoo, with a strict time schedule that could be varied only by the corps commander, failed to meet the demands of trench warfare, for promising opportunities passed all too quickly and primitive communications made the passage of formal orders a slow business, enabling the defender to repare his failure quicker than the attacker could reinforce his success.

The battlefield repays study. On the right of the Meerut Division's start line, at the La Bombe crossroads, where the D947 running north out of La Bassée crosses the D171, stands the Indian Memorial. Designed in the Indo-Saracenic style by Sir Herbert Baker, who shared the planning of New Delhi with Sir Edwin Lutyens, it includes panels bearing the names of the 4,843 officers and men of the Indian army killed in France and Belgium but who have no known graves. The Indian corps was in France for only a year – in December 1915 it was sent to Mesopotamia, whose climate was more appropriate than that of Flanders. During this time it lost 32,727 officers and men. Indian battalions had few British officers and felt their loss keenly: shortage of officer reinforcements was another reason for withdrawing the Indian corps from France.

From the La Bombe crossroads, drive north-east along the D171 into Neuve Chapelle, and turn left where signed to Neuve Chapelle Farm Cemetery. This stands on the British front line for the Neuve Chapelle attack in the sector held by 2nd Lincolns (to the north) and 2nd Royal Berkshires (to the south), the two attacking battalions of 25th Infantry Brigade, 8th Division's right-hand assault brigade. It contains the graves of several members of Princess Louise's Kensington Regiment (13th London Regiment), the one Territorial battalion in the brigade. Retracing one's steps into Neuve Chapelle, rebuilt after its almost complete destruction in subsequent fighting, then turning left on the D171 towards Fauquissart brings one (past a crucifix, which survived the battle) to the first crossroads, on the site of the German front line. It was here that the two surviving machine-guns of 11th *Jäger* stopped the Middlesex and Cameronians, the left-hand assault battalions of 8th

Division. A little further on, on the right of the road as it swings north-east, stands the the Moated Grange, prominently marked on British trench maps, its name somehow more redolent of Crécy or Agincourt than the Western Front. Although destroyed and rebuilt it retains its original plan, as well as the moat that gave the place its name.

The Germans struck at Ypres while the Allies were debating a joint offensive. Late on the afternoon of 22 April they launched poison gas against French Territorial and Algerian troops in the northern part of the Ypres salient between Langemarck and Steenstraat. On the right of the threatened sector stood the Canadian Division, left-hand formation of 2nd Army. Panic-stricken Algerian troops were seen running behind the British lines, coughing up blood and foam, and soon the 75s of the French divisional artillery, which had been firing steadily, fell ominously silent. The Germans had torn a five-mile gap in the Allied lines, but were ill prepared to exploit success. The Canadians, to their eternal credit, stood fast and their 3rd Brigade, in reserve, counter-attacked into Kitchener's Wood, just east of St-Julien. They lost it after heavy fighting, but an extemporised British detachment, known as Geddes's Force from the name of its commander, moved into the gap, and although the Germans mounted another major effort on the morning of 24 April, gaining more ground, they were unable to make much of their initial triumph.

The huge bite taken out of the salient by the gas attack gave the British command pause for thought. Smith-Dorrien recommended withdrawal to the GHQ Line, which ran just east of Ypres, giving up most of the salient. He was already on bad terms with French and this advice, together with suspicions that he had 'failed to get a real "grasp" of the situation' convinced French that Sir Horace should go: Sir Herbert Plumer of V Corps was made responsible for operations around Ypres and on 6 May he took over 2nd Army. He too advised withdrawal to a line closer to Ypres and this time the Commander-in-Chief was prepared to listen. Despite French opposition, the British withdrew, in early May, to a line running from Hill 60 through Sanctuary Wood to the Menin Road at Hooge, up to Frezenberg and the junction with the French east of St-Julien.

The ground over which the Second Battle of Ypres was fought is well marked. A mourning soldier rises out of stone to form the Canadian Memorial at St-Julien; looking out across the fields to the west of the road gives a little-changed view of the ground over which the first gas clouds rolled on 22 April. At Steenstraat, the junction between French and Belgian troops, a large cross – erected on the site of a more lavish memorial blown up by the Germans in 1941 – commemorates the first gas victims.

The gas attack coincided with fighting on the southern flank of the salient. Just over four kilometres south-west of Ypres a small but prominent hill stands between the cutting carrying the railway line to Comines and the minor road from Zillebeke to Zandvoorde. The feature, Hill 60, is not a natural hill at all, but was formed from spoil dug out of the railway cutting in 1860 and gives wide observation over the centre-south sector of the salient. It was taken from the French in December 1914 and plans were almost immediately made to recapture it. Work on tunnelling beneath the hill was begun by the Royal Monmouthshire Royal Engineers and was continued by the newly formed 171st Tunnelling Company Royal Engineers. Five mine chambers were dug and filled with explosive, and on the morning of 17 April the mines were blown. One hundred and fifty Germans were killed by the explosions and British infantry took the hill. Successive counter-attacks made little progress, but on 5 May the Germans at last wrested back the hill with the aid of gas. Hill 60 remained in German hands till 7 June 1917 when a massive mine, filled with 53,500 pounds of ammonal, was exploded beneath it as part of 2nd Army's Messines Ridge offensive.

Hill 60 is well served by monuments. At its entrance stand memorials to the 1st Australian Tunnelling Company, its plaque bearing bullet damage from the Second World War, and the 14th (Light) Division, a New Army formation consisting largely of rifles and light infantry battalions. The division first saw action at near Ypres in 1915, and thereafter it fought on the Somme and at Passchendaele, losing more than 37,000 men (well over twice its establishment strength) killed, wounded or reported missing in the course of the war. There is a memorial to Queen Victoria's Rifles (9th Battalion The London Regiment) on the hill itself, for it was an officer of the QVR, Second-Lieutenant G. H. Woolley, who won the first Victoria Cross awarded to a Territorial, for his gallantry on the night of 20–21 April 1915. The original memorial was unveiled in 1923 by General Sir Charles Fergusson, with Woolley, by then ordained, as the officiating clergyman. The memorial was destroyed by the Germans in 1940 and the present, somewhat smaller, structure uses some of the original stones.

The ground is pockmarked by shell holes and mine craters, and a pillbox, built by the Germans and modified by the British (the Germans tended to use wooden shuttering and the British corrugated iron), stands on the edge of a large crater. Across the railway, on private ground amongst the trees, is another 1917 crater, on a feature called the Caterpillar. Supplies were taken to the British front line by a light railway running up the old line to Comines, and were unloaded at the Dump, just west of where it runs beneath the Klein Zillebeke–Verbrandenmolen road. The banks of the cutting were once honeycombed with shelters – some of them occupied after the war by British soldiers who had stayed on in Belgium, and traces of these can still be seen. For many years there was an interesting if disorganised museum at Hill 60, but its place has now been taken by a tearoom and restaurant.

There is a pleasant walk along the northern bank of the Ypres–Comines Canal towards the N365. It leads to a cemetery named, from its site on earth dug out to make the canal, as Spoilbank Cemetery. The cemetery was started in February 1915, and contains officers and men killed holding this part of the salient, many of them casualties of what was euphemistically described as 'trench wastage', the inevitable day-to-day losses from shelling, mortar fire and snipers.

Any cemetery spreads its own rings of sorrow and this is no exception. Lieutenant-Colonel Binny, Major Freeman and Captain Lyons were the commanding officer, second in command and adjutant of 10th Royal Welch Fusiliers, killed when a shell hit their dugout on 3 March 1916, and lie side by side. Stewart Binny had been a regular officer in the 19th Hussars and had won the DSO in the Boer War. He had retired before the war but returned to service in 1914, first as a railway transport officer, then as deputy director of rail transport and latterly, despite his cavalry background and age of forty-four, as CO of an infantry battalion. Not far away lie the brothers Lieutenant J. Keating and Second-Lieutenant George Keating, both long-service regulars commissioned from the ranks of the Cheshire Regiment, and killed when 2nd Cheshires attacked The Bluff, a prominent feature on the north bank of the canal midway between the cemetery and Hill 60, in February 1915.

A fortnight after the loss of Hill 60 in May 1915, 1st Army mounted another attack on Aubers Ridge. The Germans were already implementing the lessons of Neuve Chapelle, deepening the barbed-wire belt, improving their front-line trenches and building a strong second line with dugouts for its garrison. Communication trenches stretched back to protect supports coming up and help contain break-ins by acting as emergency fire trenches. Between 700 and 1,000 metres behind the front trench system, a line of concrete pillboxes provided a rallying point.

The British answer to improving German defences was to mount a more ambitious attack. Along the Rue du Bois, between Port Arthur and Chocolat Menier Corner (the junction of the Rue du Bois and the road to La Quinque Rue, so called because of an advertising hoarding) I Corps and the Indian Corps were to assault on a front of 2,400 metres, while IV Corps would attack on a 1,500-metre front opposite Fromelles. The two prongs would meet on top of Aubers Ridge. At five on the morning of 9 May 600 British guns drenched the German positions below Aubers Ridge with a fire whose impact concealed uncomfortable facts. There were no more guns per yard of trench than at Neuve

Chapelle, but defences were already far stronger. Worn gun barrels and indifferent ranging in poor visibility made the fire inaccurate, and the Germans were waiting for the assault. When it came, the result was predictable. The British, advancing in dense lines, fell in heaps. Few reached the German wire and fewer still got into the German front trench. Repeated efforts fared no better: by the end of the day the two divisions assaulting the Rue du Bois lost 6,340 men and at Fromelles the 1st Division lost 4,680.

Driving north-east up the Rue du Bois (D171) from Neuve Chapelle gives the visitor an excellent feel for the battlefield. Aubers Ridge – a poor thing almost anywhere else in the world, but here dominating the flat country around it – rises gently to the east, with Aubers and Fromelles at its top. The Layes Brook, less boggy now than it once was, runs parallel with the road, across the British and Indian line of attack. Numerous German bunkers have survived – the effort of demolishing these concrete warts on the landscape is greater than the value of the land they stand on. Most post-date the fighting of 1915. A first line covers the Layes Brook, either on the brook itself, like the bunker on the Piétre–Mauquissart road, or just behind it, like the rash of concrete where the Petillon–Fromelles road ascends the ridge. The German second line is on the reverse slope, in and around Aubers. There is a multi-storey bunker in the Rue d'Houdringue in the village, and alongside the D141 Aubers–Fromelles road stands a bunker which, so local tradition asserts, was occupied by Gefreiter Adolf Hitler.

Aubers Ridge was attacked again in July 1916 by the 5th Australian and 61st South Midland Divisions to divert German attention from the Allied offensive on the Somme. Fromelles was the scene of the first Australian attack in France, and VC·Corner Cemetery, in the low ground to its north-west, reached by taking the little D22 from Fromelles and following the signs to the Australian Memorial, is the only all-Australian cemetery there. It has no headstones: the screen at the rear bears the names of 1,299 Australians killed in the battle and the unidentified bodies of 410 of them lie beneath the lawn. On a line of German bunkers captured by the Australian 14th Brigade is the Australian Memorial Park, with a moving statue, 'Cobbers', showing a soldier carrying a wounded mate to safety. Nearby is a memorial to Sergeant Pilot Kenneth Walter Bramble, Royal Air Force Volunteer Reserve, who was shot down in July 1941 and lies in the communal cemetery at Merville, where Second World War burials are interspersed with those of the First. Le Trou Aid Post Cemetery, not far away on the little D175, contains the grave of Brigadier-General Arthur Lowry Cole, whose ancestor had commanded one of Wellington's divisions and who fell commanding the 25th Infantry Brigade on 9 May 1915.

Closer to Fromelles a crucifix commemorates Lieutenant Paul Kennedy, of 2nd Battalion The Rifle Brigade, and three of his brother officers killed on 9 May 1915. 2nd Rifle Brigade and 1st Royal Irish Rifles, both in Lowry Cole's Brigade, seized about 250 metres of German trenches in this area: the ground could not be held and had to be given up the next day. Commanding the bombers (the term then used for grenade throwers) of 2nd Battalion The Lincolnshire Regiment in the same attack was Lieutenant Eric Osborne Black. In the small but beautifully maintained museum at Fromelles, the creation of the tireless Martial Delabarre, is a prismatic compass, marked 'E. O. Black, Oriel College Oxford', which has been hit with a rifle bullet while open so that it could be used for taking a bearing. The Lincolns' regimental history tells us that Lieutenant Black had broken into the German position and was killed, probably clutching this very compass, as he led his survivors back.

The Aubers Ridge area was the scene of further heavy fighting in 1917. In April 1918, during the second phase of the German spring offensive, the Battle of the Lys, the entire British front line in the Aubers area was overrun. Just along the La Bassée road from the Indian Memorial at Neuve Chapelle is a Portuguese cemetery, with a shrine to Our Lady of Fatima across the road. Portugal, proverbially Britain's oldest ally, sent an expeditionary corps to France in 1915. Just under 2,000 Portuguese officers and men lie here, some of their bodies brought in from other sectors of the front. There is something especially poignant about the cemeteries of Allied soldiers – there are also Russians and Italians buried in France – who were sent to the Western Front as a demonstration of alliance solidarity. For many years the Portuguese cemetery was dismally run-down and, though more care has been taken of it of late, there is still something dismal and forlorn about the place. By April 1918 the Portuguese Corps had been in the line too long. Its 1st Division was withdrawn just before the Germans attacked, leaving its 2nd Division, badly overextended, on a major breakthrough sector. It should not be a matter for surprise that when the attack came on 9 April the division crumbled rapidly and its collapse enabled the Germans to push forward six miles on the first day of the battle.

Off the Aubers Ridge battlefield proper, but nearby, on the D171 Béthune road from Neuve Chapelle, is the Le Touret Memorial to the Missing, commemorating more than 13,000 of those killed in the fighting in the area before September 1915 who have no known graves. The panel devoted to the Grenadier Guards reads like a page from *Debrett* and demonstrates, yet again, that whatever our view of Edwardian society, its leaders were prepared to pay with their blood for the privileges they enjoyed.

The derisory results of the battle of Aubers Ridge encouraged French to leak information on the shortage of artillery ammunition in an attempt to sidestep blame for failure, undermine the position of Lord Kitchener, Secretary of State for War, and try to prevent the diversion of resources to the Dardanelles. The resultant 'Shells Scandal' contributed to the formation of a coalition government with David Lloyd-George as Minister of Munitions. There was certainly widespread agreement in the British high command that ammunition was the key to success, and by 14 May GHQ was writing in terms of 'deliberate and persistent' attacks, which would forfeit surprise but achieve results by 'gradual and relentless' wearing down of German defences.

In mid-May 1st Army tried to put these principles into practice by attacking between the Rue du Bois and Festubert. This time the bombardment lasted three days and, though it destroyed much of the German front line, many machine-gun positions survived and the dugouts of the second line were largely undamaged. The infantry attack began on the night of 15–16 May and overran parts of the front line, but the Germans formed a new line running through strongpoints to the rear, and the battle ceased on 27 May with a maximum of 1,000 metres gained on a front of 2,700 metres, for a total of over 16,500 British and Indian casualties.

Sir John French was under pressure to co-ordinate his attacks in Flanders with French thrusts in Artois, and the Battles of Aubers Ridge and Festubert had coincided with French attacks north of Arras. There the ground is dominated by Vimy Ridge, approached up a gentle slope from Arras and the Scarpe, but tumbling down abruptly to the Lens–Douai plain. Its northern end drops into a valley sheltering the village of Souchez, before climbing on to the Lorette spur. The modern A26 runs along the back slopes, slipping through the hollow between the ridge and Lorette, and going on to cross the Scarpe near Fampoux, where the ridge melts into the marshy valley. The uplands are chalk interleaved with clay, and it is easy to dig tunnels, which require little revetment. There were already tunnels beneath the ridge – dug by chalk burners and used by Huguenots in times of persecution – before the soldiers came in 1914, and far more under the town of Arras, whose *boves*, begun in the tenth century, are a veritable rabbit warren.

The Germans took Vimy Ridge during the 'Race to the Sea', and in October 1914 they improved their position by seizing Neuville St-Vaast, Carency and Notre-Dame-de-Lorette, establishing a line running just west of the Augustinian abbey at Mont St-Eloi. In May 1915 the French attacked Vimy Ridge and Lorette behind a lightning bombardment similar to that at Neuve Chapelle, fired for the same reason: there was insufficient ammunition to maintain a heavy fire for long. After a costly battle the French secured Lorette and were firmly established on the back of the ridge, but Souchez remained in German hands.

The human consequences of this struggle are still evident. The cemetery at Notre-Dame-de-Lorette contains the graves of 20,000 Frenchmen, the ossuary the bones of another 23,000. Above it stands a lighthouse-like tower, 5.2 metres high, which provides a magnificent view over the northern end of the Vimy battlefield. The original oratory on this spot had been built in 1727 by Nicolas Florent Guilbert, a painter from the nearby village of Ablain St Nazaire who had visited Loretto in Italy and been cured of an illness. It was destroyed in 1794 during the Revolution, rebuilt in 1815 and turned into a little chapel in 1880. Work on the tower was begun in 1921, when Marshal Pétain, hero of Verdun, laid the foundation stone, and it was opened in 1925. The construction of the Byzantine-style basilica was inspired by the bishop of Arras, who laid the first stone in 1921, although the building was not at last consecrated till 1937: the bishop himself lies buried in the basilica. Also commemorated there is Louise de Bettignies, an Oxford-educated Frenchwoman who was condemned to death in 1915 for spying for the Allies – as Alice Dubois she had reported on German movements through Lille. Reprieved, possibly as the result of international outcry at the shooting of Nurse Edith Cavell, she was imprisoned at Sieburg in Germany and put in solitary confinement for encouraging other prisoners not to work for the Germans. In September 1918 she died in hospital in Cologne and was exhumed after the war for burial in the family vault at St Armand les Eaux, about ten kilometres north-west of Valenciennes.

Walking straight towards the basilica from the car park takes the visitor past the grave of Emile Barbot, *général de brigade* by rank and, during the desperate fighting of early 1915, commander of the 77th Infantry Division at Souchez. He was mortally wounded on 10 May, not far short of his sixtieth birhday, and his fine statue, on the southern edge of Souchez village, shows him striding out in his greatcoat and distinctive floppy beret of the *Chasseurs Alpins.* In the same row lie Anatole and Edmond de Sars, father and son, killed in 1914 and 1940 respectively.

In the summer of 1915 Vimy Ridge continued to feature

in Joffre's plans. He proposed to pursue the same tactics as he had in the spring, with a two-pronged thrust driving in against the shoulders of the German salient. The French 10th Army was to advance across Vimy Ridge into the Douai Plain, and Joffre was anxious that the British should assault on its left to widen the front under attack and shield its flank. The British were less sure. The ground in the Loos–La Bassée sector was unpromising, for the Germans had turned the mining villages into a formidable position. When Sir John French visited the Lorette spur on 12 July to look down on the proposed attack sector he was not impressed by what he saw, but he was aware that his grip on command greatly depended on Joffre's support and eventually gave way. At first he proposed to use artillery only, but, told that this was unacceptable, agreed to mount a major offensive at Loos.

The attack was to take place between the suburbs of Lens on the right and Givenchy on the left. The German line ran parallel with the Lens–La Bassée road. The first line, on gently rising ground, went past the fringes of Auchy, over a hummock strongly fortified by the Hohenzollern Redoubt, across the Vermelles–Hulluch road near the Bois Carré, then along the forward edge of the shallow Lone Tree Ridge to the Cité St-Pierre dormitory suburb. Experience of earlier offensives encouraged the Germans to pay attention to their reserve line, sited some three kilometres behind the front, from the edge of La Bassée, past Haisnes and Hulluch, then curling back towards Cité St-August before swinging in to Lens. It was on a reverse slope and could not be engaged from the main British observation post, the huge slag heap at Annequin on the La Bassée road (N41). There were a number of intermediate strongpoints and the Hill 70 Redoubt, near the junction of the Loos road and the Lens–La Bassée road, dominated the shallow valley running down from Loos towards Hulluch. The Germans made excellent use of the slag heaps and pitheads in the area, and enjoyed good observation from the double winding gear known as Tower Bridge near the Vermelles road.

The 1st Army planned to assault with two corps up.

Rawlinson's IV Corps, on the right, was to attack with 47th
(London) Division between the French left and the
Lens–Béthune road, then 15th (Scottish) Division, and
finally 1st Division, its left on the corps boundary on the
Hulluch–Vermelles road. The right-hand division of Hubert
Gough's I Corps, the 7th, attacked from the Vermelles road
to just south of the Hohenzollern Redoubt. 9th (Scottish)
Division faced the redoubt itself and 2nd Division took the
line out to the Indian Corps boundary just north of the La
Bassée Canal. Behind 1st Army, under GHQ's command
until released, was XI Corps, containing the newly formed
Guards Division and two 'New Army' divisions, 21st and
24th. The balance of forces favoured the attackers, for there
were some 10–11,000 Germans facing the 75,000 men of the
assaulting divisions. The odds were improved by the British
decision to use gas and 5,500 cylinders were issued to the
attacking corps.

Successful use of gas depended upon the wind and at 5.15
on the morning of 25 September Haig, watching the smoke
of his senior ADC's cigarette drifting gently north-east,
ordered that it would be used as planned. At 5.40 a.m. the
artillery, which had been firing steadily since 21 September,
stepped up to rapid fire, and gas hissed out into no man's
land. The infantry clambered from their trenches at 6.30
a.m. and went forward into the drizzly mist, peering through
the talc-covered eyeholes in their primitive gas masks. The
results of the first day's fighting were mixed. The 47th
Division took the *double crassier* slag heaps and formed a
hard shoulder on the right. The 15th Division cleared Loos
and crossed the main road near the Bois Hugo, but swung
too far south, opening a gap between itself and 1st Division,
which nevertheless reached the La Bassée road. The 7th
Division seized the Quarries, and 9th Division took the
Hohenzollern Redoubt and wrested a handhold on the
German second line at Haisnes. Only 2nd Division, on
the extreme left, failed completely. Although these gains
were not derisory, they cost over 15,000 men. The German
second position was almost untouched and reserves arrived
by railway to bolster it up.

At about midday on the 25th French decided to put XI Corps at Haig's disposal to exploit initial success, but it started some six miles behind the original front, traffic control was inefficient and orders arrived late: the corps was not ready to attack till dawn on the 26th. Its soldiers had never been in action before, and were already tired, hungry and soaked. In mid-morning they were sent forward towards the German second line between the Bois Hugo and Hulluch. There had been an understanding that the two New Army divisions would only be used against an enemy who was 'smashed and retiring in disorder', but their objective was intact and stoutly held. The twelve attacking battalions, just under 10,000 strong when they moved forward into the killing ground between Loos and Hulluch, lost over 8,000 officers and men in less than four hours. That was not the end of the day's misfortunes. The Germans had recaptured the Quarries during the night and attempts to recover them failed: Major-General Sir Thompson Capper, who had commanded 7th Division since it came out in 1914, was killed as he went forward on foot.

On the 27th the Guards Division attempted to succeed where the 21st and 24th Divisions had failed, attacking Chalk Pit Wood and the Bois Hugo. The Guards took Chalk Pit Wood and reached the main road, but the German second position still loomed in front of them. Over the next few days the battle ground on with local operations during which the Germans recaptured the Hohenzollern Redoubt. It was on 4 November that 1st Army admitted that it had no hope of continuing the offensive. It had cost 2,466 officers and 59,247 men killed, wounded and missing, and had inflicted just over 20,000 casualties on the Germans.

The best starting point for a visit to the Loos battlefield is Dud Corner Cemetery and Memorial on the busy N43 just north-west of Loos-en-Gohelle. Its observation platform gives a good view over the whole of the battlefield, with Lone Tree Ridge running off north-north-east and the Loos–Hulluch valley to its east. The Vermelles–Hulluch road crosses the middle distance to the north, with the white cross of St Mary's ADS Cemetery an easy point of reference. ADS stands for Advanced Dressing Station, the second link in the chain of medical evacuation. Wounded first received treatment at their Regimental Aid Post and were stetchered back to an ADS before being moved back, usually by ambulance, to a Casualty Clearing Station. At the ADS, alas, many were already beyond help or could not usefully be evacuated and a number of Commonwealth War Graves Commission cemeteries had their origin in the extemporised graveyards behind overflowing dressing stations. On a clear day the position of the Hohenzollern Redoubt due north of Dud Corner and on rising ground a little south of Auchy can also be seen. To the south two post-war slag heaps mark the site of the old *double crassier*: the truncated cones of the originals can be seen close to the bases of the present heaps.

Dud Corner commemorates 20,589 British missing. There is a major-general amongst them: three divisional commanders, Major-Generals Capper (7th Division), Thesiger (9th Division) and Wing (12th Division) were killed in the battle. Also commemorated on the memorial is Lieutenant John Kipling, Irish Guards, killed at the age of eighteen at Chalk Pit Wood, only son of the poet Rudyard Kipling. John was very short-sighted and was only accepted for service because of his father's influence. Kipling was broken by the young man's death and visited France in an unavailing effort to find his grave. It was on his suggestion that the words 'THEIR NAME LIVETH FOR EVERMORE' (Ecclesiasticus 44:14) are inscribed on the Stone of Remembrance, which stands in all large CWGC cemeteries, and it was also on his recommendation that the graves of unknown soldiers bear the inscription 'KNOWN UNTO GOD'. Research on the War Graves Commission's records led to a grave in St Mary's ADS cemetery (plot VII D 2), previously marked to an unknown officer of the Irish Guards, being identified as that of Jack Kipling, though opinion is divided as to the validity of the attribution.

The Vermelles road bisects the battlefield. The track running north-east towards Haisnes from the roadside shrine at the first crossroads west of St Mary's ADS gives a

good view of the area of the Quarries and the German trench
system called the Pope's Nose, which features in Robert
Graves's *Goodbye to All That*. Walk up the track for 900
yards and you are 600 yards due west of the Pope's Nose, and
1,500 yards due south of the Hohenzollern Redoubt. The
latter was a massive earthwork dug into a mining spoil heap
and the ground is still noticeably higher than its
surroundings. The site of the Hohenzollern can be reached
from Cité St-Elie, just north of Hulluch, but when I last
visited the place it had become a huge unofficial rubbish
dump and the experience was not rewarding.

The battlefield of Loos is a bare and depressing place. On
a rainy day, with fog shrouding the killing fields between
Vermelles and Hulluch, it is not hard to imagine the pipes
shrieking through smoke, gas and mist as the Scots fought
for Loos, and the hapless New Army infantry of XI Corps,
dressed by the right, marching into what the Germans called
the *Leichenfeld von Loos* – the Loos corpse field.

The French offensive fared no better. General d'Urbal's
10th Army attacked between Vimy Ridge and Beaurains,
south of Arras, with a remit to achieve 'a strategic breakthrough capable of producing decisive results'. Although the
attack was delivered with determination, its results were so
small that the French official account speaks admiringly of
comparative British success at Loos. Well aware of the
situation at Loos, the French did their best to offer support
by switching the weight of the thrust to their left, on the
immediate right of the British attack, but this, too, failed,
and the results of the great Allied offensive in Artois were a
derisory gain and thousands of casualties.

Sir John French was one of the casualties of Loos. The
issue of the reserves whipped up a furore and his powerful
political supporters could not save him. He resigned his
command in December and went off, elevated to the peerage
as Viscount French of Ypres, to be Commander-in-Chief of
Home Forces. For most of his time in France he had lived in
a lawyer's house at No. 37 Rue St-Bertin in St-Omer, moving
up to a forward command post in what is now the Place
Général de Gaulle in Hazebrouck when major operations

were afoot. For the Battle of Loos he established himself in the Château Philomel at Lillers, even closer to the front. However, St-Omer was on his mind as dismissal loomed and he quipped to his friend Lord Esher that he might take the title 'Lord Sent-Homer'.

Douglas Haig took command in France. Lieutenant-General Sir Launcelot Kiggell came out as Chief of the General Staff, while Robertson, French's CGS, returned to England to be Chief of the Imperial General Staff, providing Haig with an invaluable ally at the War Office. The directive issued by Kitchener on Haig's assumption of command emphasised that the defeat of the Germans by the Allied armies was the object of the commitment of British troops to France, and Haig at once took steps to improve GHQ's relationship with GQG. He inherited an inter-Allied agreement, concluded at Chantilly in early December 1915, that there would be a general offensive as soon as possible, with 'wearing-down' operations in the interim. In February Haig agreed in principle to relieve the French 10th Army and mount a major offensive between Arras and the Somme. This was not his first choice of battlefield: he preferred Flanders, believing that a thrust towards the German railhead of Roulers (Roeselare) offered better prospects.

The Germans had plans of their own. Falkenhayn's attack on Verdun began on 21 February 1916, impelling Haig to speed up relief of the 10th Army, and in March the British extended to their right, taking over the French line as far as Maricourt on the Somme, eleven kilometres west-south-west of the sleepy little town of Albert. The organisation of the BEF was changed to meet its increasing size and wider responsibilities. By June the 2nd Army (Plumer) held the Ypres–Armentières sector to its boundary with 1st Army opposite Fromelles. 1st Army (Monro, then Horne) continued the line to the southern foot of Vimy Ridge, 3rd Army (Allenby) held it as far as Gommecourt and 4th Army (Rawlinson) took it on to its junction with the French. A new Reserve Army was formed under Gough and this later took the title 5th Army.

The German attack on Verdun not only made it

imperative for the British to take some weight off the French; it also reduced French commitment to what was to have been the major Allied effort of the year, leaving the British as major partners in the offensive, with an initial eighteen divisions attacking along a thirty kilometre front as opposed to eleven French divisions attacking over twenty kilometres The fact that the two armies had to advance side by side left little choice of attack sector, but the area allocated to the British was unpromising. Albert, once known as Ancre from the river of the same name, which flows through it to join the Somme at Corbie, had a population of just under 9,000. It had been a pilgrimage centre, although the number of pilgrims had never come up to local expectations, and the basilica of Notre-Dame-des-Brebières, topped by a gilded statue of the virgin, seemed large for a town of its size. A Roman road runs north-east to Bapaume, twenty kilometres away, squarely across a chalky escarpment, sprinkled with large woods and shouldering large, hedgeless fields of wheat and sugar beet.

The villages of the ridge have become part of British military history. From north to south Serre, Beaumont Hamel, Thiepval, Ovillers-la-Boisselle, La Boisselle, Fricourt and Mametz all stood in the front line, and the strongpoints they sheltered made good use of the finger valleys poking up into the western face of the escarpment. It was a natural defensive line, villages and redoubts supporting one another with flanking fire. An intermediate line of detached strongpoints stood just behind the front trench system, and a second line ran from Grandcourt on the Ancre, through Mouquet Farm, behind Pozières and on to Longueval. A third line, some eight kilometres behind the front, had been begun in February 1916 and ran in front of Flers and Le Sars. The front trench system consisted of a minimum of three trenches, each 150–300 metres apart, with dugouts driven deep into the chalk. True, this formidable position had disadvantages. The front system was on a forward slope and the chalky ground meant that it was impossible to conceal digging. German doctrine required lost ground to be taken by rapid counter-attacks, and this encouraged regimental commanders to place two of their three battalions in or near the front line, and the third battalion in the intermediate and second lines. The preliminary bombardment would thus fall on an easily observed target densely packed with troops.

A heavy bombardment was an essential ingredient of the British plan. It was a plan that sprang from blurred thinking, for Haig and Rawlinson expected different things from it. Haig hoped for a breakthrough, but Rawlinson was less confident. He favoured the 'bite and hold' concept, occupying the shell-torn ground with infantry and lacerating the German counter-attack when it came. Haig objected to Rawlinson's first proposals because they did not go far enough and urged him to consider 'the possibility of pushing our first advance further than is contemplated in your plan'. Rawlinson could scarcely refuse the invitation, but never had much confidence in GHQ's belief that Gough, whose Reserve Army (later retitled Fifth Army), waiting behind Fourth Army, would speedily press through to exploit his breakthrough.

As we move from operational principles to tactical method, the picture is no more edifying. Given the amount of air reconnaissance, the existence of the German second and third lines was no surprise. However, these could not be engaged with properly observed fire, and the 18-pounder batteries, deployed to fire on the German first line, lacked the range to reach deep into the position. Thus, whatever the success of an assault on the German first line, the attackers would find themselves faced with deep trenches and uncut wire as they advanced across the ridge. If Haig's decision to attack on the Somme can rightly escape criticism on the grounds that inter-Allied relations demanded it, British failure to identify the problem posed by deepening defence was, with the evidence still whitening on the wire in front of Hulluch, little short of professional incompetence.

A misunderstanding of the effects of artillery and the potential of the infantry of the New Armies was inherent in the British plan. Rawlinson hoped that the week-long

bombardment, during which his gunners would fire
1,500,000 shells, would totally destroy the German front
position. This assumption failed to reckon with the strength
of German dugouts – no surprise, surely, for a superb
captured example was being used to train British troops in
the Touvent Farm sector opposite Serre. It also failed to note
that there was less artillery per yard of front attacked than at
Neuve Chapelle and that in terms of heavy guns the French,
attacking south of the Somme, had double the ratio used by
the British. Finally, a combination of faulty fuses and shot-
out gun barrels led to a high proportion of dud shells:
perhaps one in three of the shells fired into the slopes above
Albert failed to explode.

Rawlinson's confidence that his gunners would reduce
the German front position to a line of shell holes, torn
sandbags and shattered corpses encouraged his formation
commanders to pass on the good news to their men. 'When
you go over the top,' a Yorkshire battalion was told, 'you can
slope arms, light up your pipes, and march all the way to
Pozières before meeting any live Germans.' The infantry
was therefore trained and equipped, not for trench fighting,
but for a largely unopposed advance to occupy ground conquered by the artillery. The question of the weight carried by
infantrymen on the first day of the Somme still generates
controversy, but it is hard to disagree with General Farrar-
Hockley's assertion that '. . . no man carried less than 65 lb.
Often additional grenades, bombs, small arms ammunition
or perhaps a prepared charge against obstacles, stretchers or
telephone cable increased the load to 85 or 90 lb.'

The fact that most of the troops taking part were
volunteers of the New Armies also influenced tactics. GHQ
believed that they did not 'possess that military knowledge
arising from a long and high state of training, which enables
them to act promptly on sound lines in unexpected
situations. They have become accustomed to deliberate
action based on precise and detailed orders . . .' We have seen
how, at Neuve Chapelle, reliance on such orders led to an
over-controlled battle with troops packed like salmon in a
pool: the same was to take place on the Somme on a larger

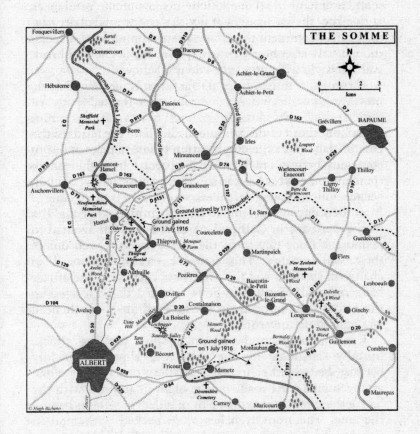

THE SOMME

© Hugh Bicheno

scale. It is hard to reconcile Rawlinson's public utterances with either his estimate of at least 10,000 wounded per day, or his diary comment that: 'I feel pretty confident of success, though only after heavy fighting. That the Boche will break and a debacle will supervene I do not believe . . .'

The high command set its face firmly against debate, however constructive. A 4th Army order warned that: 'All criticism by subordinates . . . of orders received from superior authority, will in the end, recoil on the heads of the critics.' A gunner officer in the 29th Division pointed out to corps headquarters that lifting the barrage from the enemy trench two minutes before the infantry attacked would fail, because the surviving defenders would have time to emerge from their dugouts in time to man their machine-guns. His own experience in Gallipoli suggested that barrage had to pound the front line until the attackers were well out of their own trenches. This acute observation produced no change in the plan and the result was entirely as the officer had predicted. Many battalions reported that the wire to their front had not been cut by the shelling, but little heed was paid to their warnings.

The bombardment began on 24 June and the attack, initially scheduled for the 29th, was slipped to 1 July because bad weather impeded artillery observation. Final arrangements were made on the night of 30 June, with the last of the infantry moving up from billets in villages behind the line. The front trenches were packed with men in fighting order, waiting while the artillery fireplan rose to its crescendo and the engineers prepared to fire the eleven mines buried under the German trenches. At 7.20 a.m. the first of the mines went up beneath the Hawthorn Ridge Redoubt, at 7.28 the other mines exploded and two minutes later the barrage lifted as the first waves of infantry clambered out of their trenches to go over the top.

We will follow the first day's battle from north to south, starting at Gommecourt, where 3rd Army was to attack with the object of diverting German attention and resources from 4th Army's efforts. The village lies behind the thick Gommecourt Wood and VII Corps attacked with one division, 46th (North Midland) north of the wood, and the other, 56th (London) to its south. They were to link up east of the village, but although the London Division made good progress, reaching the Gommecourt–Pusieux road, with a party of Queen's Westminsters pushing on to within 400 metres of the junction point, the 46th Division was less fortunate.

Gommecourt Wood New Cemetery, on the Fonquevillers road (D6) stands in no man's land. Five hundred metres to the north-east is Sartel Wood, and the German trenches lay along its forward edge, dominating the shallow valley between it and Fonquevillers. The wire in front of them was up to forty metres deep, firmly held on iron pickets which – here as across the rest of the Western Front – have stood the test of time well enough to be used in fences and farmyards today. The front system consisted of three trenches, each up to three metres deep, with communication trenches running back to the Kern Redoubt just south-east of Gommecourt Château and a switch line to enable the sector to be sealed off even if the village was lost. On its northern end the line bulged out, taking in a small copse still visible in the valley bottom, which housed a redoubt called Schwalben Nest by the Germans and 'The Z' by the British.

What happened to the Midlanders as they advanced into the valley at 7.30 a.m. was to be repeated, with variations, along much of the British line and the details of the 46th Division's fate will serve as a microcosm of events elsewhere that bright, bloody morning. The assaulting battalions moved into their jumping-off trenches down communications trenches running back to the northern end of Fonquevillers. The work on these headquarter dugouts and miles of signalling cable had been seen by the Germans, and the exposed position of Gommecourt made it an obvious target for attack. Accordingly, the 2nd Guard Reserve Division was held behind Gommecourt, ready to reinforce its defenders, the 55th Reserve Infantry Regiment.

The bombardment had done serious damage to the German first position, blowing in sections of trench and collapsing dugout entrances, and hitting Biez Wood, which hid a battery in direct support of the 55th Regiment. At 7.30 a.m. the two forward brigades, 137th on the right and 139th on the left, began to form up in assault formation in no man's land, covered by smoke put down by trench mortars. As the wind blew gaps in the smoke, German sentries observed the dense lines in front of them and summoned the surviving garrison of the front line from its dugouts. The wire had only a few gaps and as troops bunched to get through them they were hewn down by machine-gun fire. Artillery had already begun to hit the British front and support trenches, and it soon brought a belt of defensive fire down across the line of advance.

A few men of 137th Brigade's first wave – 1/6 North Staffordshire and 1/6 South Staffordshire – reached the front line, but were soon killed or captured, and the fire falling on communications trenches disrupted the advance of the follow-up battalions, 1/5 South Staffordshire and 1/5 North Staffordshire. 138th Brigade, consisting of Territorial battalions of the Sherwood Foresters, did better, despite vicious flanking fire from the Schwalben Nest. Elements of the two leading battalions, 5th and 7th Sherwood Foresters, entered the front trench and struggled on beyond it, but the Germans swiftly re-established themselves and mopped up the isolated parties of Foresters who had penetrated their position. Survivors, wounded and unwounded, lay up in shell holes, grenaded from the German trenches and sniped if they showed themselves.

It is a measure of the division's efforts that five of its battalion commanders were killed or wounded – Lieutenant-Colonel C. E. Boote of 1/6 North Staffs is buried in Gommecourt Wood New Cemetery. The Medical Officer of 1/5 Sherwood Foresters, Captain J. L. Green, was hit while searching for wounded in no man's land and was eventually killed as he dragged a wounded officer back to the British lines. He was awarded the Victoria Cross, one of the nine won that day, and lies in Fonquevillers Military Cemetery. Although Lieutenant-General Snow of VII Corps told his men that their attack had been partially successful because it had pinned troops to Gommecourt, the venture was regarded as a failure by participants.

The 46th Division's plan, rigid and inflexible, had no chance of success against uncut wire and intact positions. The policy of bringing supporting troops forward through communication trenches, which were soon blocked with wounded, meant that they arrived too late to take advantage of the few gaps in the German front line: the division was simply less well prepared for the operation than some of its counterparts further south. Its commander, Major-General the Honourable E. J. Montagu-Stuart-Wortley, was relieved of his command after the battle. But the Midlanders rose above their miseries. They went on to become an excellent division and we shall meet them later, breaking the Hindenburg Line in 1918.

The villages of Serre and Beaumont Hamel were the objective of VIII Corps, left-hand formation of 4th Army. They differed from the fortress-villages south of the Ancre in that they were in dead ground to the attacker, with the German front line running along crests just in front of them. Serre was attacked by the 31st Division, mainly Pals' Battalions from the industrial towns of Yorkshire and Lancashire. A track running north from the Serre–Mailly– Maillet road (D919) takes the traveller up to Railway Hollow Cemetery, lying in a gully with the preserved woodland of Sheffield Memorial Park to its front. Running along the edge of the wood is the remnant of forward trench used by 11th East Lancashires (The Accrington Pals), who attacked with 12th York and Lancaster (The Sheffield City Battalion) on their left. For once, the staid prose of the *Official History* rises to the occasion when describing the 31st Division's attack:

> The extended lines started in excellent order, but gradually melted away. There was no wavering or attempting to come back, the men fell in their ranks, mostly before the first hundred yards of No Man's Land had been crossed. The magnificent gallantry, discipline and determination displayed by all ranks of this North

Country division were of no avail against the concentrated fire of the enemy's unshaken infantry and
artillery, whose barrage has been described as so
consistent and severe that the cones of the explosions
gave the impression of a thick belt of poplar trees.

The division made negligible progress, and amongst its hard-
hit units the Accrington Pals lost 585 officers and men, and
the Sheffield City Battalion 512. About halfway along the
track joining Sheffield Memorial Park to the Serre road the
crosses of sacrifice in four British cemeteries mark the high-
water mark of the attack, for many of the dead were buried
close to where they fell. On the road itself are two large
'concentration cemeteries', Serre Road Nos 1 and 2,
containing the bodies of men killed on and after 1 July.
Many were interred there when the area was cleared by the
British in 1917 and more were added after the Armistice
when some smaller cemeteries were emptied. There are
7,127 British and Commonwealth burials in Serre Road No.
2, a dreadful 4,944 of them unidentified. Two Sheffield
brothers, Frank and William Gunstone, lie in Luke Copse
Cemetery. In Queen's Cemetery lies Lieutenant Stanley
Bickersteth of the Leeds Pals, a clergyman's son: his family
correspondence is in the evocative *Bickersteth Journals*.
Lieutenant M. W. Booth (his Christian name, unusually,
was 'Major') lies in Serre Road No. 1. He was a Yorkshire
cricketer who had twice played for England: his sisters left
his room just as it was and lit a candle for him every night.

Between Serre and Beaumont Hamel stands Redan Ridge,
crossed by a track that runs up from opposite Serre Road No.
1 Cemetery. One of its two main defence works, the
Heidenkopf (roughly where the Cross of Sacrifice in Serre
Road No. 2 Cemetery now stands) was taken, but the Redan,
halfway on to Beaumont Hamel, remained in action all day
and those troops who crossed the ridge were eventually
mopped up by German troops freed by the failure of British
attacks on both flanks.

Beaumont Hamel was the objective of the 29th Division,
composed in great part of regular battalions, which had been

in garrison overseas when war broke out. The division had fought in Gallipoli, where it had already earned the right to its nickname 'incomparable'. Its path on 1 July can be charted all too simply. On its left, the direct route to Beaumont Hamel (the modern D163) was dominated by Hawthorn Ridge, which the Germans had secured with the Hawthorn Ridge Redoubt. The mine beneath it was fired at 7.20 a.m., but the Germans reached its crater before two platoons of 2nd Royal Fusiliers, sprinting for the same objective. A sunken lane lay midway between the British jumping-off trenches and the German front line. There is a memorial to 1/8th Battalion The Argyll and Sutherland Highlanders, part of 51st Highland Division, which took the village in November 1918, close to where the sunken lane meets the D163 just west of Beaumont Hamel. From its mound bushes surrounding the mine crater on Hawthorn Ridge, on the slope across the road, stand out prominently and the line of the German front trenches can be seen 400 yards due east, along the tree line on the edge of the village. 1st Lancashire Fusiliers attacked it across the sunken lane as part of 29th Division's first wave and had pushed two companies, together with the brigade trench mortars, forward into it under cover of darkness. At zero hour the leading companies were machine-gunned almost as soon as they left the cover of the road, and Corporal George Ashurst, moving up to the road with the commanding officer and the two follow-up companies, ran towards it through what was, all too literally, a hail of fire:

> Miraculously I breathlessly reached the sunken road, practically leaping the last yard or two and almost diving into its shelter. Picking myself up and looking around, my God, what a sight! The whole of the road was strewn with dead and dying men. Some were talking deliriously, others calling for help and asking for water.

The Lancashire Fusiliers' attack ended, like so many others, on the German wire, and a few fusiliers lie where they fell, in Beaumont Hamel Cemetery, which marks the wire's forward edge. Failure to capture Hawthorn Ridge Redoubt was fatal to the battalion's chances, for it was not simply engaged by the garrison of the trenches straight ahead, but by the Germans who had occupied the mine crater on the ridge and who now enjoyed an easy shoot into the Fusiliers' right flank.

The right flank of 29th Division's attack extended over the ground now preserved as the Newfoundland Park. Here the German trenches ran just in front of a sharp-sided little valley known as 'Y Ravine'. 1st Royal Inniskilling Fusiliers and 2nd South Wales Borderers, in the first wave, were stopped well short of the ravine, and 1st Border Regiment and 1st King's Own Scottish Borderers, following up, fared no better. The usual catalogue of misunderstandings led to two more battalions attempting the impossible and just after 9 a.m. the Newfoundland Battalion advanced through the wreckage of the previous attacks. With a determination that almost defies description by the written word, the Newfoundlanders went straight on into a searing fire, and had the melancholy distinction of losing 715 officers and men, the heaviest loss incurred by any battalion that day. Their memorial, a great bronze caribou, roars out, in grief and rage, towards Y Ravine across the grassy hummocks of old trenches and distinctive corkscrew barbed-wire pickets. A visitor centre near the entrance to the park tells us much about the tight-knit little maritime communities that sent off their men with the Newfoundlanders.

Although the custodian's desire to prevent visitors from damaging the park by uncontrolled access, for this is a favourite tourist spot, has made the place rather over-controlled, it is still worth walking past the 'danger tree', the remains of a tree that stood in no man's land, to Y Ravine Cemetery, then swinging north to the German front line. A superb kilted highlander commemorates 51st Highland Division's success in November, and nearby is Hunter's Cemetery, which, as the War Graves Commission's description says, is actually 'a great shell hole' in which forty-one dead highlanders were buried. Remember that the battlefield as we now see it does not simply mark the fighting of 1 July 1916. Frenchmen fought and died there before British and Newfoundlanders arrived, and a trench still visibly creasing the no man's land of 1914 actually dates from much later in the war.

Across the marshy valley of the Ancre, the 36th (Ulster) Division had concentrated in Thiepval Wood (Bois d'Authuille) opposite its first objective, the Schwaben Redoubt, an imposing defence work dominating the ridge-line between Thiepval and the Ancre valley. The division, largely composed of pre-war members of the Ulster Volunteer Force, had a spirit and character all of its own. There were more than a few orange sashes about that morning and the old cry of 'No surrender' stirred the blood as warmly on Picardy chalk as it had in the streets of Belfast. At 7.30 a.m. the two leading brigades attacked and took the Schwaben, with some men pushing on as far as the German second line, where they were caught by their own barrage, crashing on by timetable. Worse still, the keep-like eminence of Thiepval, away to the Ulstermen's right, was still in German hands and its machine-guns covered the open ground across which the supporting brigade had to pass. As the day wore on the division lost almost all the ground it had gained, although it still had 750 metres of German front trench to hand over to a relieving formation when it was pulled back that night, having lost just over 5,000 men. The Ulster Division's memorial, a replica of Helen's Tower in the Marquess of Dufferin and Ava's park at Clandeboye, County Down, where the division had trained, stands in the middle of the attack sector. The Schwaben Redoubt stood just east of it and the ground in the area is so unstable owing to underground workings that the gravestones in Mill Road Cemetery, just beyond the memorial, are laid flat to prevent them falling over.

It is often hard, when looking at 'preserved' trenches, to get a real feel for trench life during the war. The trenches on Vimy Ridge, with their concrete sandbags, are too neat, while those in Newfoundland Park, atmospheric though they are, are in contrast too softened by the passage of time.

In Thiepval Wood a section of the Ulster Division's trench has been excavated by professional archaeologists, its contents carefully recorded and conserved, and the trench itself then restored to a condition that the men of 1916 might recognise. The custodian of Ulster Tower leads tours of the trench and it is well worth visiting, for it is too easy, on this evocative battlefield, to spend too long looking at memorials and too little time trying to make sense of the lanscape itself.

Thiepval, then one of the strongest points in the German line, is now crowned by Lutyens's brick-and-stone Memorial to the Missing, which bears the names of 73,412 officers and men missing on the Somme in 1916–17. It is now approached through a visitor centre that includes a bank of computers so that visitors can use the admirable CWGC website to track down relatives who might be commemorated on the memorial or buried in the area. Like all memorials to the missing, this is organised by seniority of regiments and corps, with individuals listed alphabetically by rank within their regiments. In some places there is a gap where a name has been filled in, because a body has been found and identified. It is easy to imagine that the men commemorated on the Memorial were blown to tatters or simply lost on this broad-shouldered landscape, but in fact many of them would have received hasty burial at the time of their death, only for the site of the grave to be lost and for their body, now without adequate identifying features, to be buried in a War Graves Commission cemetery as an unknown soldier. The Commission does its best to provide such information as is available and the headstones of some men whose full details are unknown bear inscriptions like 'An officer of the Great War', or 'A corporal of the Great War, Royal Fusiliers'. Non-commissioned personnel wore brass shoulder titles, giving the designation of their regiment or corps, fastened to their epaulettes, and these often survived even when uniform and equipment had perished and identity discs had disappeared.

There is something uncomfortable, almost brutal, about the Thiepval Memorial. It takes the form of a giant arch because it was first intended to bestride a road at the far end of the Somme battlefield, marking the furthest point of advance. It was then to be placed on the Albert–Bapaume road, but that site too was rejected by the French authorities. Its present location, just to the south of the site of Thiepval Château, on ground taken, in late September 1916, by 18th Division, attacking along the high ground from the south, was the third choice. The union flag flies from one side and the tricolour of France from the other, and in front of the Memorial lie 300 Commonwealth soldiers and 300 French soldiers, making the point.

On 1 July 1916 the fortified village of Thiepval and the Leipzig Salient that curled out along the ridge line to its south were the objectives of the 32nd Division. No impression was made on Thiepval itself, although misleading reports that British soldiers had entered it contributed to the misfortunes of the Ulster Division, whose reserve brigade went forward in the expectation that its right flank was secure. The Scots unit attacking the Leipzig Redoubt was formally 17th Battalion The Highland Light Infantry but rejoiced in the unofficial title the Glasgow Commercials. Its brigadier, J. B. Jardine, had been an observer in the Russo-Japanese war, and knew that it was important to minimise the gap between the lifting of the barrage and his troops' arrival in the enemy trenches. The Glasgow men crept forward to within twenty or thirty yards of the German front line while the bombardment was still going on, and 'in one well-organised rush' overran the Leipzig Redoubt before the Germans had emerged from their dugouts. The Germans had recovered from the shelling and the British remained in possession of the redoubt throughout the day, but it was the only success on this part of the front. 'It was said, with some truth,' observed the *Official History*, 'that only bullet-proof soldiers could have taken Thiepval on this day.'

III Corps advanced astride the Albert–Bapaume road. On its left, 8th Division attacked on a wide front opposite Ovillers la Boisselle, its right flank pushing up the long re-entrant of Mash Valley, with the La Boisselle spur looming on its right. It made little progress and one of its battalions,

2nd Middlesex, lost 623 of its soldiers in Mash Valley, chiefly to flanking fire from the spur, a fate which so grieved its commanding officer that he later committed suicide. The fate of 34th Division, aiming for Sausage Valley on the other side of the spur, was even crueller. The configuration of the ground was such that one of its brigades had to advance not from the British front line at the foot of the slope, but from the Tara–Usna line, nearly two kilometres behind the British front line. The front-line battalions achieved some success, partly because of the damage done by mines, including the huge Lochnagar Mine, whose crater, now owned by an Englishman, remains a poignant memorial. The left brigade pushed up towards La Boisselle and the right brigade took the Heligoland Redoubt, on the crest a kilo-metre north-east of Bécourt. But the rear brigade, four battalions of Tyneside Irish, marched over the open ground east of the Tara–Usna line, keeping step to a single big drum in the centre of the brigade. Such was the intensity of machine-gun fire that it arrived at the German front line with only a handful of survivors, but they pressed on: isolated parties reached the village of Contalmaison. The fate of the Tyneside Irish brigade helped push the 34th Division's casualties up to the highest suffered by a British division that day. No less than 6,380 of its members were hit, including one of three brigade commanders and seven out of twelve commanding officers. The divisional com-mander, Major-General 'Inky Bill' Ingouville-Williams, was killed by shellfire on 22 July.

The 21st Division attacked Fricourt, an exceptionally well-fortified village, from the west. Although three mines removed part of the German front line, progress was poor, and few soldiers got beyond the first enemy fire trench. 10th West Yorkshire lost its commanding officer, second in command and adjutant, with nineteen other officers and 688 men. South of Fricourt the front line turned to run eastwards and in this sector results were far more encouraging, partly because the character of the ground changes: gone is the wicked sequence of spurs, standing like natural bastions above the re-entrants they dominate, which had proved so damaging in the north. Mametz was attacked by 7th Division, whose commander had anticipated the effects of German shellfire and kept his front and support trenches clear of troops until shortly before the attack. Although British losses were heavy, the garrison had been badly shaken by the bombardment and the village was taken – and retained.

This little victory was not without episodes of extreme poignancy. Due south of Mametz, 8th and 9th Devons attacked from Mansell Copse, on the high ground just south of the Albert–Maricourt road. Captain D. L. Martin had made a model of the position and predicted that his battalion would be hit by the German machine-gun dug into the base of the crucifix in Mametz Cemetery, across the main road. When the Devons crossed the crest line that morning, Martin's prophecy was fulfilled: the machine-gun had survived the bombardment and killed him and many of his comrades. They lie in Devonshire Cemetery, whose entrance bears a stone inscribed:

> THE DEVONSHIRES HELD THIS TRENCH
> THE DEVONSHIRES HOLD IT STILL

East of Mametz, 18th Division made good progress, reaching the Montauban Alley trench system parallel with the Mametz–Montauban road and a little to its north. On the division's right Captain W. P. Neville, commanding a company of 8th East Surreys, had issued a football to each of his platoons and the company attacked behind footballs kicked into no man's land. Neville himself was killed and lies buried in Carnoy Military Cemetery.

On the extreme right, 30th Division's progress was even more spectacular. The bombardment had been unusually effective, in part because the French, with their greater number of heavy guns, had added their weight to it. A single machine-gun caused difficulties, firing from 18th Division's sector into the Manchester and Liverpool battalions on the division's left, hitting all the leading company commanders, but eventually a Lewis gun team dealt with it. Montauban, badly knocked about, was occupied by late morning, and the brickworks, once a formidable strongpoint but literally flattened by Allied artillery, was secured by midday. The brickworks' chimney, rebuilt after the war, still stands, close to Bernafay Wood, entered by British patrols on 1 July. There is a dignified memorial to 30th Division's pals' battalions from Manchester and Liverpool opposite the church in Montauban.

The French, attacking on the British right, had done altogether better. General Balfourier's XX 'Iron' Corps went over the top at the same time as its British neighbour – the right-hand British commanding officer, Lieutenant-Colonel Fairfax, and the left-hand French *chef de bataillon, Commandant* Le Petit, went forward arm in arm. South of the Somme, the French achieved surprise by attacking two hours later: the Germans did not believe that the Allies would be able to extend their attack that far south. The French also profited from the liberal use of heavy guns: eighty-five heavy batteries had engaged 7,300 metres of trench for nearly eight days. But it was their tactics that set the French firmly apart from their British comrades-in-arms. The results of set-piece attacks in 1914–15 had encouraged them to let attacking lines break down into small groups, ducking and weaving amongst shell holes and trenches, supporting one another by fire. By nightfall the French had taken most of their objectives and 3,000 prisoners.

The overall results of the day were deeply disappointing. Although there had been substantial gains on the right, the German second line was intact and north of Mametz the Germans recaptured most lost front-line trenches by nightfall. These meagre gains had cost the British 57,470 officers and men, 19,240 of them killed and another 2,152 missing. It is difficult to calculate German losses with any certainty, but an estimate of 8,000, including the known figure of 2,200 prisoners, cannot be far from the truth. 1 July 1916 was the British army's bloodiest day, and it ushered in a long and costly battle as hope of quick breakthrough turned into the reality of obdurate attrition.

Over the next two days British plans evolved. Gough was

put in charge of the two northern corps, VIII and X, and rapidly decided that an advance on 2 July was out of the question. On the 3rd attacks on Thiepval and Ovillers failed, and XV Corps failed to exploit a promising advance in front of Contalmaison. That night XV Corps at last inched forward to occupy the southern fringe of Mametz Wood, while XIII cleared Bernafay Wood, on the extreme right of the British line. Both sides regrouped, bringing up fresh divisions, and on the night of 4–5 July Gough's right boundary was moved south of the Albert–Bapaume road. Here he faced the largely intact German front line, while Rawlinson prepared to assail the German second position.

The great wedge of Mametz Wood stood like a redoubt just in front of the German second line: its capture was a prerequisite to any assault on Longueval Ridge beyond it. The Quadrangle strongpoint beyond its south-west edge was taken on 5 July, but the wood itself, its tangle of undergrowth thickened by branches in full leaf blown down by shelling, proved a formidable obstacle. The brunt of the attack was borne by the 38th (Welsh) Division, which had been fought to a standstill by the time it was relieved by the 7th and 21st Divisions on 12 July, the day the wood was at last cleared. One of the battlefield's most striking memorials stands above the track connecting Mametz with the Longueval–Contalmaison road, the D20. A defiant Welsh dragon, barbed wire crunched up in a mighty paw, snarls out towards Mametz Wood. Closer to the road is Flat Iron Copse Cemetery, set up on the site of a dressing station, for this little valley – Happy Valley to the south, but Death Valley to the north – remained an important route to and from the front line. In the cemetery lies Corporal Edward Dwyer of the East Surreys, who had won his VC at Hill 60, outside Ypres, in April 1916, and three pairs of brothers, two of them from Welsh battalions.

Rawlinson was learning. On the night of 13–14 July he launched III, XV and XIII Corps in a night attack against the Longueval Ridge, breaching the German second line on a front of 6,000 metres from Bazentin-le-Petit to Longueval itself. The aptly named High Wood – Bois des Fourcaux –

was within Rawlinson's grasp and its capture would have enabled him to interdict Germans moving down the Bapaume road to defend Pozières, and supports coming from Flers to assist with the defence of Delville Wood. The opportunity was missed and although the British reached the edge of the wood that evening, it was not to be cleared for another two months.

'Ghastly by day, ghostly by night, the rottenest place on the Somme' was one contemporary description of High Wood. Today it gives the visitor a feel for the central part of the Somme battlefield. Memorials commemorate the 47th London Division, which took the wood on 15 September, the Cameron Highlanders and the Black Watch, and the 9th Highland Light Infantry – the Glasgow Highlanders. The crater of a mine blown on 3 September lies just in the eastern edge of the wood and a concrete pillbox skulks amongst the trees facing the main road. To its south-east, through Longueval, lies Delville Wood. The village and the southern fringes of the wood were entered, during the 14 July attack, by 9th (Scottish) Division, which included a South African Brigade. The battle swayed to and fro throughout July and August, and it was not until 3 September that the wood was at last secured. Its rides are now named after streets in Edinburgh and Cape Town, and the South African Memorial in its centre has been sensitively expanded into a museum. Just north-west of the museum is the only tree that survived the battle, scarred but growing strongly.

The Delville Wood–High Wood line ran just to the rear of the initial German second position, but its garrison, making good use of the cover provided by the woods, held off a succession of attacks in July and August, in conditions made even more hellish by unseasonably rainy weather. Gough continued the pressure on his front, using the newly arrived Australians against Pozières and Mouquet Farm. Three Australian divisions – 1st, 2nd and 4th – fought in the sector, losing 22,826 officers and men. The 5th Division was, as we saw earlier, engaged in a diversionary operation at Fromelles, in the Aubers sector, where it lost another 5,533 men. Pozières, still visibly dominating the centre of the ridge, with a television mast providing a useful point of reference when the village itself is out of sight, was taken by the Australians whose memorials can be seen on the south-western edge of the village and on towards Bapaume, on the crest where Pozières Mill stood and where it was said that Australian dead lay thicker than at any other spot.

In his excellent book, *Australians on the Somme: Pozières 1916*, Peter Charlton suggests that the fighting around Pozières had more than a narrowly military importance. Most Australians were not disposed to reflect on the problems confronting generals waging industrialised war on a vast scale. They were, as one officer wrote, tired of 'the British staff, British methods, and British bungling . . .' The disillusionment that resulted from the appalling casualties suffered for so little apparent purpose made it hard for Australians to reconcile nationalism with imperialism thereafter. 'If Australians wish to trace their modern suspicion and resentment of the British to a date and a place,' declared Charlton, 'then July–August 1916 and the ruined village of Pozières are useful points of departure. Australia was never the same again.'

On 29 July Robertson, Chief of the Imperial General Staff, warned Haig that 'the Powers that be' were showing disquiet over huge losses and small gains. Haig replied that he proposed to wear down the Germans by steady pressure, grasping any chance of a breakout.

The dogged attrition of August and early September saw Haig under increasing political pressure to make recognisable gains. He decided that another push would be made in mid-September and as part of it he proposed to use what he termed 'a rather desperate innovation' – the tank. The first tanks, part of the Heavy Section, Machine Gun Corps, arrived at Abbeville in early September, and were taken by train from their training area to a railhead near Bray.

Before the main offensive began 4th Army improved its position by taking Guillemont, Ginchy and Faffémont Farm on its right. Capture of the latter enabled the French to put fresh momentum into their own advance, recommenced on 3 September, and by the 14th they were up on the

Bapaume–Péronne road at Bouchavesnes. Need to reorganise prevented them from joining the British attack on the 15th, but their heavy guns made a valuable contribution by reaching across into the German flank.

On 15 September the main attack began at 6.20, but an hour earlier a single tank had assisted infantry in clearing the last Germans from a pocket of resistance in Delville Wood, earning the distinction of being the first tank ever to see action. We will consider the battle from left to right. Reserve Army allocated its tanks to the 2nd Canadian Division, which used them to spearhead its advance on the Sugar Factory just short of Courcelette. The start point for this attack is marked by a memorial – an obelisk with four model tanks at its base – on the main road at Pozières. The Canadians took the factory, though without the assistance of the tanks, which either broke down or moved too slowly to keep up with the infantry. On their right, 15th (Scottish) Division of III Corps took Martinpuich with the help of half the corps' eight tanks, providing a fulcrum for the successful Canadian advance on Courcelette.

47th Division's attack on High Wood got off to a bad start largely because III Corps had decreed that the four tanks accompanying it would pass through the wood, despite the objections of tank officers, who feared that their machines would belly on tree stumps. Three duly ditched and the fourth reached the German support line where it was hit by a shell. The infantry made no progress, despite what the *Official History* terms 'reckless bravery'. At midday, however, after a trench mortar battery had put 750 bombs into the wood in seven minutes, the Londoners tried again and this time the Bavarian defenders had reached the limit of their endurance.

To the east of High Wood, XV Corps advanced on Flers, and although fire from High Wood, not yet cleared by the Londoners, caused difficulties for the New Zealander Division – whose obeslik between High Wood and the Flers road marks their line of advance – the tanks proved decisive. When the attack stalled on the edge of Flers, four tanks rumbled into the village and the infantry surged in behind them. On the far right, XIV Corps made some progress towards Lesboeufs, but the tanks made little effective contribution and the Guards Division halted short of its objective in a wilderness of murdered nature that made navigation excruciatingly hard. Amongst its dead was Lieutenant Raymont Asquith of 3rd Grenadier Guards, son of the Prime Minister, H. H. Asquith. He was mortally wounded between Lesboefs and Ginchy, and is buried in Guillemont Road Cemetery, with an inscription from Shakespeare's *Henry V* on his headstone: 'Small time, but in that small most greatly lived this star of England.' Close by rests Lieutenant the Hon. Edward Wyndham Tennant of 4th Grenadiers, killed a week later. The officers were related (Tennant's aunt was Asquith's stepmother) and both were members of that charmed circle known as the Souls.

The day ended with 4,100 metres of the Germans' third line in British hands, but as long as Lesboeufs and Guédecourt held, there was no chance of exploiting these gains. Although any prospect of a breakthrough had now gone, the battle continued as the British sought to consolidate their grasp on the whole ridge. In late September XV Corps, assisted by the French on its right, took Morval, Lesboeufs, Combles and Guédecourt. In the last-named village another caribou marks the site of a position taken by the Newfoundlanders. On a clear day, the Thiepval Memorial can be seen by looking due west across eleven kilometres and tens of thousands of lives.

The Reserve Army was also on the move in late September, fighting the Battle of Thiepval Ridge to take Thiepval and the Schwaben Redoubt – both 1 July objectives. In early October the attacks edged north-east and northwards, with 4th Army pushing up to Le Sars and the Butte de Warlencourt, and the Reserve Army – renamed 5th Army on 1 October – working its way up to dominate the high ground above the Ancre. Beaumont Hamel was at last taken, by 51st Highland Division whose memorial, a kilted soldier, we have already seen in Newfoundland Park. Beaucourt fell to the Royal Naval Division, but Serre, bloody Serre, remained in German hands.

In mid-November Allied leaders met at Compiègne to discuss strategy. There was outline agreement on another combined offensive in 1917, but a meeting of Allied government representatives in Paris revealed alarm at the casualties of 1916 and suspicions that there might be more promising areas of operation than the Western Front. In December Joffre was gently propelled upwards, becoming a marshal of France but losing all influence. His successor was General Robert Nivelle, who had made his reputation at Verdun. In Britain, Asquith was replaced as prime minister by Lloyd-George. The latter was already deeply concerned about casualties and over the next few months this concern would influence events in France.

Historians still cannot agree on the butcher's bill for the Somme. Differences in casualty reporting procedures meant that German official figures are certainly too low, though some have accused the official historian, Sir James Edmonds, of inflating the total to an unreasonable 660–680,000. It is safest to say that German casualties were in the region of 600,000, and the British and French roughly similar, with the British suffering about two-thirds and the French one-third of this total. Haig's supporters point out that even if he did not achieve a breakthrough, he had done lasting harm to the German army. Ludendorff acknowledged that it was 'completely exhausted', and a survivor wrote, 'The Somme was the muddy grave of the German field army.'

This damage had been inflicted at a horrifying price. The burden of the Somme fell disproportionately on the New Armies and especially on those Pals' Battalions whose destruction spread rings of sorrow through close-knit communities. The loss was also qualitative. The most dangerous rank on 1 July was captain, and as bad weather and diminishing confidence robbed some units of their élan in the weeks and months that followed, officers sacrificed themselves to breathe fire into exhausted men. Almost any British cemetery on the Somme makes the point graphically, with subalterns and captains in profusion, and a strikingly large number of lieutenant-colonels, commanding officers of battalions and, in the tribal world of the regimental system, very important men.

In a German cemetery, in contrast, there are fewer officers, partly because the German army had a smaller ratio of officers to men and used senior NCOs in appointments that would have been filled by officers in the British army. The German cemeteries at Fricourt and Rancourt on the Somme show the price paid by these individuals, denied an officer's status but sharing his risks. Here lies an *Offizierstellvertreter*, there a *Feldwebel-Leutnant*: in the next row a *Fahnenjunker*, and behind him a *Wachtmeister*.

Scarcely had the fighting drawn to a close than the Germans were analysing its results. Their deductions fitted into the pattern of reappraisal that had followed the elevation of Hindenburg and Ludendorff to the high command. A pattern of new defensive lines was sketched out. *Flandern* was to run from the Belgian coast, along the Passchendaele Ridge, behind Messines and on to Lille. *Wotan* ran behind the Loos–Vimy battlefield and on through Quéant to Sailly; its central section, east of Arras, was known to the British as the Drocourt-Quéant Switch. *Siegfried* cut off the Noyon salient, from east of Arras to St-Quentin, Laon and the Aisne east of Soissons. The *Hunding* and *Michel* lines ran off to the east, behind the Champagne and Verdun battlefields. Work began first on the *Siegfried* Line, for its completion would enable the Germans to shorten their line by almost twenty-five miles, and some 65,000 labourers started work in late October. Most were Russian prisoners of war, their presence recalled by the Russian headstones in many Western Front cemeteries.

On 4 February 1917 OHL ordered withdrawal to the *Siegfried* Line, known to the Allies as the Hindenburg Line, staged over the period February–15 March. The operation was code-named *Alberich*, after the spiteful dwarf in the Nibelungen saga. It was an appropriate choice for, despite protests by senior officers, including Crown Prince Rupprecht of Bavaria, army group commander in the north, the land given up was devastated with ferocious thoroughness. Bridges were destroyed, roads mined, villages and orchards levelled. The keep of the fine thirteenth-century castle at Coucy le Château was blown up, although the town's medieval walls remain intact. Vengeful cunning was displayed in the siting of booby traps. Attractive items of equipment were connected to explosives and delayed-action fuses set off charges long after the Germans had left.

The Hindenburg line embodied the concepts resulting from the experience of 1915–16: Colonel Fritz von Lossberg, deputy chief of the Operations Section of OHL, played a leading role in their development. The pamphlet *Conduct of the Defensive Battle*, issued in December 1916, proclaimed the official doctrine of defence in depth. Where the ground permitted it, the outpost line lay on a forward slope, and was lightly held, giving advance warning of attack and enabling artillery observers to engage enemy concentrations. The main trench system, usually consisting of three distinct trenches, lay on a reverse slope, 500–1,000 metres from the outpost line. The battle zone ran back from this, its rearward edge marked by the artillery protection line, a double trench line some two kilometres to the rear. As time went on, fresh layers of defence were added further back. Care was taken that only a small proportion of troops would be in the outpost line and vulnerable to the enemy's field artillery. A three-battalion regiment would have a battalion forward, holding the outpost line and providing reserves for local counter-attacks, a battalion in the trenches and strongpoints of the battle zone, and a reserve battalion in the rearward zone. Counter-attack divisions were posted behind the front in readiness to assail an exhausted enemy who had gasped his way into the labyrinth.

German withdrawal caught the Allies off balance. Nivelle, the new French commander-in-chief, had oversold a spring offensive to politicians and soldiers alike. The story of its tragic failure on the Chemin des Dames above Rheims will be told in a later chapter, but its consequences bear all too closely on what happened in Flanders in the summer of 1917.

For the moment, though, let us turn our attention to the British component of Nivelle's offensive. This was originally intended to comprise three distinct attacks. Horne's 1st Army would assault Vimy Ridge; Allenby's 3rd would attack east of Arras, while Gough's 5th slogged on towards Bapaume. German retirement not only deprived Gough of an objective, but made offensive action on the southern end of the British line difficult because of the need to get troops and equipment over the desert of the old Somme battlefield. The offensive therefore consisted of only two of the proposed attacks, launched by Horne and Allenby.

The battle had mixed results. The Vimy attack was the responsibility of the Canadian Corps, fighting its first battle as a unified corps. Its preparations were thorough. There were 480 18-pounders, 138 4.5-inch howitzers, and 245 heavy guns to support the attack, giving roughly twice the concentration of gun to yard of front available on the Somme. Piping was laid to supply water – five gallons per man per day for 80,000 men and ten gallons per horse per day for 36,000 horses.

The most distinctive characteristic of the preparatory phase was the digging of tunnels from the western slope of the ridge to the front line. Mining operations began there in 1915, leaving a number of craters around the opposing front lines. The tunnels used by the Canadians in 1917 were bigger than the tunnels dug for mining. There were twelve main tunnels in the Canadian sector, with a network of smaller tunnels and dugouts running off them. The main east-west shafts were two metres high and at least one metre wide. Lit by electricity, they carried light railways and telephone cables, and permitted troops to reach the Canadian lines with safety. A section of Grange Tunnel, one of the longest, is open to the public and visitors are taken round, appropriately enough, by young Canadian guides. However, a visitor to the Grange is only scratching the surface of the tunnel system, literally miles of which lie beneath the chalk. These tunnels are now most unsafe, not merely because of the dangers of gas and cave-ins, but because of unexploded mines – chambers packed with explosive – one of which was discovered in 1988 and made safe by Royal Engineers.

Canadian preparations included an element missing from the lavish orchestration on the Somme nine months before. Over 40,000 maps were issued and troops were briefed on the details of German defences, so that, despite centralised planning, there was room for low-level initiative if there were local hold-ups. The phases of the advance were marked by four report lines – black, red, blue and brown – but they were simply intended to chart the progress of the advance, whose successive objectives were real features – trenches and strongpoints – on the ground.

The Corps, under Lieutenant-General Sir Julian Byng, was to attack with its four divisions side by side. On the right, the 1st Canadian Division, with its start line just north of Ecurie, had the near edge of Farbus as its final objective. The 2nd Division, attacking from Neuville St-Vaast, was to pass through Les Tilleuls to the Bois du Goulot, while the 3rd Division, also debouching from Neuville, was to clear the Bois de la Folie to reach Petit Vimy. On the extreme left was the 4th Division, going for Hill 145, the high ground now topped by the Canadian Monument. The ridge was not deemed suitable for tanks, but eight were made available to assist 2nd Division with the capture of Thélus. Shortly before the offensive started, Horne put the British 5th Division under Byng's command: Byng sent a brigade of it to strengthen his 2nd Division and kept the rest in reserve. Once the ridge was safely in Canadian hands, the corps cavalry would begin the process of exploitation, reinforced, as required, by cavalry divisions from GHQ reserve.

The battle began at 5.30 on the sleety morning of Easter Monday, 9 April 1917, with a hurricane bombardment that cut the remaining German wire and knocked out most batteries supporting the ridge's defenders. The 4th Division's 11th Brigade, attacking towards Hill 145, encountered an intact strongpoint and lost heavily. The delay enabled defenders to emerge from cover and the division's attack stalled short of its objective. It was the only significant check. Elsewhere the Canadians boiled forward, taking the crest line and objectives beyond it, but the cavalry exploitation, perhaps predictably, failed to materialise. Over the next few days the Canadian hold on the ridge was strengthened by the capture of the whole of Hill 145 and the Pimple feature to its north.

Loss of the ridge shook the Germans profoundly. The Allies now enjoyed excellent observation over the Douai Plain, and had captured 4,000 prisoners and fifty-four guns. Prince Rupprecht confided to his diary that he doubted if there was any point in continuing the war and even the stolid Hindenburg found Easter Monday a day of 'much shade, little light'.

The effect on the Canadians was as dramatic. Byng shortly went off to command 3rd Army, while Currie, commander of 1st Division, became corps commander in his stead – the first ever Canadian lieutenant-general. But for most Canadian participants, and to their families and friends across the Atlantic, the victory was a proud demonstration of nationhood: it is not too much to say that Canada became a nation at Vimy. The battlefield has been preserved, perhaps too well. Trenches, revetted with cement-filled sandbags, mark the Canadian and German outpost lines, and craters gape amongst the pine trees. The Memorial, designed by Walter S. Allward of Toronto, bears the names of 11,285 Canadians who were declared 'missing, presumed dead' in France. Of the 619,636 Canadians who served abroad during the war, 66,655 were killed. The contribution of the Canadians was seldom less than distinguished, from their first appearance amidst the gas at Second Ypres, through the battle for Courcelette on the Bapaume road in 1916, and on to the victorious advance of the Canadian Corps in August 1918.

So much for the northern flank of the British attack. Things went less well to the south, where 3rd Army was pushing out of the suburbs of Arras towards Monchy-le-Preux and Fampoux. Allenby's task was made harder by the fact that the exits from Arras were constricted and German defences were strong. The old front line ran through Neuville Vitasse and up to Roclincourt, where it climbed Vimy Ridge. Behind it ran the artillery protection line of the Hindenburg system, hooking back from Tilloy to Athies and

Point du Jour. A third line, the Monchy Riegel system, ran just west of Monchy, which village dominated the centre of Allenby's attack.

Allenby's men attacked, like Byng's, at 5.30 in the morning. The westerly wind blew sleet into the defenders' faces and an effective counter-battery programme crippled their artillery support. Progress was good north of the Scarpe, where XVII Corps took Fampoux and secured Point du Jour Ridge. It was more patchy on VI Corps' front south of the river. The 12th and 15th Divisions took Observation Ridge, the long slope running north of Tilloy towards the Scarpe, now midway between Arras and the A1. They crossed it in time to catch German artillery in action in Battery Valley below it, and captured sixty guns. Then the old problem of infantry-artillery co-ordination reared its ugly head and although Orange Hill, part of the next ridge line towards Monchy, was taken, the attack halted on the uncut wire in front of the Monchy Riegel system. On the right flank, Neuville Vitasse fell and the 14th Division took Telegraph Hill, two kilometres east of Beaurains, but progress here was generally disappointing.

Nevertheless, the considerable gains made on the 9th, added to Canadian success on the ridge, left 3rd Army in a position to do serious damage to the Germans, possibly compelling a withdrawal to the Drocourt–Quéant switch. However, on the 10th an attempt to exploit success with cavalry failed, with horsemen caught by machine-gun fire amidst flurries of snow. On the 11th Monchy fell to the 37th and 15th Divisions, aided by three tanks. The cavalry came up again, this time to fight as infantry in the defence of Monchy. 8th Cavalry Brigade lost its brigadier, Charles Bulkeley-Johnson, to small-arms fire, and Lance-Corporal Harold Mugford of the Essex Yeomanry earned the VC helping to hold the village, turned into a shambles in the most literal sense of the word as German artillery fire hit the cavalry's horses, which could not be got back to safety.

The fighting dragged on till 14 April, when the 29th Division, which had replaced the exhausted 37th, sent its 88th Brigade against Infantry Hill, a wooded feature just short of the Bois du Sart, on the track running from the north-east edge of Monchy towards Boiry Notre Dame. The attackers, 1st Essex and the Newfoundland Battalion, took the wood easily, but were immediately caught in a pincer-like counter-attack that destroyed most of them and swept on towards Monchy. Of the brigade's other battalions, 2nd Hampshires had a company in the village – effectively its only garrison – and the rest on Orange Hill, and 4th Worcesters were holding the line to the south-west.

Lieutenant-Colonel Forbes-Robertson, the Newfoundlanders' commanding officer, who had not been permitted to accompany the attack, learned of the disaster from survivors and led a handful of men – including his signals officer, provost sergeant and a straggler from the Essex – into the attack's jumping-off trenches and held the Germans off. The Hampshires' commanding officer heard of the peril and brought the remainder of his battalion forward. The Germans had sealed off Monchy with a box barrage, but the Hampshires saw that one battery was firing slightly off line and slipped in through the gap. Monchy was saved. Forbes-Robertson was recommended for the Victoria Cross, but was not awarded it: happily another act of heroism brought him his VC next year.

Monchy contains a memorial to the 37th Division, three bronze figures of soldiers in fighting order. On top of a bunker stands a Newfoundland caribou, the third we have seen: we shall see the fourth and last before long. Monchy Cemetery, on the minor road west of the village, gives a good view of the slope crossed by 2nd Hampshires on the 14th. The Boiry track, up near the church in the north-east corner of the village, offers excellent observation to Infantry Hill. Due north of Monchy, across the Scarpe and between the villages of Plouvain and Roeux, its slopes crossed by the A26, is Greenland Hill. On 23 April a series of attacks began towards this objective, with the Canadians taking Arleux, to its north. The battle reached its zenith in early May, with fruitless and unsubtle fighting for Fresnoy, eventually recaptured by the Germans, and Roeux, held by the British.

Further south, there had been two distinct battles, both

involving the Australians, at Bullecourt. On 11 April the 4th Australian Division attacked with ineffective tank support, and its 4th Brigade, which went forward with 3,000 men, suffered a shocking 2,339 casualties. Amongst these were 1,200 prisoners, about one-third of all Australians captured during the war. Second Bullecourt, which began on 3 May, saw 2nd Australian Division attack to the east of Bullecourt while 62nd (West Riding) Division attacked and eventually, though heavily reinforced, took the village itself. Casualties – Australian, British and German – were heavy: 62nd Division lost over 4,000 officers and men. There was friction, by no means one-sided, between the Australians and the British and, as Jonathan Walker demonstrates in his history of the battle, *The Blood Tub*, significant backbiting amongst Australian generals.

A disused railway line, much of it on a wooded embankment, runs south of Bullecourt, enabling the visitor to follow, for some 4,000 yards, the jumping-off line for 4th Australian Division for the first battle. There is a good museum in the village, and in front of the church are memorials to the British 7th, 58th and 62nd Divisions, and a bronze slouch hat for the Australians. East of the village a bronze digger, one of the finest statues on the Western Front, presides over the ground assaulted by the left-hand brigade of 2nd Australian Divison during 2nd Bullecourt.

The Battle of Arras cost the British and Canadians 150,000 men and the Germans only 20,000 fewer. It demonstrated yet again that the exploitation hinged on command and control: if decisions were made at corps or army headquarters the orders they issued arrived too late. Co-ordination of effective artillery fire in the advance remained a problem: laying line from observers to batteries took time, and while coloured flares could provide rough and ready communication, they were no substitute for the systematic adjustment of targets by trained observers.

Part of the answer lay in acknowledging that the chain of command could not function coherently in battle with the communications then available, and to recognise this by delegating decision-making, allowing local commanders to act within the framework of their superior's general plan. This concept, often known by the term *Auftragstaktik*, was, increasingly, employed by the Germans. It was certainly no panacea, for during the early stages of the Arras–Vimy fighting the Germans mishandled their counter-attack formations, and early in the battle a junior staff officer exercised his initiative disastrously by ordering the evacuation of the Feuchy–Wancourt line, exposing Monchy. It is fashionable to celebrate the excellence of Australian and Canadian formations, and we should guard against too easy an acceptance of the view that all troops from the Dominions were brave and innovative, and all British troops leaden and hidebound. But a major reason for the success almost consistently enjoyed by Australians and Canadians was their more flexible attitude, stemming from a closer relationship between leaders and led than was often the case in the British army. Peter Charlton gets to the heart of the matter when describing Brigadier-General John Gellibrand, who commanded the 6th Australian Brigade on the Somme and at Bullecourt. Regular officer turned fruit-grower, Gellibrand

> set himself only slightly apart from his soldiers. He wore the same uniform (by choice not affectation) and endured the same hardships. His headquarters functioned more as a co-operative venture than a military hierarchy with the commander sitting back and discussing the problems with the staff and with his unit commanders; at the same time he was not above intervening personally to ensure that his standards were being met and his orders carried out.

The results of the Arras–Vimy battle were substantial when set alongside the utter failure of Nivelle's offensive in Champagne, a disillusionment so sharp that it broke the heart of a French army already worn to a thread by heavy casualties, poor man-management and successive disappointments. From 25 April it began to show signs of mutiny. It took Nivelle's successor, General Philippe

Pétain, several months to restore his army to full reliability. In the meantime the weight of the war fell upon the shoulders of his allies.

The precise details of Haig's knowledge of the parlous state of the French army and its impact on his planning for the coming months must remain matters of conjecture. It is clear, however, that the French collapse was a major motive behind his desire to resume offensive operations in Flanders. Moreover, he had long believed that Flanders offered better prospects than Artois or Picardy, and only the need to conform to French wishes had drawn him south. German submarines, many based at Ostend and Zeebrugge, were sinking large numbers of Allied merchant ships and in November 1916 the Cabinet's War Committee had declared that it deemed no single measure more important than the expulsion of the Germans from the Flanders coast. Although America had entered the war in April 1917, it would be some time before she could make her presence felt and in the meantime Russian weakness would enable the Germans to switch more troops to the Western Front.

In the strategic circumstances of late spring 1917, therefore, standing on the defensive on the Western Front appealed neither to Haig nor to his political masters. Plans for a Flanders offensive had been elaborated since late 1916 and were well developed even before the Arras–Vimy operation was mounted. There were, in essence, to be three thrusts: firstly, an attack on Messines Ridge to clear the British right; a drive from the salient on the axes Passchendaele–Roulers and Staden–Thourout; and an amphibious hook along the Flanders coast. It had been intended that Rawlinson should command the main operation from Ypres, but in May the task was given to Gough of the 5th Army, at forty-seven the youngest of the army commanders. The *Official History* was to maintain that 'he was less inclined than an older man or a foot soldier to the tactics of "wearing down" battles and trench warfare'.

The curtain-raiser for the offensive was in very different hands. The commander of 2nd Army, 'Daddy' Plumer, and his chief of staff Sir Charles Harington knew the ground well and had a reputation for sound planning. The Messines Ridge offensive was meticulously prepared and was to employ three corps, X, IX and II Anzac, each attacking with three divisions in line. Plumer had three divisions in reserve and XIV Corps, in GHQ reserve, was just behind him. Seventy-two tanks were allocated to the attack and nineteen mines, with nearly a million pounds of high explosive between them, were dug beneath the German lines. The fireplan was comprehensive, and paid attention to the need to deal with German batteries and to break up counter-attacks. A total of 2,266 guns and howitzers, 756 of them medium or heavy, supported the attack. In order to ensure air superiority over the battlefield, a complete Royal Flying Corps brigade of 300 machines was put at Plumer's disposal, outnumbering German aircraft by two to one.

Although the timing and methods of the offensive came as a surprise, Rupprecht had foreseen a British move against Messines Ridge, and suggested that it should be forestalled by withdrawal to the German third line, the Warneton Line. But neither this suggestion, nor that of withdrawal to the *Flandern* Line even further back, appealed to the commander of the German 4th Army, who elected to offer battle on the strong but exposed ground of Messines Ridge.

Because of the high water-table the Germans were unable to dig as many dugouts as they wished, and had instead built concrete shelters, which could withstand the direct hit of a 6-inch shell. During the preparatory phase shelling soon knocked the earth covering off the pillboxes, and on the bright May afternoons, as the sun sank behind the British observers, the pillboxes shone out as perfect targets. From 26 May to 6 June over three and a half million shells were fired on the Messines Ridge front. The first line was almost totally destroyed, but many of the pill-boxes of the second and Warneton lines remained intact. Front-line divisions were mauled before the battle started and the commander of *Gruppe Wytschaete* – the German corps level of command in the defensive battle – unwisely replaced them with counter-attack divisions, bringing fresh divisions up to take over from the latter. This decision, which later cost him his job, resulted in the well-trained counter-attack divisions being worn down in the trenches, and the new and unpractised divisions arriving on the battlefield too late to help.

At 3.10 on the morning of 7 June the mines went up with what was the loudest man-made sound created thus far in history, and the whole of 2nd Army's artillery opened fire at a rapid rate, lashing German forward lines and reaching out to blanket battery positions with gas and high explosive. The mines did fearful damage: on the following morning Harington visited a dugout and found four dead German officers sitting at a table with no mark of a wound on any of them: they had been killed by concussion. The front line was quickly overrun. II Anzac Corps barrelled on into Messines, taking the ruins after close-quarter fighting. To its left, IX Corps cleared Wytschaete Wood (Kapelleriehoek on modern maps), but X Corps had a hard fight for the Damm Strasse, the minor road running from Wytschaete to Hollebeke. Nevertheless, by five o'clock the British were in possession of their second objective, the front trench of the German second line.

There followed a two-hour pause while the British consolidated and prepared to attack the rear trench of the second line along the back of the ridge. The bombardment resumed at 7 a.m. and the attack went on. Wytschaete itself was taken by Munster Fusiliers and Royal Irish, assisted by tanks. By 9 a.m. the British held the whole of the ridge from the Douve to east of Mount Sorrel. It was only then that problems arose. The attackers, who had so far suffered fewer casualties than had been expected, were closely packed on the ridge and the expected counter-attacks failed to materialise. They were easily shrugged off when they came, but much time was lost and the troops on the ridge, wielding pick and shovel in their 'grey-back' shirts, were hard hit by long-range machine-gun fire. Despite this, the advance to the third objective, the Oosttaverne Line, was generally accomplished by nightfall. The long pause had allowed the Germans to get most of their guns away – only forty-eight of the expected 120 were captured. The Spoil Bank, up by Hill

60 on the canal, remained impervious to attack and it took the Australians much hard fighting to clear the Blauwepoortbeek valley north-north-east of Messines: it was not secured till 14 June. Messines was a clear victory. The ridge had been taken, over 7,000 Germans captured and perhaps another 15,000 killed or wounded. 2nd Army's casualties totalled 24,562, most of them suffered by II Anzac Corps.

Unlike so many other Flanders 'ridges', Messines Ridge looks worthy of the name. Seen from Kemmel Hill it rises square against the skyline, with the villages of Wytschaete and Messines prominent on it. At the top of Kemmel Hill stands a memorial to French soldiers who died in Belgium and there is an ossuary, containing the remains of over 5,000 Frenchmen, on its western slopes. Many of these soldiers were killed when the Germans took the hill in April 1918: it was eventually recaptured at the end of August. At the Lettenberg, not far from the ossuary, at the junction of Kattekerkhofstraat and Lokerstraat, are British dugouts which were cleared in 2004 and are now open to the public. The Belvedere Café, built on the site of a similar structure destroyed in 1918, has a tower that gives a good view over the battlefield: it is easy to see why the café was used by senior officers and artillery observers alike. On the Ypres road out of Kemmel is a memorial to the US 27th and 30th Divisions who fought there in August–September 1918.

Crossing Messines ridge from the north on the N336, the traveller passes through St-Elooi. This village was captured by the Germans in 1915 and thereafter was the scene of heavy fighting, particularly in March–April 1915. Two craters of mines exploded on 27 March lie just north of the Hollebeke road out of the village and there are two more craters – one from March 1916 and the other from June 1917 – in the V formed by the Messines and Warneton roads. Two pillboxes are visible to the east of Messines on the edge of the village: one is a typical German structure, the other British, with the marks of its 'wriggly-tin' shuttering, looking for all the world like a concrete Nissen hut.

Wytschaete, along the N365, has a memorial to the 16th

Irish Division, which took the village in 1917. The ridge is
still gouged by mine craters. The most easily reached is the
Spanbroekmolen crater, the 'Pool of Peace', just south of the
Wytschaete–Kemmel road. Work on the mine was begun on
New Year's Day 1916 and a gallery 521 metres long was
driven up from the British lines. This was the largest of the
Messines mines, containing 91,000 pounds of ammonal,
whose detonation left a crater seventy-six metres wide and
twelve metres deep, extending total devastation out to a
diameter of 131 metres, and killing by its blast and debris
over a much wider distance. Two of the mines prepared for
the battle were not fired and their locations were subsequently lost. One blew up in July 1955 during an electrical
storm, but the remaining one still remains packed with
explosive somewhere beneath the fields.

Messines is rich in memorials. On the N365 to its north
stands a monument to the London Scottish, the first
Territorial battalion to fight there during First Ypres. The
'Four Huns' bunker stands beside the road closer to the
village. The imposing church of St Nicolas in Messines was
rebuilt to replace that totally destroyed in the battle, but its
crypt is original. In it rests Adela, wife of Count Baldwin V
of Flanders, and mother of William the Conqueror's wife
Matilda. The town hall, also rebuilt, contains a small war
museum, open during office hours in the summer. The New
Zealand Memorial Park, south of the village, houses not
only a white stone monument to the New Zealanders,
inscribed with their characteristic tribute 'FROM THE
UTTERMOST ENDS OF THE EARTH', but also two
pillboxes, part of the rear defences of the German first line.
Down the Wulvergem road is the New Zealand Memorial to
the Missing, with the names of missing New Zealanders
inscribed on the base of a Cross of Sacrifice.

Just south of Messines on the N365 is the Island of Ireland
Peace Park, to my mind one of the most important
memorials on the Western Front. When Plumer's men took
the ridge, 16th (Irish) Division, mostly Roman Catholics
from the south of Ireland, went into battle alongside 36th
(Ulster) Division, composed largely of Ulster Protestants.

Although officers and men from the two divisions differed sharply in both religion and politics, they were often conscious of sharing a common Irish identity. Mortally wounded that day was Major Willie Redmond, a nationalist MP and brother of John Redmond, leader of the Irish parliamentary party. Aged fifty-six, he slipped away from a staff job to rejoin his old battalion, 6th Royal Irish, and was mortally wounded during its advance. John Meeke, a stretcher-bearer in 36th Division, saw him fall, tended his wounds and helped carry him to safety. Redmond died in a 36th Division dressing station and now lies in a lonely grave near the entrance to Locre Hospice Cemetery.

Many of the nationalists who had enlisted in the British army did so in the belief that their conduct in the war would advance their cause, but the Easter Rising of 1916 left them exposed and after the war they were reviled as men who had betrayed their country. It is only in the last decade or so that the wounds have begun to heal. The Irish regiments disbanded when the south of Ireland became independent in 1922 now have regimental associations once again and I am proud to wear my Royal Dublin Fusiliers tie, its broad red and blue stripes, traditional fusilier colours, divided by a thin green line to make clear the regiment's Irish lineage.

The Irish Peace Park is part of this process of reconciliation. Although strictly speaking it is not on quite the right part of the battlefield, for the Irish divisions attacked further north, it would be wrong to allow academic carping to obscure the greater principle. At the centre of the little park is an Irish round tower, built partly from stone partly from an old barracks at Tipperary, and partly from a workhouse near Mullingar. The sun illuminates its interior on the eleventh hour of the eleventh month each year, the time at which the Armistice came into force. The park was inaugurated by President Mary McAleese on 11 November 1998 and in her speech she emphasised the tragedy of Irish soldiers in the Great War: 'They fell victim to a war against oppression in Europe. Their memory then fell victim to a war for independence at home in Ireland.' The memorial strives not to redress the balance, but to recognise the sacrifice, and does so very well indeed.

Ploegsteert Wood – inevitably known to the British as Plugstreet – lies further south along the same road, across the Douves three kilometres from Messines. The greater part of the wood was held by the British for most of the war but was captured by the Germans in April 1918. It contains an assortment of pillboxes, and in and around it lie a number of British cemeteries. At Hyde Park Corner, a road junction at the wood's northern end, is the Ploegsteert Memorial to the Missing, bearing the names of the 11,447 soldiers missing in the Ploegsteert–Armentières–Fromelles sector. In early 1916 Lieutenant-Colonel Winston Churchill, who had lost political office as a result of the failure of the Gallipoli campaign, which he had championed, commanded 6th Royal Scots Fusiliers in the Ploegsteert sector. His battalion headquarters, a farmhouse just south of the little road that runs along the wood's southern edge, from Ploegsteert to Le Gheer, has long gone, although its well-head still stands in the pasture. In Lancashire Cottage Cemetery, on the road, lie some of the soldiers killed during Churchill's time in command.

Despite the success of the Messines Ridge attack, Haig found it hard to convince the politicians that he should press on with the Flanders offensive. Eventually the War Cabinet agreed to it, on the understanding that it would be called off if gains did not live up to expectations. The British had, in the meantime, been concentrating for the attack, with Gough setting up his headquarters in Lovie Château, just north of Poperinghe, and four corps – II, XIX, XVIII and XIV – coming under his command. 2nd Army was stripped of much of its artillery, and other guns were brought in from Artois and Picardy to give Gough 725 heavy and medium, and 1,422 field pieces. Three tank brigades, each with seventy-two tanks, were put at his disposal. The coastal operation, once intended to come under Gough's command, was now the responsibility of Rawlinson's much-reduced 4th Army, whose old front line on the Somme was gradually taken over by 3rd Army.

In late June 1st Army feinted towards Lens and Lille, and made small gains. This continued in mid-August, when the main battle was raging at Ypres, and the Canadian Corps made a useful advance front of the old Loos battlefield. Hill 70, on the Lens–La Bassée road, was captured and five German divisions were cut up in the process. Yet the Germans were anything but a spent force. On 10–11 July they mounted an attack in the Nieuport sector, punishing its defenders so severely that the coastal offensive had to be postponed until the main attack had made some progress. In the event, the coastal operation never took place and the troops destined for it found themselves committed to the main Flanders offensive.

The outline plan for Third Ypres was for Plumer to take the outposts of the Warneton Line, threatening Lille and pinning German reserves to the southern sector. Gough would clear the Gheluvelt plateau and Pilckem Ridge, and exploit north-eastwards to the line Thourout–Couckelaere before advancing towards Bruges and the Dutch frontier. On his left, Anthoine's French 6th Army would support him by attacking towards Bixschoote. The *Official History* acknowledges that, while 'no special difficulties were expected as regards the ground', the land-drainage system, based on the network of little *beeks*, was fragile. As the battle went on, the British created their own obstacle in front of them: 'shell craters and mud, and the Ypres mud was the consistency of cream cheese. One sank in it.'

Inconsistencies in planning contributed to British misfortunes. Gough hoped to make a rapid advance on the first day and spread his resources evenly to achieve this. Haig thought that he would be better advised to swing hard at the Gheluvelt plateau and then, having seized it, to exploit to his left. Misled by information from his own intelligence section, Haig believed that he could afford to engage the Germans in a 'wearing down' battle and that any one of his blows might result in a sudden enemy collapse.

The attack came as no surprise and the Germans responded to its preparations by strengthening their deep defensive fabric. Colonel von Lossberg arrived to be Chief of

Staff of 4th Army and ensured that the defensive layout
embodied all the lessons learnt to date. The *Flandern* I Line
ran from Lille, through Wervicq on the Lys, and on through
Becelaere and Broodseinde. Behind it stretched *Flandern* II
and III, the latter going from Menin to Westroosebeke and
beyond. There were three distinct systems in front of
Flandern I. The third line, *Wilhelm*, ran from Zandvoorde to
Zonnebeke, the second, *Albrecht* – with the bulk of field
batteries in it – on a reverse slope on the line of the
Steenbeek, and the lightly held front line curled round the
salient where the ebb of Second Ypres had left it. There was
the usual emphasis on soaking up the momentum of the
attack with strongpoints and providing counter-attack
divisions – one to each pair of front-line divisions – to deal
with the overextended attacker.

The attack began on 31 July. 2nd Army achieved
immediate success, taking several positions on the forward
edge of the Warneton Line, including the village of
Hollebeke. 5th Army also made good progress, especially on
its left, securing the gentle swell of Pilckem Ridge and
reaching a line a little short of Langemarck in the north,
through St-Julien and Frezenberg to Shrewsbury Forest
south of Hooge. There were two disappointments. The first
was that progress towards Gheluvelt plateau was unsatisfactory; the second that counter-attacks recaptured a large
segment of ground in the centre.

It began to rain on the evening of the 31st and continued
to do so for the next three days. The combination of the
downpour and heavy shelling – over four and a quarter
million shells had been fired by the British – turned the
battlefield into a quagmire and tanks were of little use
thereafter. On 10 August II Corps failed to take Gheluvelt
plateau and on the 16th it renewed its efforts with no better
result. On its left XIX Corps had no more success, trudging
out of the morass of the Hanebeek and Steenbeek valleys to
take some of its objectives before losing them to counter-
attacks. XVIII Corps made small gains in front of St-Julien
and XIV Corps reached the outskirts of Langemarck. Over
the next few days further attempts on Gheluvelt plateau

failed, convincing Haig to put Plumer in command of the central attack. On 24 August he ordered 2nd Army to take over II Corps' sector and to extend its own attack well south of the Menin Road to force the Germans to disperse their effort.

On 27 August 5th Army tried again. Men marched up in atrocious weather and spent ten hours in mud and water up to their knees waiting for the assault. It was too wet to light the smoke candles, which each man carried, and the going was so heavy that the barrage soon got ahead of the plodding infantry, who fell in swaths to machine-gun fire. Faced by this ghastly evidence, Haig decided to suspend operations by 5th Army until the 2nd could intervene. The month's fighting had cost him 68,000 casualties and had resulted in fourteen divisions being so badly bruised that they were taken out of the line to refit.

The unsatisfactory results of the battle, coupled with the visible collapse of Russia and French exhaustion, persuaded Lloyd George that it would be better to abandon it and husband resources for the following year. He later modified this view, suggesting that the effort should be diverted to the Italian front, where the Austrians seemed to be in serious difficulties. Haig argued that withdrawing troops from Flanders would free Germans for use elsewhere and it was only by maintaining the pressure in an area to which the bulk of the British army was already committed that the Germans could be beaten. Although Haig was not instructed to abandon the battle, Lloyd George had reason to hope that indirect pressure would force him to do so. The British army was running short of manpower: drafts arriving in France in the summer of 1917 were sufficient only to compensate for normal wastage, not to make good the losses in Flanders.

In late August Haig modified his plans in an attempt to achieve gains commensurate with these losses. 2nd Army was to mount a series of methodical attacks with limited objectives, while 5th Army, ploughing across the swampy desolation of the Steenbeek valley, would follow suit. 5th Army would eventually be in a position to secure the Passchendaele–Staden sector as planned and, if the Germans broke under the relentless pressure, it could exploit the collapse. Changes in tactics and organisation mirrored this approach. The infantry adopted a new, flexible attack formation, with a loose skirmish line covering small attacking groups and mopping-up parties moving up behind them. Artillery liaison was improved, and the engineers made heroic efforts to establish some sort of infrastructure in the wilderness of mud and shell holes.

On 20 September the attack recommenced with the Battle of Menin Road. 2nd Army attacked with three corps up. On its right, IX advanced in front of Klein Zillebeke, taking all its objectives and reaching the Bassevillebeek. In the centre, X also did well, though it failed to take the Tower Hamlets spur on the *Wilhelm* Line, just south of Gheluvelt Wood. On the army's right, I Anzac Corps secured the *Wilhelm* Line, taking Glencorse Wood and Nonne Boschen, and biting deep into Polygon Wood. The southern element of 5th Army's attack went almost equally well. V Corps crossed the Hanebeek and approached Zonnebeke; XVII Corps got well into the *Wilhelm* Line, though its advance along the Hanebeek valley towards Gravenstafel was slow. Only XIV Corps, on the left flank, forward of Langemarck, failed to make significant gains.

The improved artillery communications – everything from dogs and pigeons to balloons, aircraft and observers in the ruins of the cloth hall – meant that German counter-attack divisions were broken up before they could reach the British infantry. The view from the tower of the cloth hall confirms the lesson today. Provided there was reasonable visibility, the Germans could not cross the Gheluvelt plateau or the western slopes of the Passchendaele Ridge without being flayed by the superior British artillery. But in 5th Army's sector the roles were reversed. The attackers were overlooked from Passchendaele Ridge and counter-attack divisions could form up under its shelter.

The second step of 2nd Army's march across the Gheluvelt plateau, the Battle of Polygon Wood, was taken on 21st September. The Australians captured the wood itself, but progress on their right was less good and the

Tower Hamlets spur was only partially secured. On the left, 5th Army mounted just a modest operation to cover the Australian flank, but succeeded in getting into the western outskirts of Zonnebeke. Counter-attacks were repulsed and the difficulty of getting supports to the front line through the British barrage was such that the Germans began to modify their own doctrine, putting more troops in the front system, permitting stronger resistance and better local counter-attacks.

The new German tactics were in use by the time the next phase of the offensive, the Battle of Broodseinde, began on 4 October. The defensive barrage was formidable and approximately one in seven of the attackers was hit as they waited behind the jumping-off tapes. Nevertheless, there were substantial gains across the front attacked. I Anzac Corps overran a large section of the *Flandern* I Line, taking Zonnebeke and Broodseinde: Windmill Hill, just north-west of Zonnebeke on the Langemarck road, was stormed by 3rd Australian Division after some large-scale bayonet fighting. II Anzac Corps pushed out towards Gravenstafel, and XVIII and XIV Corps advanced abreast of Poelcappelle, taking the village. Over 4,500 Germans were captured, and Ludendorff bleakly acknowledged that 'the battle . . . was extra-ordinarily severe, and we only came through it with enormous losses'.

The weather, which had set fair during the late summer, broke in early October and the next attack, the Battle of Poelcappelle, on the 9th, took place in heavy rain. Little ground was gained, though again the German losses were heavy. Brigadier-General Charteris, Haig's intelligence chief, went forward to watch the battle and pronounced it 'the saddest day of the year. It was not the enemy but the mud that prevented us from doing better . . . I got back very late and could not work, and could not rest.' The New Zealand Division was especially hard hit in its attack on the Gravenstafel spur, losing 2,700 men on 'New Zealand's blackest day'. On 12 October, in the First Battle of Passchendaele, the Australians breasted the ridge line south of Passchendaele but were swept off it by fire from the reverse slope. The terrible weather caused another pause and on the 26th the Second Battle of Passchendaele saw the Canadians seize a portion of the *Flandern* I Line near the headwaters of the Raavebeek, but the 5th Army, on the Canadian left, made no progress. In the dying jerks of the battle, on 30–31 October, and 6 and 11 November, the Canadians renewed their struggle for the ridge. The battle ended with Passchendaele in their hands.

The hideous conditions in the salient, the dispatch of troops to shore up the Italian army after the disaster at Caporetto and the fact that an offensive was poised at Cambrai induced Haig to call off the battle. It had cost the British some 245,000 men. The Germans admitted to 217,000, but the British Official Historian, Sir James Edmonds, estimates that their losses might have been as high as 400,000. Critics maintain that Edmonds was over-generous in his assessment of German casualties in an effort to show British strategy in general, and Haig in particular, in a better light. Although it is impossible to establish German casualties with precision, the Germans regarded them as unacceptably high. General von Kuhl, Chief of Staff of Rupprecht's Army Group, called the campaign 'the greatest martyrdom of the World War', and admitted that it had 'consumed the German strength to such a degree that the harm done could no longer be repaired'.

Historians will continue to haggle over the merits of Third Ypres or, to use the name burnt into popular memory, Passchendaele. Some will show that it prevented the Germans from demolishing the tottering French and reduced their strength by attrition: they will also declare that opponents of the Western Front can point to no really satisfactory alternative theatre of operations. Others will survey the tiny gains – a short morning's walk will take you comfortably from the British start line to Passchendaele – and ask whether this slip of Flanders soil justified the expenditure of so much life. They will also examine GHQ's relationship with the army commanders, to suggest that at Third Ypres, as on the Somme, there was no clear logic linking what Haig wished to achieve and the means used to achieve it.

Nowhere can the human cost of Passchendaele be better appreciated than from Tyne Cot British Cemetery and Memorial on Passchendaele Ridge. It contains nearly 12,000 graves and the names of almost 35,000 men are inscribed on the curving panels at the rear of the memorial. It was once thought that it was so called because the bunkers resembled Tyneside cottages back home, but it is now clear that the name, then spelt 'Tyne Cott' was in use from 1916 and actually pre-dates the bunkers: it was probably a nickname given to a farmhouse, for there were Marne, Seine and Thames farms in the area. The great cross stands on a German bunker and there are others, remnants of the *Flandern* I Line, in the cemetery and the nearby gardens. Around the central cross sprawl the graves of soldiers buried when the bunkers were used as British dressing stations: the neatly ranked rows of headstones mark the resting places of soldiers brought to the cemetery from the surrounding battlefield.

Tyne Cot gives a panoramic view over the field of Third Ypres. Standing at the entrance, with your back to the cemetery, you look straight down the valley of the Hanebeek towards St-Julien with the traffic of the N369 trundling through it, a little under five kilometres away. The British line on 15 July was just on the far side of this village: when the battle opened it ran just over the low crest – Pilckem Ridge – two kilometres beyond. Slightly to the right of St-Julien is a windmill, built on the site of the *Totenmühle* – mill of death – used as an observation post by the Germans until British gunners brought it down. Almost directly behind it is Langemarck, lost during second Ypres and recaptured on 16 August 1917. It contains a large German cemetery, which includes the mass grave of nearly 25,000 Germans and a row of pillboxes of *Flandern* I. The difference in style between British and German cemeteries is striking, and a visit to Langemarck makes the point well. The place is dark and sombre, from the red granite of the entrance building to the low teutonic crosses dotting the greensward. While British graves bear individual headstones, low tablets, flush with the ground, mark German graves, often with more than eight men commemorated on each.

Further to the right is Poelcappelle, taken by the 11th and 14th Divisions, with the assistance of tanks, in October 1917. In it is a memorial to the French air ace Georges Guynemer, shot down in the area in September: it is topped by a stork, the emblem of his squadron. On the far left from Tyne Cot is Zonnebeke, with the valley of the little Zonnebeek running up towards it, taken by the Australians on 4 October. Polygon Wood lies two kilometres beyond it. The wood was not named because of its shape, but because it contained a *polygone*, the practice ground of a Belgian artillery school: the *butte* within it was the butts of a rifle range. On the skyline to the left the twin pimples of Kemmel Hill and the Scherpenberg, to its right, break the horizon. Half-left rise the towers of Ypres. They are only nine kilometres away, but that short step from the top of Passchendaele Ridge, behind Tyne Cot, to Ypres itself, represents the full depth of the Ypres battlefield. Until their victorious advance in the summer of 1918 the Allies never got beyond Passchendaele – and the Germans never took Ypres.

On a bright day the battlefield holds no menace, with complacent cows amongst the lush grass and the A19 scything along past Black Watch Corner and over the blink-and-you've-missed-it Frezenberg Ridge. The little museum at Sanctuary Wood, south of Hooge, has the usual packed display cases and some photo viewers containing numbing photographs of the salient as it was. In cold or wet weather the trenches behind the museum, shallow and slope-sided with age, begin to touch the emotions. Strip away the trees and think of living there, amidst the stink of high explosive, decomposing corpses and chloride of lime. It is small wonder that thousands died in the salient, not from shell splinters or rifle bullets, but from utter exhaustion, in bone-chilled despair amongst the shattered timber and yawning shell holes.

The prosperous agricultural landscape, neat houses and orderly cemeteries mask the horrors of the salient. Even the place names on contemporary maps – Mouse Trap Farm and Tin Houses, Lower Star Post and Church End – have a cheery, bucolic ring. The difference between cartographer's theory and soldier's practice was not lost on Lieutenant-Colonel Graham Seton Hutchison, who wrote: 'God knows what cynical wit christened those splintered stumps Inverness Copse or Sanctuary Wood. Who named that stinking quagmire Dumbarton Lakes? And who ordained that those treacherous heaps of filth should be known as Stirling Castle or Northampton Farm?'

For the ninetieth anniversary of Passchendaele, in 2007, a visitors centre was opened just behind Tyne Cot cemetery. Its panoramic windows give an excellent view across the battlefield and there is a wide assortment of memorabilia of those who fought in that blighted landscape. It is now possible to walk along the line of the old railway from Zonnebeke up towards Tyne Cot, and there are helpful boards which relate today's landscape to that endured by the combatants. Whizzing cyclists are, however, almost as much of a danger now as whiz-bangs were in 1917. The inimitable Franky Bostyn, who runs the first-rate Memorial Museum in Zonnebeke Château, to my mind quite the best of the museums in the salient, was lead author of the 2007 *Passchendaele 1917: The Story of the Fallen and Tyne Cot Cemetery*, which is required reading for those who seek to put the cemetery into context. I was painfully struck by the fact that the skeletons of four members of the Royal Field Artillery, buried together in 28 H 18, were found in a sandbag. The Zonnebeke museum holds the keys to one of the most evocative spots on the battlefield, the German dugout at Cryer Farm, just north of the Menin Road between Hooge and Veldhoek, at what my grandfathers' generation called Clapham Junction, where the pavé track towards Reutel and Beselare leaves the N8 in the shadow of Nonnebosschen. It is named for Second-Lieutenant Bernard Noel Cryer of 7th London, killed taking the place on 15 September 1917. He has no known grave and is commemorated, with so many of his comrades, on the Menin Gate.

It is a relief to turn from Flanders mud to the downs of the

Cambrésis. Here the British had followed up German withdrawal to the Hindenburg Line to occupy a line running from St-Quentin, past Cambrai and swinging in towards Monchy. The ground is similar to the Somme, sand and loam on chalk. The valley of the Escaut cuts the Cambrésis from north to south and through it runs the St-Quentin Canal, joining the Escaut and the Oise. As the memorial plaque on the canal tunnel at Bellicourt proudly proclaims, the canal was built during the First Empire. It is a remarkable feat of engineering, running in a cutting eighteen metres wide at the top and ten metres at the bottom south of Cambrai, but flattening out nearer the town. Running more or less parallel with the St-Quentin Canal, and about eight kilometres west of it at Cambrai, is the Canal du Nord, linking the Somme and the Sensée. It is of more recent construction than the St-Quentin Canal: in 1914 it was fully dug but not yet filled with water. The Canal du Nord cuts hard into the Havrincourt Ridge west of Cambrai and at its deepest is twenty-seven metres below the surface.

Cambrai, its reputation once founded on the production of good-quality linen, lies on the St-Quentin Canal, with the prominent mass of Bourlon Wood to its west. Known to the Romans as Cameracum, it has a long and distinguished history. The earnest François de Salignac de la Mothe Fénelon, scion of a noble family from Périgord, was its archbishop in 1695–1715. Salignac-Fénelons feature prominently in French army lists and it is a sad little irony that young Bertrand de Salignac-Fénelon, second-lieutenant in the 236th Infantry, killed at Mametz in December 1914, is commemorated in the chapel at Rancourt, behind the Somme battlefield and in old Fénelon's archdiocese.

Cambrai's war damage has been sensitively effaced. Its cathedral was torn down after the Revolution, but the fine entrance to the archbishop's palace still leads off the Place Fénelon. St Géry's church, on the site of the temple of Jupiter Capitolinus, has a striking baroque rood-screen in red and black marble. In the nearby Place Aristide Briand is the imposing town hall, built in 1786 and rebuilt after the

First World War. It is probably a blessing to discover that most of Vauban's fortifications were demolished in the nineteenth century. His citadel stands in the leafy public gardens between the Boulevard Vauban and the Avenue Villars. There is a well-preserved seventeenth-century gate once incorporated in Vauban's defences, the Porte de Paris, on the Péronne road out of the town, and three medieval round towers cover the line of the Escaut where the Boulevard Dupleix spins round the northern fringe of the town centre. Cambrai forms a hub with the spokes of Roman roads radiating from it: the D939 to Arras, the N43 to Le Cateau, and the N30 to Bapaume.

In 1917 German engineers strengthened a position that was naturally formidable. The outposts of the Hindenburg Line – disconnected lengths of trench and individual strong-points – ran just west of the Canal du Nord, crossing it at Havrincourt to swing east in front of Ribecourt and thence pick up the line of the St-Quentin Canal. The Canal du Nord is still in operation, and the depth of its cutting between Havrincourt and Hermies shows just what a formidable obstacle it was. The front trench system – *Siegfried I* – lay on a reverse slope, one to two kilometres behind it, its trenches 3.65 metres wide and three metres deep to make them impassable to tanks. Four rows of wire formed an obstacle ninety metres deep, slanted out in great wedges to channel an attacker on to strongpoints. The support system, also well dug and wired, ran three to four kilometres further back, and the *Siegfried II* system was further back still, stretching north from Beaurevoir to Crêvecour, on behind Masnières to Noyelles, where it branched with one spur running from Cantaing up to Bourlon Wood and another up past Sailly on the Arras road.

There were good reasons for attacking this robust defence. Breaking the Hindenburg Line at Cambrai would outflank its north-western run towards Monchy and the capture of Cambrai would seriously disrupt German communications. The ground had not been fought over, its water-table was low and it offered good going to tanks. It was held by the British 3rd Army, now under Byng, victor of Vimy, who

succeeded Allenby in June 1917 when the latter was sent off
to command in Palestine.

The Tank Corps, with its energetic commander Brigadier-
General Hugh Elles and brilliant Chief of Staff Lieutenant-
Colonel J. F. C. Fuller, saw glittering opportunities on the
Cambrai front. Separate schemes came together in the
summer of 1917. Fuller envisaged a large-scale hit-and-run
raid with the aim of getting tanks on to the German gun
line, and Brigadier-General Tudor, commanding the artillery
of 9th Division, had developed a plan for marking targets by
survey rather than adjusting them by fire, so that the barrage
would be sudden and accurate. To avoid lengthy wire-
cutting, Tudor suggested that tanks should crush the wire.
3rd Army favoured an offensive. GHQ would not grant
authority for it as long as Third Ypres squelched on, but in
late October Byng was given qualified permission to
proceed.

The plan was one of limited liability. 3rd Army was to
break the German line by a *coup de main* spearheaded by
tanks, push cavalry across the St-Quentin Canal, seize
Cambrai, Bourlon Wood and the Sensée crossings, then
advance north-eastwards. If the early results were not
impressive, the operation would be stopped after forty-eight
hours. 3rd Army had five corps, two of which, Pulteney's III
and Woollcomb's IV, would attack between Bonavis Ridge
and the Canal du Nord. III Corps, with 216 tanks, would
break in as far as the St-Quentin Canal at Masnières and
Marcoing, where the Cavalry Corps would pass through it.
IV Corps, with 108 tanks, would overrun the Flesquières
Ridge and secure Bourlon Wood. Over 1,000 guns would
support the attack, using Tudor's system of silent adjustment. Tanks and infantry carried out combined training,
using drills which enabled tanks to breach the wire and,
using bundles of brushwood called fascines, cross trenches,
while infantry mopped up behind them and formed
springboards for the next phase of the attack.

Concentration was carried out in great secrecy. Tanks
moved up to railheads by night, then crawled into their
assembly areas – Gouzeaucourt, Villers Guislain, Dessart

Wood, Villers Plouich and the huge Havrincourt Wood. They left these after dark on 19 November and edged cautiously, in bottom gear with their engines just ticking over, to the forming-up line. The advance began at 6.10 on the morning of the 20th, and the tanks and infantry were just visible from the German outpost line when the barrage came down. The combined effect of the tanks and the sudden bombardment was too much for most of the defenders, and by 11.30 a.m. both corps had overrun the outpost line and main battle line on almost all the attack frontage.

Only at Flesquières was there a major setback. The village was attacked by the 51st Highland Division aided by the tanks of D and E Battalions of the Tank Corps. The divisional commander, Major-General 'Uncle' Harper, had strong views on the use of tanks and those supporting his division moved further ahead of the infantry than was the case elsewhere. Many were hit as they crossed the ridge and came within view of the gun line on its reverse slope. Haig's dispatch later paid tribute to a German artillery officer who had manned his gun single-handed until killed. In fact, sound German tactics played a greater role than individual valour and the story of 'the gunner of Flesquières' is largely myth. Lieutenant-General von Watter, commanding the 54th Division, holding this sector, had a brother who had encountered British tanks on the Somme and warned him what to expect. He trained his gunners to deal with them by pulling their 77mms out of gun pits and taking on tanks with direct fire. As the tanks lumbered over the crest they were at a temporary disadvantage and the bold handling of German guns paid dividends. If there was a particular hero at Flesquières that day, it may have been either Lieutenant Müller or Sergeant-Major Kruger of the 108th Field Artillery Regiment, both of whom played distinguished parts in facing the tank attack.

The results of the afternoon were more disappointing. Flesquières still held out. II Corps was up on the canal – and across it at Marcoing – but the tank 'Flying Fox' had tried to cross the Masnières bridge and crashed through it. The cavalry had been ordered up and some had been in action, but it had proved impossible to get them across the canal in strength. Tank losses had been heavy. Of the 378 fighting tanks available that day, 179 were out of action, sixty-five of them destroyed by direct hits and the others either ditched or broken down. 51st Division had lost twenty-eight of its tanks to the guns behind Flesquières Ridge.

On the 21st, in worsening weather, III Corps went firm on the line it had reached, while IV Corps resumed its attack on Bourlon. It is not hard to see why. Bourlon Wood lowers down upon the battlefield: German observers on its forward edge brought fire to bear on the northern half of the field. A line just short of the wood would be untenable: IV Corps had either to take the wood, or to fall back on the next defensible line, the Flesquières Ridge. That evening Haig decided to persist with the offensive, despite the need to release troops for Italy, believing that the Germans were 'showing a disposition to retire'. The scene was set for a long and debilitating battle for Bourlon Wood.

In the wood itself and in Fontaine-Notre-Dame, on the Bapaume road at its foot, 23 November saw bitter fighting. It is hard, driving through the decidedly ordinary Fontaine, to form an impression of the ferocity of the battle. The Germans, elements of the 46th and 50th Infantry Regiments, fought with determination against 51st Division and its supporting tanks. Some even rushed the tanks and hung on to their guns to prevent them from being elevated, while their comrades maintained a brisk fire from barricaded houses: B Battalion lost ten of the thirteen tanks it committed. Over the next few days the battle went on relentlessly, ending with the British in possession of most of Bourlon Wood. Although reinforcements, including the Guards Division, had arrived, they had been fed into the furnace of Bourlon and largely consumed, so that Byng found himself holding a wide salient with few fresh troops available and his tank strength badly depleted.

By this time the Germans had recovered from their initial shock and were redeploying to pinch out the salient. On 27 November Ludendorff met Rupprecht, and Marwitz of 2nd Army, in the latter's headquarters at Le Cateau, and that evening Rupprecht issued orders for a counter-attack. It began on the morning of 30 November and, although the British had suspected that an attack was coming, the Germans achieved a remarkable degree of tactical surprise.

In the south, they broke in on the boundary of III and VII Corps, taking Villers Guislain and cutting up into III Corps' flank. They entered Gozeaucourt, overrunning several howitzer batteries and coming close enough to the head-quarters of the 29th Division to bring it under machine-gun fire. A sharp counter-attack by 1st Guards Brigade regained Gozeaucourt, but Gonnelieu was lost. With it went four batteries of artillery, though not before most guns had expended all their ammunition. The survivors of one battery, an officer and five men, fought on for two hours, then withdrew taking their sights, breech blocks and wounded with them. There was a desperate battle for Les Rues des Vignes, where Captain R. Gee, staff-captain of 86th Infantry Brigade, won the Victoria Cross for a Homeric struggle in which he recaptured the village almost single-handed. In the north the Germans made no effort to achieve surprise and attacked the junction of VI and IV Corps on the Bourlon–Moeuvres front with dogged persistence, but failed to gain much ground.

Byng assessed that the Germans were certain to resume their attack on 1 December and ordered a counter-attack early that morning. It was too crudely cobbled together to have much chance of success. In the south, British and Indian cavalry regiments of 4th and 5th Cavalry Divisions made a series of mounted charges between Epehy and Villers Guislan. One example speaks for all. The 6th Inniskilling Dragoons, accompanied by a section of mounted machine-gunners, took the Villers Guislain road out of Epehy (D24). They crossed the railway and came under constant machine-gun fire, and the leading squadron and machine-gunners reached the sugar beet factory just short of Villers Guislain. Here they were engulfed, and the rest of the regiment wheeled about and regained Epehy. The Inniskillings lost six officers and 108 men – the machine-gun section, with its two officers, fifty-three men and four guns, was a total loss. The Guards Division made some progress towards Gonnelieu and 4th Grenadiers actually got into the village but were unable to hold it.

When the German attack got under way later that day the 29th Division was hard pressed to hold Masnières and Les Rues des Vignes, and gave up the former at nightfall. There was little movement on the Bourlon–Moeuvres front that day and on the 2nd there was much less activity, although La Vacquerie was lost. On the 3rd Marcoing was heavily attacked, and the Newfoundland Battalion again suffered severely: its caribou on the N44 north of Masnières, offering a fine view out to the north and north-west, commemorates its efforts.

On the morning of 3 December Haig visited Byng at Albert and decided that holding what remained of the salient would be prohibitively costly. Orders were issued for withdrawal to the Flesquières Line, which took place on the night of 4–5 December and left the British holding part of the old Hindenburg support system around Flesquières, overlooking the battlefield. Work began on converting the German defences to face eastwards and at the same time, across the whole of the British front, the defensive layout was modified to reflect German practice, with forward, battle and rear zones.

Both sides settled in for the winter and mulled over the results of Cambrai. From an initial success so considerable that the church bells had been rung in England for the first time in the war, Cambrai had turned into a reverse made all the more painful by the early gains. Some elements of German tactics – short bombardment, penetration by small assault groups, and the blinding of defensive fire by cutting communications between observers and guns – were to be essential ingredients of the spring offensive in 1918. The balance sheet of the battle was even. German attacks between Epehy and the Péronne road had gained them about as much ground as the British retained around Flesquières. Both sides had lost a little in excess of 40,000 men, but the British lost 158 guns to the Germans' 145.

The Memorial to the Missing on the N30 Cambrai–
Bapaume Road at Louveral, about eighteen kilometres to the
west of Cambrai, is an excellent place to begin a visit to the
battlefield. The reliefs carved on the walls facing the
entrance are of particular interest. The work of C. S. Jagger,
who had served as an infantry officer, they show vignettes of
trench life from an unheroic perspective. Just to the south of
the main road is the village of Hermies, in whose British
Cemetery lies Brigadier-General R. B. Bradford VC MC,
killed at the age of twenty-five, commanding the 186th
Brigade near Bourlon. Havrincourt Château, three kilometres east, across the motorway, was the headquarters of
the German 2nd Army during the Somme battles, and in
1917 the village and château were incorporated into the
Hindenburg Line. Cemetery Alley trench, connecting the
front and support systems, ran on the line of the D15
Havrincourt–Flesquières road. One kilometre beyond the
cemetery the line of the old railway cuts obliquely across
the road. This marked the divisional boundary between 51st
Highland to the east, and 62nd West Riding to the west: the
latter's memorial obelisk is on the eastern edge of
Havrincourt. The British cemetery east of Flesquières
provides a good feel for the reverse slope of the ridge, from
which the gunners of 108th Field Artillery engaged the tanks
coming over the slope. To the north-west is Orival Wood, its
little cemetery containing the graves of many Highlanders,
including the poet Lieutenant E. A. Mackintosh MC.

Bourlon Wood blocks the northern horizon. Approaching
through Anneux and along its western edge takes one up the
Siegfried II system. There is a Canadian Memorial in the
wood, reached through Bourlon village: it commemorates
the Canadian Corps' capture of the hill on 27 September
1918. Going on into Cambrai one can imagine the Guards
Division attacking in the drizzle on 27 November, with no
opportunity for reconnaissance. An Irish Guards battalion,
under the future Field Marshal Alexander, struck up into
Bourlon Wood, navigating by compass bearing, biting deep
into the wood to reach their objective. Leaving Cambrai via
the N44 towards St-Quentin the traveller passes the

Newfoundland caribou, before crossing the canal at Masnières on a mundane span replacing the bridge wrecked by 'Flying Fox'.

Philippe Gorczynski, proprietor of the charming Hôtel Beatus, 718 Avenue de Paris, Cambrai, had long believed that it should be possible to locate one of the tanks that had been abandoned and subsequently buried in a shell hole. In November 1998 digging began in a field outside Flesquières and a tank, still in astonishingly good condition, was quickly found and eventually recovered to a barn on the D89 Ribecourt road. Research has identified it as D51 'Deborah', commanded by Second-Lieutenant Frank Heap. 'Deborah', moving ahead of the Scots infantry of 153rd Brigade, who were checked by fire from Germans concealed in the ruins of Flesquières, passed through the village and was eventually knocked out by direct hits from a field gun, which had been hauled out of its pit. Four of the crew were killed (they were probably reinterred in Flesquières British Cemetery after the war) and Heap led his survivors back through the village to safety, earning a Military Cross in the process. M. Gorczynski was also well to the fore in the construction of a memorial in Flesquières, unveiled on the ninetieth anniversary of Cambrai, which shows tank tracks followed by the footsteps of the infantry.

On 11 November 1917 Ludendorff met a select group of officers at Mons. They included Major – soon to be Lieutenant-Colonel – Georg Wetzell and an infantry captain, Hermann Geyr. The meeting's aim was to discuss how best to seek a decision in the west before the Americans arrived in strength, and while Wetzell was OHL's chief of operations for the Western Front, Geyr was a leading theorist on offensive operations and an author of the forthcoming pamphlet *The Attack in Position Warfare*. It concluded that the Germans could afford one large-scale offensive in 1918 and should pre-empt American arrival by attacking in February or March. Beating the French might not drive Britain out of the war, but defeating the British would unhinge the northern part of the front and provoke a French collapse. Ergo, concluded Ludendorff, 'we must beat the British.'

The tactics employed were the distillation of experience gleaned from several sources. A French infantry officer, Captain André Laffargue, had written *The Attack in Trench Warfare* in 1915: he advocated deep penetration of enemy defences by small groups of the attackers who threw their opponents off balance. His ideas had limited currency in his own army, but were absorbed by the Germans. German experimentation with storm troops – *Stosstruppen* – was also important and the success of Captain Willy Rohr's *Sturmbataillon* was avidly studied. Finally, in Colonel Georg Bruchmüller the Germans had an artillery expert of the first order. On the Eastern Front he had developed tactics based on the sudden application of fire throughout the depth of the enemy position and favoured the use of gas shells, which did not damage the ground like high explosive, enabling attackers to move more freely. All this – and German experience of being surprised by the British at Cambrai – was encapsulated in *The Attack in Position Warfare*. This envisaged an attack prepared by a lightning bombardment, carried out by storm troops, equipped with flame-throwers, mortars and machine-guns, driven deep into the enemy position, bypassing strongpoints, which would be dealt with by follow-up units.

Not only did the Germans succeed in developing a new doctrine: they organised and trained units capable of implementing it, by combing the weak elements out of chosen battalions. The incorporation of formations from the Eastern Front helped make Ludendorff's fighting machine large as well as efficient. The terrible strain of the war had brought tsarism crashing down and the Russian Provisional Government, which had at first tried to continue the war, was controlled by the Communists from November 1917. In February 1918 Russia agreed to suspend hostilities, enabling the Germans to switch troops to the west: on 3 March the treaty of Brest-Litovsk formally brought Russia out of the war. Ludendorff also squeezed men out of industry, and by late March he had three and a half million on the Western Front. Of his 192 divisions, sixty-nine were concentrated along the

ninety-six-kilometre stretch of front between Arras and La
Fère.

Emphasis on this sector was the result of a thought
process which, while logical as far as it went, contained the
seeds of disaster. Ludendorff's stated aim was to beat
the British and, by driving them out of the war, engender the
collapse of France. Yet it was difficult to reconcile this with
the tactical realities of the Western Front. Ludendorff
believed that attacks on both sides of Ypres – plans code-
named *St George I* and *St George II* – could link up at
Hazebrouck and swing right to pocket a substantial chunk
of the British army. But the British had given an object
lesson in the difficulties of attacking in the salient: Flanders
mud was impartial and would impede Ludendorff's movements just as it had Haig's. Ludendorff therefore rejected the
northern option and looked further south, to Arras–Vimy.
The plans for an attack here – *Mars* and *Valkyrie* – would be
imperilled by the British hold on Vimy Ridge, even harder to
attack from the east than from the west, and by the cluttered
country to its north.

Ludendorff then looked at the familiar territory between
the Sensée and the Oise. His attacking formations could
concentrate in the Hindenburg Line, and the wasteland of
the Somme battlefield would hamper the British and provide
excellent opportunities for German infiltrators. The offensives in this sector bore the code name *St Michael*. General
Oskar von Hutier, a successful practitioner of attacks on the
Eastern Front, commanded the 18th Army, on the German
left, mounting *St Michael III*. Despite his considerable
reputation, von Hutier's role was largely subsidiary: his left-
hand corps would take the Crozat Canal, securing the
attack's left shoulder, while the remainder of his army
advanced on either side of St-Quentin. Marwitz's 2nd Army
would mount *St Michael II* across the Cambrai battlefield
and make for Albert, while Below's 17th Army, attacking
with *St Michael I* on the German right, would aim for
Bapaume. If early objectives were secured, the Germans
would swing right, driving in on the right rear of the British
armies. At this juncture *Mars South* would be mounted by

Below's right wing against Arras south of the Scarpe and later, with the weather improving, *St George* could be sprung at Ypres, subjecting the British to a concentric attack they would have little hope of resisting.

The concentration of attacking forces was, as the British *Official History* recognises, 'a gigantic problem which was solved with complete success'. Yet there remained an integral flaw in the plan. Ludendorff's desire to achieve tactical success led to his attacking in the sector where the British enjoyed most depth. It was some fifty kilometres from the German front line to Albert, and a further westward advance of fifty kilometres would still leave the Germans well short of the coast. In the north, however, an initial advance of fifty kilometres from Ypres would put the Germans within comfortable striking distance of the Channel ports and place the Béthune–Hazebrouck–St-Omer area, crammed full of supply depots, under their control.

The possibility of German attack had long been considered by the Allies. At the tactical level the British planned to meet it by adopting deep defence based on German doctrine, with Forward, Battle and Rear Zones. Unfortunately the concept, embodied in a GHQ instruction of December 1917, was not properly thought through. It encountered much consumer resistance and there was precious little time to carry out the work on rearward defence lines upon which it depended: moreover, the southern part of the British line had only recently been taken over from the French and there were the customary complaints that it was in a shocking state.

There was evidence, as early as February 1918, that the Germans intended to attack 3rd and 5th Armies, and an ambiguous GHQ instruction told Gough to meet an attack in his present position but to consider falling back to prepare a counter-attack. The absence of overall command and a central reserve were serious problems, and were well identified by the Allied Supreme War Council. But nothing effective was done. Neither Pétain nor Haig could spare men for a reserve, although each agreed to provide support if the other was attacked, and this left Pershing's Americans, still

far short of their anticipated strength, as the only candi-
dates. No overall commander was appointed, and matters
were not improved by friction between Pétain and Foch. The
latter was now Chief of the French General Staff, France's
military representative on the Supreme War Council, and
enjoyed the confidence of the French Premier Clemenceau.

The storm broke on 21 March. At 4.40 that morning the
concentrated German artillery shook the earth between the
Sensée and the Oise with the impersonal malice of a natural
disaster. At 9.40 the German infantry advanced, looming
through fog which masked the fire of British machine-gun
posts, pushing on with bomb and bayonet to drive iron
wedges right through the Battle Zone before nightfall.
Usually the British had put too many men – and too many
of the all-important machine-guns – in the Outpost Zone, at
the mercy of the hurricane of fire and the first infantry
assault. Command and control had been gutted by the
bombardment, and survivors were fighting blind, encircled
by follow-up troops, or reeling in the chaos of the Battle
Zone. Much of 3rd Army's Battle Zone south of the Scarpe
had gone, though a haunch of it remained in front of
Flesquières, and 5th Army had lost a thick wedge of territory
north of Ham.

The second day of the attack achieved less impressive
results. 17th Army ran into stiffening resistance south-west
of Arras and although 2nd Army was halfway to Péronne, it
was 18th Army that made the running, coursing out along
the Crozat Canal in the face of a defence so ramshackle that
Rupprecht suspected he might have caught Gough in the
process of withdrawal. Gough was clearly in difficulties.
That morning he declared, 'In the event of serious hostile
attack corps will fight rear-guard actions back to forward
line of the Rear Zone, and if necessary to rear line of Rear
Zone.' What was afoot evidently was, to those gasping for
breath beneath its torrent, a 'serious hostile attack'. The
Rear Zone existed on maps but not on the ground: one
hard-pressed officer found the only evidence of it 'an
immaculately patterned mock trench'.

The gap between the theory of lines on maps and the

dreadful reality of an army in retreat widened on the 23rd. Haig ordered Gough to hold the line of the Somme 'at all costs', but Péronne was abandoned that day and Hutier was over the river. Haig was becoming increasingly concerned at his plight. The French had already sent the six divisions agreed in discussions, and these managed to fill the gap that opened on the British right flank. But on the 23rd Haig asked Petain to concentrate another twenty divisions around Amiens, only to discover that Petain was himself worried about a forthcoming German attack in Champagne and could not spare them. Petain announced that he would try to prevent the Allied armies from being split, but in his diary Haig admitted the possibility that 'the British will be rounded up and driven back into the sea!'

On the 23rd 3rd Army evacuated the Flesquières salient, so that the front bellied out like a filling sail from Arras, past Péronne, to Chauny and Noyon on the Oise. Ludendorff's sense of purpose was not equal to the opportunity. In new orders that day he sent 17th Army north-west, 2nd due west, and 18th south-west, capitalising on tactical success, but not contributing to the operational purpose, which alone could justify the efforts his soldiers were making. Already the offensive was beginning to run out of steam: sheer fatigue and the difficulty of maintaining supply was slowing up even the aggressive von Hutier.

This was scant comfort to Pétain and Haig. The former, preoccupied by the need to cover Paris, made it clear that his own Reserve Army Group would have to fall back from Montdidier to the south-west if attacked and was not to be cut off with the British. Haig felt that he was 'confronting the weight of the German Army single-handed'. On 26 March he met his army commanders in the little Picardy town of Doullens, though it is doubtful whether any of them had much interest in the macaroons for which it is famous. Gough was not present and old Plumer muttered that Byng, too, ought to have been with his army, before character-istically offering to send fresh divisions south to relieve those ground down in the fighting. Haig ordered Byng to anchor his right flank in front of Bray-sur-Somme, south of

Albert, and emphasised the importance of covering Amiens.

Haig was then summoned to the town hall, where President Poincaré was waiting with a French delegation, together with Lord Milner, a member of the British War Cabinet, and Sir Henry Wilson, Robertson's successor as Chief of the Imperial General Staff. In response to a question from Clemenceau, Haig announced that he proposed to defend Amiens and hold north of the Somme, although he had placed elements of the 5th Army below the Somme under French command. This was Pétain's cue to declare that little of 5th Army now remained. Foch was undaunted by his colleagues' gloom and burst out: 'We must fight in front of Amiens, we must fight where we are now. As we have not been able to stop the Germans on the Somme, we must not now retire a single inch.' Haig declared that he would happily follow Foch's advice if Foch would consent to give it. There was a brief discussion amongst national groups and a draft agreement was read, charging Foch with co-ordinating the Allied armies in front of Amiens. This did not go far enough for Haig and the finalised agreement concluded that Foch should be given command of Allied armies on the Western Front. It was a momentous decision: not because it made an immediate difference to the battle – Foch could only produce eight divisions at Amiens and then only by early April – but because of its implications for the future. The meeting took place in the town hall's council chamber, which retains the furniture used on 26 March 1918 and is open to visitors.

Although the Doullens meeting brought little relief to a 5th Army which was still falling back – the Germans were in Albert that day – there was light on the horizon. The pace of battle had slowed as exhaustion and lack of supplies bore down on the Germans. Captured dumps were raided with a single-minded determination that testified to German privation. Ludendorff pumped at the chain of command like a driver running out of fuel, changing the axis of advance yet again and, on 28 March, unleashing *Mars. Mars* got nowhere and *St Michael* at last ground to a halt. Ludendorff's March offensive had taken over 90,000 men and 1,000 guns, and overrun more territory than all the costly Allied offensives of the past three years – but it had demonstrably not won the war.

In April Ludendorff tried again. *St Michael* had consumed so many of his troops that *St George* was out of the question and an emasculated alternative, named *Georgette* by the waggish Wetzell, was launched on the morning of 9 April. We saw, earlier in this chapter, how the brunt of this attack fell on the Portuguese north-west of La Bassée. Over the next weeks it hammered a deep dent into the British lines, from Givenchy on the La Bassée Canal, up to Merville on the Lys, ebbing round the Mont des Cats before running west of Kemmel Hill, and inside a much-reduced Ypres salient – with all the agonising British gains of 1917 gone – to the edge of Houthulst Forest. Foch was reluctant to move reserves to bolster the sagging front until the crisis had passed, but towards the end of the month he produced three divisions and used these to replace five British divisions, badly clawed in March and April, which were sent off to 'the sanatorium of the Western Front' on the Chemin des Dames. We shall meet them again in a later chapter. The end of *Georgette* saw the Germans tantalisingly close to the Mont des Cats and Mount Cassel, and thus outflanking the Ypres salient and the Yser position. Once again a decision had eluded Ludendorff and once again his finite resources had been reduced.

On 24 April *St Michael* gave a dying twitch. The Germans had produced a tank of their own, the A7V *Sturmpanzerwagen*, and they used thirteen of these monsters to attack either side of Villers-Bretonneux, on the Roman Road from St-Quentin to Amiens. They broke the British line with little difficulty and duly captured the town. Just west of Villers-Bretonneux the main road skirts the Bois d'Aquenne and then, after the D168 turning to Cauchy, the Bois de l'Abbé. A section of British tanks was lying up in the wood on 24 April, their crews suffering ill effects from mustard gas, which had been fired into the wood the day before and still hung about in the damp undergrowth.

The section moved out to engage some of the German

tanks that had swung south of Villers-Bretonneux. The first two tanks out were Mark IV Females. Their machine-guns were useless against the armour of the A7Vs, whose 57mm guns drove them back into the cover of the wood. The remaining tank in the section was Second-Lieutenant Frank Mitchell's Mark IV Male. Mitchell's gunners found Leutnant Biltz's 'Nixe', the leading German tank, no easy target, because their eyes were inflamed by mustard gas, and armour-piercing machine-gun fire filled the inside of the tank with sparks and flying splinters. Eventually Mitchell halted to give the gunner in his left sponson a better shot, and was rewarded by seeing the white puffs of three direct hits on 'Nixe', and the German tank was duly knocked out. For many years it was believed that Mitchell had engaged 'Elfriede', which turned over while trying to cross a steep bank, but it now seems clear that his adversary was in fact 'Nixe'.

Later the same morning a British reconnaissance aircraft spotted two German battalions making for the village of Cachy, just south of the Bois de l'Abbé. It dropped a message to seven British whippet light tanks, which immediately advanced and caught the Germans in the dead ground west of the village. The whippets drove through the two battalions, machine-gunning as they went, then turned at the bottom of the slope to repeat the performance, destroying the German units as a fighting force. That night three brigades, one British and two Australian, counter-attacked Villers-Bretonneux, and had cleared the town by dawn on 25 April. *St Michael* was dead beyond dispute.

In May Ludendorff tried again, attacking the French 6th Army in the Chemin des Dames, in a battle described in a later chapter. The attack pushed a new malignant swelling into the Allied lines, but the Americans were now present in strength, arriving, as Henri de Pierrefeu was to write, like a transfusion of blood to reanimate the mangled body of France. Fresh American troops helped limit German exploitation of victory on the Chemin des Dames, and a final offensive, launched between Noyon and Montdidier in an effort to link up the salients produced by earlier attacks, produced the familiar story: tactical success but no long-term strategic gains.

By the early summer of 1918 the Germans had shot their bolt. The transfer of troops from the east and the combing-out of men from the ancillary services could not conceal the fact that, with hundreds of thousands of Americans arriving in France, Germany was losing the manpower battle. Moreover, the burden of casualties had fallen dispro-portionately heavily on storm troop units. Other resources were also running low. The Allied blockade had already had a marked effect, sapping morale on the home front and reducing the flow of raw materials. The simplest battlefield archaeology makes the point effectively. Copper driving bands encircled shells, to bite into the gun's rifling and produce a gas-tight seal between shell and barrel. Sometimes these can still be found on shells, like the omnipresent shrapnel shells, which look for all the world like a rusty beer can, and sometimes they turn up as bent strips of verdigris lying in furrows. But occasionally the familiar copper is replaced by a non-ferrous alloy, a sure sign that the shell it encircled was produced in a German factory gasping for raw materials. Ludendorff had forged a formidable weapon that spring, and by summer he found its blade nicked and dull.

The riposte was coming. The recapture of Villers-Bretonneux had left the Australians holding a line running below the low ridge that stretches from Le Hamel, near the Somme, through Vaire Wood. Australian positions were overlooked from the ridge and the Australian Corps Commander, Lieutenant-General Sir John Monash, a civil engineer by profession and one of the two non-regular officers to reach this level of command in the British and Dominion forces during the war (the other was Currie the Canadian), was anxious to take it. It was a measure of his businesslike approach that he succeeded first in overcoming Australian reluctance to work with tanks, a consequence of the Bullecourt venture in 1917, then in planning a battle which involved not only ten of his own battalions but also part of the US 33rd Division.

Monash attacked in the pre-dawn of 4 July – a date chosen

out of deference to his American comrades-in-arms. His gunners had shelled Le Hamel and Vaire Wood at the same time every morning for a week, so the bombardment that accompanied the attack caused no great concern. The Australians and Americans were into the German first line before its defenders realised what was happening. Fighting tanks dealt with intact strongpoints, and supply tanks brought up wire and pickets to help secure the ground gained. The well-handled little battle contained two significant firsts. It witnessed the first use of resupply from the air: ammunition was parachuted into Le Hamel and Vaire Wood. And it saw the first US National Guard Medal of Honor, won by Corporal Thomas Pope. But perhaps more important was the fact that Monash's plan – with its emphasis on the co-ordinated effect of artillery, armour, infantry and air power – was circulated as a training document. The British had at last institutionalised the learning process.

Later in July Ludendorff attacked between Rheims and Soissons, made little progress and was counter-attacked so briskly that he abandoned plans for another stab at the British. It was with the tide beginning to run in Allied favour that Rawlinson suggested to Haig that his 4th Army would be strong enough to attack on the axis Amiens–Péronne if the Canadians were brought down from the Vimy sector to reinforce him. Foch warmly supported the proposal, although he demanded that the date of the planned offensive be advanced to 8 August.

It is a quirk of the British character that poignant defeats or hard-won victories attract an interest denied to well-deserved but cheaply bought success. Thus the battle of Amiens attracts far less comment than the Somme or Passchendaele, though its results were arguably greater than those of any other British offensive during the war. Although planning was detailed, there was none of the 'big push' puffing which had preceded the Somme battle. Staffs discussed the battle on a need-to-know basis, and divisional commanders learned of their involvement only a week before the attack. Foch had placed Debeney's French 1st

Army, on the British right, under Haig's command for the attack, but Haig deliberately avoided visiting Debeney until the attack had started so as not to excite comment. The Canadians were pulled out of the line on Vimy Ridge and, while most of them moved south, a few went up to Ypres, where their ill-concealed presence helped create the impression that the British offensive would be in the north. Low-flying aircraft helped mask the sound of Rawlinson's 534 tanks – 342 of the new Mark Vs, seventy-two whippets, and 120 supply tanks. The 2,070 guns supporting the attack were allowed to adjust their targets under such strict control that there was no increase in the volume of fire.

Discussions between Rawlinson and Haig mirrored their disagreement before the Somme. Haig thought that Rawlinson's initial objective was too modest: if surprise was achieved, it might be possible to push cavalry and whippets deep into the German rear. Although Rawlinson was weaker than he had been two years before, with fifteen infantry divisions and the Cavalry Corps, he was not attacking a well-prepared defensive position; he had plenty of tanks; and, by no means least, he had learnt hard lessons.

Maintaining surprise proved difficult. It had been decided that the Canadians would be kept out of the line till the last moment, when they would move up into the sector south of Villers-Bretonneux. The Australians took it over from the French in late July and the British III Corps filled in around Morlaincourt on the Australian left to permit this. The Australians, entirely characteristically, had put in a destructive trench raid the night before departing. This jolted the Germans into replying with a jab of their own, which caught the 18th and 58th Divisions in the middle of a relief and threw them back with loss. There was concern as to III Corps' fitness for the coming attack and wider fear that the Germans were aware of what was afoot.

The attack began at 4.20 on the morning of 8 August. It was a morning of Picardy mist, burnt off, as the day went on, by brilliant sunshine. On the right and in the centre the Canadians and Australians achieved complete surprise, tanks and infantry coming through the fog behind a creeping barrage. Many Germans quickly surrendered, but elsewhere machine-gunners and artillerymen stuck to their posts as the tanks closed in. At Marcelcave, five kilometres south-east of Villers-Bretonneux, six German machine-guns held up the Australians until a single Mark V dealt with each in turn. In the Canadian sector, nine of the ten tanks of the Tank Corps' 1st Battalion were knocked out by field guns firing at point-blank range.

On the right and in the centre these checks were temporary and, as the defence crumbled on the Amiens–St-Quentin road, the whippets of the 17th Battalion moved through to reap the harvest. They first shot up a column of transport on the main road, then, reaching the crossroads three kilometres east of the British cemetery at Le Bois du Sart, one element swung north into Proyart and destroyed more transport, while the other drove south into Framerville, catching a corps staff at lunch. Further south, the whippet 'Musical Box', detailed to escort the cavalry, first destroyed a German battery holding up some Mark Vs, then sped on towards Chaulnes, machine-gunning infantry and transport until it was knocked out far behind the German lines.

Between them the Canadians and Australians captured nearly 13,000 Germans and over 300 guns for the loss of some 6,000 men. Progress on their flanks was more disappointing. III Corps made heavy weather of its initial objective, the Chipilly spur, standing like one of Vauban's ravelins above the Somme, enabling the Germans to enfilade the Australian left flank above Le Bois du Sart. Debeney's Frenchmen, attacking on the Canadian right, also moved slowly and when Haig visited Debeney that afternoon he found him nonplussed by his soldiers' lack of dash. Yet for all the disappointment on the flanks, and the Cavalry Corps' inability to achieve a breakthrough into open country, the results of 8 August were impressive. The German *Official History* saw the battle as 'the greatest defeat which the German army had suffered since the beginning of the war . . .', while Ludendorff hailed it starkly as 'the black day of the German Army'.

The battle slowed down over the days that followed. On 9 August the US 131st Infantry, fighting under III Corps, took the Chipilly spur, but Rawlinson, to Monash's irritation, was reluctant to throw himself wholeheartedly into open warfare. Resistance was stiffening, many of the tanks used on the first day had been knocked out or broken down and the Allies were now approaching the moonscape of the Somme battlefield.

The Australian contribution to the battle of Amiens is well remembered. On a hilltop midway between Corbie and Villers-Bretonneux stands the Australian National War Memorial commemorating nearly 11,000 Australians who fell in France but have no known grave. The Lutyens-designed memorial is topped by a thirty-metre-high tower, whose lantern contains a helpful orientation table. The excellent views offered from the memorial made it useful as an observation post in 1940, when the French held the area, and it was badly damaged by German fire: there is a bullet hole in the bronze door of the register of names. In Villers-Bretonneux, which celebrates ANZAC Day each year, is the Sir William Leggatt Museum, housed in a school building some 200 metres south of the Mairie, given to the children of the town by Australian children whose fathers and brothers had been killed there. There is a major Australian memorial in Le Hamel, with extensive orientation panels.

The events of April–August 1918 can be traced in the woods and fields around Villers-Bretonneux. By leaving the main N29 on the Cachy road just east of the town the visitor passes the Bois de l'Abbé and the field to its south-west, which was the scene of 'Nixe's' misfortune. The two German battalions caught by the whippets on the same day had moved through Cachy and were attacked on the farmland west of the village. Many of the villages in the Santerre plateau to the east contain Commonwealth War Graves Commission cemeteries with the graves of Canadians killed on 8–9 August and a memorial stone just off the D934 near Le Quesney commemorates the Canadian achievement.

North of the Amiens–St-Quentin road the battlefield of Le Hamel–Vaire Wood appears to good advantage. The Chipilly

spur lunges down to its east and in Chipilly itself is a
memorial to the 58th Division, part of the hard-pressed III
Corps. Corbie, birthplace of Ste Colette, contains the fine
abbey church of St Peter, badly damaged in 1918. On the D1
Bray road, about three kilometres from Corbie, is the spot
where the German air ace, Baron Manfred von Richthofen,
crashed on 21 April 1918. It remains uncertain whether he
was shot down by Captain Roy Brown of the Royal Flying
Corps or an Australian machine-gunner. He was commemorated by a memorial in Glisy airfield, on the eastern
outskirts of Amiens, but it seems to have been stolen some
years ago. The Baron's remains enjoyed an uncertain repose.
They were first buried, with military honours provided by
an Australian firing party, at Bertangles, north of Amiens.
After the war they were reburied in the German military
cemetery at Fricourt, on the Somme, and finally, in 1975,
they were moved to the Richthofen family vault at
Wiesbaden.

Amiens, ancient capital of Picardy, was an important
trading centre in the Middle Ages and its locally grown woad
was much in demand as a plant dye. Traditional textiles,
velvet and wool, are still produced there, although they now
vie with tyres, car parts and electrics. Choderlos de Laclos,
author of *Les Liaisons Dangereuses*, was born in Amiens and
the novelist Jules Verne passed much of his life there. The
city was badly damaged by shellfire in 1918 and its centre
was burnt during the German invasion of 1940.
The massive Cathedral of Notre-Dame was the subject of
some nineteenth-century 'restoration' by Viollet-le-Duc but
miraculously escaped the fire of 1940. Its nave is the highest
in France, and amongst the cathedral's many monuments is
a plaque to Lieutenant Raymond Asquith, son of the Liberal
Premier Herbert Asquith, killed on the Somme and buried at
Guillemont.

In mid-August 1918 the Allies had turned their back on
Amiens and were poised for the next assault. Foch, vigorous
as ever, pressed Haig to attack with his 3rd and 4th Armies
as soon as possible, but, strongly influenced by the views of
his army commanders, the British Commander-in-Chief

declined to be hustled and it was not until 21 August that Byng attacked north of Albert. He had little to bite on, for the Germans had already withdrawn to an intermediate line, but a counter-attack was firmly repulsed. Rawlinson kept pace, recapturing Albert, and on 23 August both armies attacked again, rolling the Germans across the Somme battlefield. Three days later 1st Army joined in, advancing to the east of Arras.

Péronne, which lay on the axis of 4th Army's advance, had been an important fortress for centuries. In 1468 Charles the Bold of Burgundy had forced Louis XI to agree to a humiliating treaty in its château. Besieged by the Emperor Charles V in 1536, it had been saved by the efforts of the townswoman Marie Fouré, who animated its defence. Its fortifications had been remodelled at the turn of the sixteenth and seventeenth centuries. One of the gates, the Porte de Bretagne, stands on the old Péronne–Cambrai road, with a bastion, protected by the River Cologne, to its south, and there is a brick bastion behind the château. On 26 June 1815, in the advance after Waterloo, the Guards light companies took the defences covering the Porte de Bretagne and the town capitulated a few hours later. In 1870 it was more obdurate and the Prussians bombarded it for thirteen days before it fell.

Péronne had been declassified as a fortress long before the First World War, but its natural position was strong, especially against an enemy coming from the west, who had to cope with the high ground of Mont St-Quentin to the town's north-west, and the Somme and Canal du Nord curling around it. The Allies had narrowly failed to reach the town in 1916, though they got as far as Biaches, just across the Somme. The Germans had carried out thorough demolitions before giving up Péronne when they withdrew to the Hindenburg Line and dug themselves in securely when they recovered the town. In August 1918 all the nearby bridges over the Somme had been blown and elements of five divisions, including 2nd Guards, held the sector.

On 30 August the Australians crossed the Somme over the damaged railway bridge at Feuillères – where the autoroute now passes over the river – already held by III Corps, and seized two bridgeheads from which they attacked the next morning. The assault had little artillery preparation and resistance was stiff. The northern prong of the attack took the village of Mont St-Quentin but was driven out by fierce counter-attacks, and by nightfall was compressed into a narrow strip between hill and river. The southern prong had stalled and troops from it had been diverted to feed the battle in the north. The fighting around Mont St-Quentin had fixed German attention on the north and on 1 September the southern thrust made unexpectedly good progress, breaking into Péronne and clearing much of the town. The Germans weakened their hold on Mont St-Quentin in an effort to recover Péronne, and the Australians secured both village and hill. On 2 September the Australian 15th Brigade crossed the Somme well south of Péronne, outflanking the Germans still holding out around the town. Byng, meanwhile, was making good progress to the north – the New Zealand Division entered Bapaume on 29 August – and it is doubtful whether the Germans would have held Péronne with both flanks turned.

The Battle of Péronne might have been avoidable, but there is no denying the Australian achievement. It is marked by a monument to the 2nd Australian Division on Mont St-Quentin, on the Bapaume road out of Péronne. The present structure, a slouch-hatted 'Digger' in fighting order, replaces an earlier statue of an Australian soldier killing an eagle, which was destroyed by the Germans during the Second World War. A few kilometres on, up the N17 towards Bapaume, in the village of Bouchavesnes-Bergen, is a statue of Foch, occasionally to be seen incongruously clutching a tricolour. In the château at Péronne is the *Historial de la Grande Guerre*, opened on 1 August 1992 and described as 'an international museum of comparative history'. In contrast to so many museums on the Western Front, in which exhibits are often crammed together without giving the visitor much idea of their significance, the *Historial* has a much more measured tone and is concerned not simply with what happended in the front line, but with life behind the lines, for soldiers and civilians too. It contains a documentation centre, with written material, photographs and printed works, available for consultation, though visits must be pre-booked.

Foch himself, promoted to Marshal of France on 6 August 1918, was hustling Allied armies along with the cry *'Tout le monde à la bataille'*. John Terraine describes this as 'vigorous opportunism' and, as Foch acknowledged in a letter to Haig, it consisted of 'an offensive fed from behind and strongly pushed forward on to carefully selected objectives, without bothering about alignment nor about keeping too closely in touch'. Given Pétain's caution and the exhaustion of his armies, it was the British and Americans whose offensives were to be 'pushed forward'. And although there were signs that German morale had begun to crack, Foch's policy of all hands to the pumps could not be implemented cheaply or easily. The fighting of August had cost the British armies over 80,000 men and in front of them stood the Hindenburg Line.

The Canadians, up in the Arras–Vimy sector once more, had broken the *Wotan* Line by a well-handled attack on 2 September, with the 63rd Royal Naval Division, which had held the Flesquières salient in March 1918, playing a prominent part. This, coupled with the loss of Bapaume and the outflanking of Péronne, compelled Ludendorff to order withdrawal to the old pre-21 March British front line. The next phase of the battle began on 26 September, when Byng's army attacked the Canal du Nord opposite Cambrai, breaking clean through the *Siegfried* positions. On the following day Plumer's men plodded through the mud of the salient to take the whole of Passchendaele Ridge. Between Cambrai and St-Quentin the Hindenburg Line remained intact. It ran parallel with the St-Quentin Canal, which had been incorporated into the defences. Between Le Catelet and the southern edge of Bellicourt the canal flows through a tunnel over seven kilometres long and this, together with a shorter tunnel from Le Haucourt to Le Tronquoy to the south, had been turned into an underground fortress, with shafts leading up to the trench systems above, hospitals, headquarters and barges used as accommodation for troops. Rawlinson's gunners bombarded the position for two and a half days and at first light on 29 September 4th Army's attack began in earnest.

Its main weight was delivered by the US 27th and 30th Divisions, attacking between Vendhuille and Ville Noire, a sector crossed by the principal canal tunnel. On their right the 46th Division was to attack south of Ville Noire, and British divisions covered the flanks. The American divisions, part of the US II Corps, were fighting under the direction of the seasoned Australian corps headquarters, but this could not compensate for their own lack of experience.

Both divisions had secured some ground before the attack proper began. On 26 September 30th Division took Quarry Wood (still a prominent feature south of the D331 Bellicourt–Hargicourt road) and a section of trench to its north. On the following day 27th Division attacked the three strongpoints of the Knoll, just south of Vendhuille, Guillemont Farm, off the little road from Bony to Ronssoy, and Quennemont Farm, off the D57 Bony–Hargicourt road, but advanced only a few hundred metres. The main attack initially fared little better. 27th Division's 107th Infantry Regiment lost 377 men killed and 658 wounded – the heaviest loss suffered by any American regiment in the entire war – assaulting up the slopes west of Bony. Part of the American problem stemmed from an excess of enthusiasm and unfamiliarity with trench fighting: parties of Germans who had been bypassed by earlier attacks emerged to machine-gun follow-up troops. There were also some difficulties in co-ordinating the American attackers with their supporting British tanks and artillery, and the Australians who were to pass through the Americans once their objectives were secured.

The task facing the 46th Division seemed even harder, for it had to cross the canal as well as fight its way through several layers of defence. We last saw the division perform disappointingly at Gommecourt, but in the intervening two years it had changed considerably. Its officers and men were battle-wise, and they approached the problem of crossing

the canal with wisdom and resource. All available crossing
aids were secured. The engineers brought up pontoons and
rafts, and soldiers borrowed lifebelts from cross-channel
steamers. Fortunately it was another foggy morning. The
Midlanders got through the German positions east of the
canal with relative ease, and were soon across the canal and
driving up the high ground on its eastern side. As the 46th
Division bit deep into the German defences, the US 30th
Division began to make better progress and this in turn
helped free the 27th to its north. By nightfall the *Siegfried*
system was in ruins, and 4,000 men and seventy guns had
been captured. It was the knock-out blow to an already
groggy German high command. On the 29th, Ludendorff
warned that only an immediate armistice could avert a
catastrophe and on 2 October his representative told
Reichstag party leaders that the war was lost.

No visitor to the battlefields of the Cambrésis can fail to
be struck by the large number of British soldiers killed in the
closing weeks of the war. Over 111,000 became casualties in
October alone, many of them caught in savage battles with
German rearguards whose soldierly determination was
stronger than their sense of self-preservation. The little York
Cemetery at Haspres, off the D955 between Solesmes and
Denain, tells a typical story. Against the back wall of the
cemetery, behind the rows of British infantrymen, many of
them young conscripts, stands a row of ten German
headstones. The soldiers they honour form, as nearly as it is
possible to judge, two detachments of the machine-gun
company of the 295th Infantry Regiment, who died on the
same day as most of the British soldiers whose burial ground
they share: victors and vanquished under the cropped turf
amongst the beet fields.

The American contribution to breaking the Hindenburg
Line is marked by a memorial on the N44 just north of
Bellicourt. This block of white marble with its symbolic
figures commemorates all American units that served with
the British armies during the war. On its western side is a
map and orientation table, which identifies key points in the
fighting. Both entrances to the main tunnel can be reached

easily, though the southern entrance, between Bellicourt and Riqueval, is the more impressive: an inscription above the entrance celebrates the tunnel's opening. On the main road nearby stands a memorial to the Tennessee soldiers of the 30th Division's 59th and 60th Brigades.

Further down the N44 an easily missed turning leads to the Riqueval bridge. This was the scene of a famous photograph depicting Brigadier-General Campbell of 46th Division's 137th Brigade addressing his men. Some are still wearing life jackets and a few have improved the shining hour by donning captured German helmets. The bridge survived the battle more or less intact, but the bare horizon in the photograph gives little clue to which side of the canal it was taken from: by matching shell damage still on the arch with that shown in the photograph, it becomes clear that the photograph was taken from the west bank. A brick monument on the eastern side of the bridge announces that it was taken by A Company of 6th Battalion The North Staffordshire Regiment and a detachment of Royal Engineers. Even at this late stage in the war, when regional recruiting had been badly bruised by casualties and the policy of posting infantry reinforcements from base depots in the Etaples area without much regard for cap badges, 6th North Staffordshire still cherished its origins and entertained the Mayor of Wolverhampton at Riqueval bridge soon after its capture. The monument was erected by the Western Front Association, and no First World War enthusiast can be unaware of the research, battlefield tours and commemoration organised by the regional branches of this commendable organisation. The 46th Division has a memorial almost opposite the Bellenglise turning on the N44. It was a tired and unprepossessing obelisk when I wrote the first edition of this book, but it has been restored relatively recently, and its groundwork well refurbished. On the high ground about a kilometre north-west of the village of Bellenglise, between canal and autoroute, is an obelisk commemorating the Australian 4th Division.

The American Cemetery at Bony is officially entitled the Somme Cemetery, though it lies some distance from battlefield and river alike. It contains the remains of 1,844 American war dead and the names of another 333 missing are inscribed on the interior walls of the Chapel-Memorial at the front entrance. British war cemeteries are reminiscent of an English country garden; German are stark and teutonic, and French are utilitarian. American cemeteries, with their white marble crosses, have a clean and dignified air. The mighty bronze gates of the Chapel-Memorial at Bony are especially impressive. Winners of the Congressional Medal of Honor, America's highest decoration, have the inscriptions on their headstones picked out in gold: two of them lie buried at Bony.

IV

Flanders, Artois and Picardy – III

THE MAGNITUDE OF Allied victory in 1918 led some to believe that the bloodletting of 1914–18 had indeed been 'the war to end wars'. But Foch had predicted that the Treaty of Versailles was merely a twenty years' armistice and the breathless events of the 1930s proved him right. Flanders, Artois and Picardy did not feature greatly in French military policy in the inter-war years. Belgium was a French ally, with defences of her own along her eastern frontier, while the mighty Maginot Line guarded the Franco-German border. There thus seemed little likelihood of war coming to the north-west. Even when Belgium declared her neutrality in 1936, thus imperilling the northern flank of the Maginot Line, little work was done to protect the northern frontier, although a few simple concrete bunkers moulder there as evidence of last-minute attempts to fortify the open flank.

Nor was there much British interest in resuming a Continental commitment. The slump of the late twenties and early thirties did not encourage a vigorous defence policy, and the strife-torn state of France made her an unappealing ally. The legacy of the Somme and Passchendaele also militated against plans which might lead to a repetition of the bloodletting of 1914–18. In May 1938 one senior officer declared: 'Never again shall we even contemplate a force for a foreign country. Our contribution is to be the Navy and the RAF.' Even when faced with clear evidence of German expansionism, it took the British government perilously long to react. Planning to send an expeditionary force to France began in earnest in August

1938 and was barely completed by the time that war broke out.

The planners' task was complicated by the fact that Britain and France had pursued different strategic policies in the inter-war years, and by Belgium's adherence to strict neutrality. Nothing could be done to help the Poles – although Hitler's invasion of Poland was the *casus belli* – and a French offensive into the Sarre in September was a low-key affair. The French Commander-in-Chief, General Maurice Gamelin, adopted the conservative policy of building up his strength in men and equipment until he was able to take on the *Wehrmacht*. In the meantime, he believed that if the Germans attacked, they could only do so effectively through Belgium. However, the Belgians refused to allow the Allies into their country before the Germans had actually launched an attack and the army Chief of Staff, General van den Bergen, was dismissed in January 1940 for removing the barriers on the French frontier. In October 1939 Gamelin ordered the commander of his North-East Front, General Georges, to move into position on the Scheldt (Escaut) from Antwerp through Ghent in the event of German attack, and the following month Gamelin decided on an even deeper advance, this time to the River Dyle.

In January 1940 the fortuitous capture, near the Belgian town of Mechelen, of a partly burned copy of *Case Yellow*, the German operations plan for the invasion of Belgium and northern France, produced a flurry of activity and convinced Gamelin that he was right to meet the German attack in Belgium. He planned to send his entire left wing – from west to east, the French 7th Army, Lord Gort's BEF, and the French 1st and 9th Armies – into Belgium, pivoting on the 2nd Army near Mothermé at the confluence of the Meuse and Semois. The plan resulted in both Allied wings being strong, the left poised for the move into Belgium and the right in the Maginot Line. But in the centre, behind the Ardennes, stood the weakest of Gamelin's divisions – with no strategic reserve behind them.

The story of the piercing of the Allied centre will be told

in a later chapter. The plan that brought it about was the brain child of Lieutenant-General Erich von Manstein, Chief of Staff of Rundstedt's Army Group A. *Case Yellow* had placed the weight of the attack to the north, with Bock's Army Group B, which was to carry out the advance into Belgium. Manstein believed that this would produce at best an ordinary victory – whereas he argued that the aim of the plan should be to force a decisive issue by land. Bad weather forced the postponement of *Case Yellow* and in February it was replaced by a new plan which placed the *Schwerpunkt* in the centre. Hitler's influence proved crucial in bringing about the change. Manstein visited him to explain his project and the Führer was impressed by its audacity – and by the fact that its decisive point was Sedan, where the Germans had won their great victory in 1870. The *Sichelschnitt* – scythe cut – Plan gave pride of place to Army Group A, now comprising forty-five divisions, seven of them armoured. Army Group B, still to advance into Holland and Belgium but no longer mounting the major effort, had twenty-nine divisions, three of them armoured, while Leeb's Army Group C, covering the Maginot Line, had only nineteen divisions, none of them armoured.

The attack began on the morning of 10 May. As the German armoured spearhead entered the Ardennes, the Allied left wing advanced, as planned, into Belgium. The American war correspondent, Drew Middleton, was not alone in noticing how strange it was for British soldiers who had fought in Belgium in the First World War to be revisiting the scenes of their youth. 'It was almost as if they were retracing steps taken in a dream,' he wrote. 'They saw again faces of friends long dead and heard the half-remembered names of towns and villages.' By dusk on 11 May the BEF was dug in in the Dyle between Louvain and Wavre. Blanchard's 1st Army was moving up to the Gembloux gap on its right, with the excellent light mechanised divisions of Prioux's Cavalry Corps covering it. On the BEF's left the 7th Army had moved up north of Antwerp and had already been roughly handled by German armour near Tilburg. Over the next few days the situation in north and centre alike

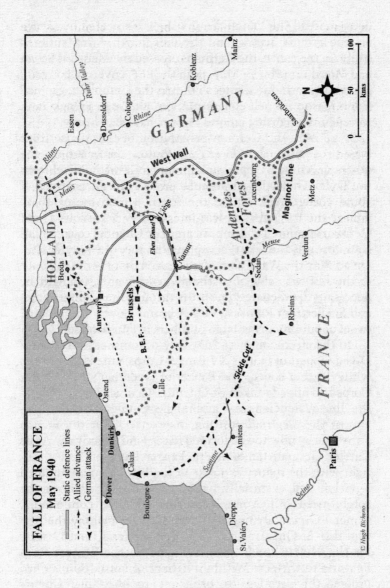

FALL OF FRANCE
May 1940

— Static defence lines
↓ Allied advance
↓ German attack

deteriorated. The Dutch army was dismembered by savage air and ground attack, and the French 7th Army suffered badly. In the centre the Germans crossed the Meuse at Sedan and Monthermé on 13 May. Both the BEF, covering Louvain, and French armoured forces holding the Gembloux gap had some reason for self-congratulation, although French tank losses in the fighting around Perwez had been heavy.

By 16 May the German penetration of the Meuse front was so serious that General Billotte, commanding the northern Army Group, issued orders for a withdrawal from the Dyle. The BEF was soon faced by hard alternatives. It could either break out to the south, which meant abandoning the Belgians and was increasingly hazardous as the German armoured columns neared the Channel coast, or fall back towards the coast, accepting the risk of encirclement. On 20 May the War Cabinet recognised that even if the bulk of the BEF was able to withdraw southwards, it might be necessary to evacuate some units through the Channel ports and Sir Bertram Ramsay, Vice-Admiral Dover, was ordered to set about procuring large numbers of small vessels.

On the afternoon of 18 May Erwin Rommel's 7th Panzer Division, part of Hoth's XV Panzer Corps, entered Cambrai, while, further south, 2nd Panzer of Guderian's XIX Panzer Corps had already taken St-Quentin. Georges hoped to hold the line Valenciennes–Cambrai–Le Catelet–St-Quentin – almost the *Siegfried* position in reverse – but the French army was by now too dislocated to respond effectively. At Le Catelet German tanks of 6th Panzer Division had a stiff fight with the robust 'B' tanks of the French 2nd Armoured Division and eventually overran the headquarters of the sorely tried 9th Army, capturing its new commander, General Giraud. By nightfall even Georges realised that the Cambrai–St-Quentin line was untenable and issued orders for the establishment of a new line on the Somme from Péronne to the sea. Within a matter of hours Billotte had ordered his armies to back on to the line Ghent–Valenciennes–Bouchain, and the following day he swung his right flank round to face south on the Sensée at Aubigny and the Scarpe at Arras.

1. This medieval cast-iron monkey, in the main square of Mons, epitomises the story of the fatal avenue. Over the centuries its head has been patted by Spaniards, Frenchmen, Austrians, British, Germans and Canadians as successive invasions have ebbed and flowed. Passing Belgians still give the creature's shiny pate a rub for luck.

2. The main gate to the citadel of Lille, now *Quartier Boufflers*, Boufflers barracks, named for the squat and energetic marshal who held it against Marlborough and Eugène in 1708.

3. The Valmy windmill, with the memorial to the battle just visible to its right, dominates this open landscape like French gunners in 1792.

4. Point of impact. The edge of the Bois Pierrot, about 1 km north of the Metz–Verdun road at Rezonville. On 16 August 1870 the regiments of von Bredow's cavalry brigade swung from column into line on the other side of the thin line of poplars in the distance. They then charged towards our viewpoint across the open ground on the left, riding through the gun-line of the French VI Corps.

5. Thor's anvil. A cupola at Loncin, one of the forts protecting Liège, smashed by German heavy guns in August 1914.

6. The rear of Fort de Leveau at Maubeuge, taken by assault in the face of determined resistance on 7 September 1914.

7. Lieutenant Colonel Emile Driant's command post in the Bois des Caures north of Verdun, in the very epicentre of the German attack in February 1916.

8. Somme skyline. Unexploded shells, the battlefield's iron harvest, alongside a track which runs northwards from Beaumont Hamel towards the Serre road. Munich Trench cemetery is just visible on the crest, and in the summer and autumn of 1916 the British attacked up this slope, their final line running roughly across its top.

9. War's hard wire. Many of the iron pickets used to support First World War barbed wire entanglements are still in use. Here, just north of Beaumont Hamel, a fence supported on these pickets keeps cattle in their field.

10. Jetsam of war. A rusty rifle lies atop a pile of empty shrapnel shells in a Somme wood.

11. Lochnagar crater, on the 1 July 1916 German front line. The village of La Boisselle, on the main Albert-Bapaume road, is 750 metres to the left. 66,000 lbs of ammonal were exploded to make this crater: the tunnel along which British miners made their laborious way ran under the plough in the left foreground.

12. A British front line trench, with a communication trench joining it in the centre of the photograph, in Newfoundland Park at Beaumont Hamel. This photograph was taken in 2007, ninety-one years after the Newfoundlanders attacked here.

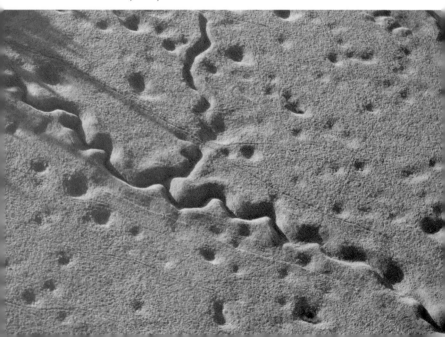

13. One of the most evocative sights on any battlefield. The 38th (Welsh) Division's memorial looking out – towards the direction of the British attack – across Death Valley to Mametz Wood.

14. A foreign field indeed. Sir Herbert Baker's Neuve Chapelle memorial to the Indian Army's officers and men who have no known graves. The column on the right is reminiscent of pillars erected in India by the Emperor Ashoka.

15. A brave man's last view. The ditch at the citadel of Arras, the scarp rising up to the right and the counterscarp, its covered way now overgrown, to the left. The Germans shot resistance workers here, and the pillar in the foreground represents the post to which they were tied.

16. Per Ardua ad Astra. The Flying Services Memorial at Arras commemorates the 1,000 airmen of or attached to the Royal Naval Air Service, the Royal Flying Corps and the Royal Air Force who died on the Western Front in 1914–18 and have no known graves. This face of the pillar, with its Canadian, Australian, New Zealand and South African badges makes the never-to-be-forgotten point that Britain relied heavily on the support of her Dominions.

Between the Somme and Douai there now gaped a hole filled only by fragments of the French 9th Army and two British Territorial Divisions, 12th and 23rd. The latter formations had been employed on lines of communication duties, an appropriate role in view of their sketchy training and limited equipment. There was a pitiful shortage of machine-guns and anti-tank rifles. 12th Division had only four field guns and although 23rd Division could boast thirteen, eleven of them could only fire over open sights and the remainder had no sights at all. Faced with the emergency of the German armoured breakthrough, 23rd Division was given the task of holding the Canal du Nord from Douai towards Péronne, where it was to join elements of the newly formed French 7th Army, replacing the original 7th Army, which had been lacerated in the fighting in Holland. The unlucky 12th Division, subject to conflicting orders from its parent headquarters, HQ Lines of Communication North in Rouen, and Army Rear Headquarters in Arras, was widely spread in Amiens, Abbeville, Cléry-sur-Somme and Doullens. In the course of an ill-structured deployment, the divisional commander, Major-General Petre, was whisked away to command Petreforce, an ad hoc force cobbled together to defend Arras. Time and energy were wasted in moves and countermoves, and on the 18th the 7th Royal Sussex suffered heavy casualties when its train was dive-bombed at Amiens.

We should not be surprised that when the Germans attacked on the 20th they made good progress. Albert, defended by 7th Royal West Kents, part of 36th Brigade, fell to 1st Panzer Division early that morning, and Amiens, blazing from repeated air attacks, was in German hands by midday, despite the spirited resistance of 7th Royal Sussex of 37th Brigade. The 35th (Queens) Brigade was scarcely more fortunate. Two of its battalions were sent by train from Abbeville to Lens. This was speedily discovered to be an error and the battalions returned to Abbeville, where they were soon in serious difficulties as German armour approached the town. 2nd Panzer, urged on by Guderian himself, took Abbeville that evening and 35th Brigade soon ceased to be an effective fighting force. Further north, Reinhardt's XLI Panzer Corps moved rather later. 6th Panzer met its first British at Mondicourt, just east of Doullens, then went on to attack Doullens, defended by 5th Buffs and 6th Royal West Kent. Two companies of the Royal West Kent held out in the town centre until evening, but by nightfall 36th Brigade was destroyed as a fighting formation. Its commander, Brigadier George Roupell, had won his VC as a lieutenant on Hill 60 in 1915: his proven valour was no substitute for orders and information, and these Rear Headquarters in Arras seemed unable to provide. Some elements of 12th Division fell back south and others became involved in the defence of the Channel ports. But six of its nine battalions had been destroyed on 20 May in an unplanned battle, which testifies to Rear Headquarters' total incapacity for commanding troops.

The fate of 23rd Division was almost equally hard. It had only two brigades, which held the Canal du Nord from Arleux on the Sensée to Ruyaulcourt, east of Bapaume. The southern brigade, 69th, was able to withdraw to the Scarpe east of Arras to form part of the BEF's southern shoulder, but 70th Brigade, trying to move south of Arras to cover the Arras–Doullens road, was caught on the move by 8th Panzer Division. One of its battalions, 11th Durham Light Infantry, was overrun in Wancourt, on the 1917 battlefield, and another, 1st Tyneside Scottish, was cut to pieces just east of Ficheux: some of its soldiers took on tanks with rifle and bayonet. By the end of the day the brigade numbered only 233 officers and men.

The remaining misfortune of the 20th overtook 137th Brigade, a West Yorkshire battalion and two battalions of the Duke of Wellington's regiment, with a battalion of King's Own Yorkshire Light Infantry attached. This was part of our old friend the 46th (now North Midland and West Riding) Division. Although this formation was in much the same state as the other two second-line Territorial divisions, its commander had more gallantly than advisedly declared it battleworthy, and two of its brigades had been ordered north where they became part of 'Macforce', under Gort's Director of Military Intelligence, Major-General Mason-Macfarlane, tasked with protecting the right rear of the BEF. 137th Brigade was following suit, in a four-train convoy, and reached Abbeville – under heavy air attack – at much the same time as the Germans. After a series of hair-raising adventures – the Duke's battalions found themselves navigating with the aid of a copy of Bradshaw's *Railway Guide to Central Europe* – all four battalions managed to make their way to Dieppe.

The panzer breakthrough had split the Allied armies, but it was not, in itself, conclusive. The German infantry were stretched out far behind, marching hard under a blazing sun down the poplar-lined *pavé*, and until they caught up there was every chance that the Allies, jabbing in from north and south, could sever the armoured spearhead from the shaft which followed it. On 19 May, the day on which he discovered that he was to be replaced by General Maxime Weygand, Gamelin suggested a combined attack with exactly this end in view. On the 20th, with the Territorials of 12th Division dying in Picardy, General Ironside, Chief of the Imperial General Staff, visited Gort at his forward command post at Wahangies, twenty kilometres south of Lille. Gort had moved there as soon as operations began, taking with him a small nucleus of his staff and leaving the remainder of it at Arras. This was not a satisfactory arrangement, and the appointment of Mason-Macfarlane to command Macforce made matters even worse. The acerbic Montgomery, deftly commanding 3rd Division, thought the scheme was 'amateur and lacked the professional touch'.

Ironside, armed with a War Cabinet order that the BEF was to make for Amiens, found Gort adamant on the point: seven of his nine divisions were fighting on the Escaut and he could not disengage them. He had already told Major-General Franklyn, commanding the strong 'Frankforce' – nominally 5th and 50th Divisions with 1st Army Tank Brigade – to support the garrison of Arras, occupy the line of the Scarpe to its east, and establish contact with the French by patrols on the 21st, but this was as much as he could do. Ironside then visited Billotte in his headquarters at Lens and, after a difficult interview in which the huge Ironside shook the Frenchman by a button on his tunic, Billotte agreed that the French would attack with two divisions towards Cambrai on the following day. Unfortunately Franklyn was not informed that his very modest single-nation venture was now an Allied counter-attack and when he met Prioux later that day he was unable to conform to the scale of operation now envisaged by the French. Prioux's corps had been mauled in the defence of the Gembloux gap and subsequent fighting, and its two light mechanised divisions were now down to some thirty-five tanks, but he nevertheless agreed to screen the right flank of Frankforce when it attacked.

Arras was evacuated by Rear Headquarters on the 19th, but the headquarters defence battalion, 1st Welsh Guards, remained there with a scratch force of light tanks. Battalion headquarters was established in the Palais St-Vaast and the approach roads were blocked. Air attacks set fire to part of the town: a plaque on the wall of the church of St Jean Baptiste commemorates a captain and warrant officer of the local fire brigade who died trying to save it from the flames, and the First World War memorial is chipped by splinters of the bombs that destroyed the station. 7th Panzer Division approached the town on the 20th. Rommel himself reached Beaurains, to its south, that day, and while trying to bring up more troops he was nearly bagged by French tanks in Vis-en-Artois, a village on the Cambrai road with its Memorial to the Missing of 1918. On the 21st German reconnaissance units probed the defences of Arras, while 8th Panzer Division forged on ahead to its south-east and 7th Panzer, with the SS *Totenkopf* motorised regiment on its left flank, hooked around the town. Armoured cars of the 12th Lancers fell back ahead of the German advance and on the 20th the Lancers' headquarters were at Mont St-Eloi, where the ruined towers of the abbey destroyed by German shelling in 1915 offered an excellent field of view.

The attack was entrusted to Major-General Martel of 50th Division, who was ordered to clear the area south of the Scarpe from the southern outskirts of Arras, including

Monchy and Pelves, to the line of the River Cojeul, as far as
Boiry on the Arras–Bapaume road. It was already evident,
from the reports of the 12th Lancers, that the Germans were
in the area in strength and Martel's force was hopelessly
inadequate for its task. He had two mobile columns, each
consisting of an infantry battalion, a tank battalion, a field
battery, an anti-tank battery and a motorcycle reconnaissance company. Most of his tanks were the slow Mark I
Infantry, mounting only a .303 machine-gun, though a few
had a .5 inch. There were also 16 Mark II Infantry tanks,
which mounted a 2-pounder gun as well as a machine-gun
and were more heavily armoured than any other tank then
in service. The Arras–Doullens road (the modern N 25), start
line for the attack, was to be crossed at two on the afternoon
of the 21st. This left insufficient time for proper orders and
gave little opportunity for tanks and infantry to marry up.

The right column, 7th Royal Tanks and 8th Durham
Light Infantry, moved down the gentle slope towards the
Scarpe, then up to the Arras–St-Pol road, where it found
some soft-skinned vehicles destroyed by the 12th Lancers
that morning. Duisans was cleared, with the assistance
of French infantry and tanks, but thereafter the infantry
and armour became separated, the tanks swinging up to
Dainville, while the infantry, leaving two companies to hold
Duisans, advanced to Warlus. By losing direction 7th Royal
Tanks missed a head-on collision with 25th Panzer
Regiment, which eventually reached Acq, north of the St-
Pol road, before returning to join in the battle. At Warlus the
Durhams were dive-bombed and German armoured attacks
grew increasingly dangerous as 25th Panzer reappeared, but
the Durhams fought on in the blazing village and were able
to withdraw, with the timely assistance of six French tanks,
in the early hours of the 22nd.

The left column had made better progress. It enjoyed
several advantages: it had to cover a shorter distance than its
comrades on the right, and the tank crews of 4th Royal
Tanks had had some chance to rest and to ensure that their
radios were working properly. However, 6th Durham Light
Infantry – whose commanding officer had fought at Arras as

a subaltern in 1917 – was well behind the armour as it
rattled off past Dainville and Achicourt, carving up
Rommel's 6th Rifle Regiment in the process. Between
Beaurains and Telegraph Hill is a shallow valley, and as the
leading tanks of 4th Royal Tanks, attacking over the same
ground as D Battalion of the Tanks Corps in April 1917,
moved into it they were knocked out by the direct fire of
Rommel's 105mm field guns firing from the crest line.

Part of the battalion making for Wailly, 7th Royal Tanks,
which had swung off course in Duisans and cut in behind 4th
Royal Tanks in Dainville, continued its advance. Here it
encountered Rommel himself, on hand at the crucial point.
He thought that it was 'an extremely tight spot' and his
ADC, Lieutenant Most, was shot dead at his side as they
encouraged field and anti-aircraft gunners. Another detachment of 7th Royal Tanks passed through Agny before being
stopped in front of Mercatel by 88mm guns, already proven
in the Polish campaign, but here making their first
appearance in the anti-tank role in the west. The advance
was now well and truly halted. By this time the two forward
companies of the Durhams had reached Beaurains, where
they were dive-bombed and shelled, before eventually falling
back as 5th Panzer Division, which had moved up from
Bouchain that day, appeared on the scene. Y Company of 4th
Royal Northumberland Fusiliers fought to a finish at a
crossroads on the Beaurains–Achicourt road, and the survivors of the attack made their way back up on to Vimy Ridge.

The action south of Arras on 21 May 1940 is sometimes
described as a counter-attack or, to bring it into line with
British military terminology of the 1980s, a counter-stroke.
We should not read too much into the event itself. Its objectives were decidedly modest and it was conceived in the
false expectation that there were no substantial German
forces in the area. The British force engaged was tiny –
especially in view of the fact that Franklyn had two
divisions at his disposal – and the operation lacked most of
the ingredients normally regarded as essential. There was no
air cover, communications were atrocious and artillery
support was derisory.

The immediate results of the action were not negligible: about 400 Germans were captured, another 200 killed or wounded and about thirty German tanks were destroyed. Its real importance, though, lay in the widening ripples of concern that spread amongst the German high command. Rommel, not easily given to alarm, believed that he was facing five British divisions. The panzer divisions, which had already reached the coast, sent strong forces back inland: the delays generated by the Arras operation enabled both Boulogne and Calais to be resolutely held, and helped make possible evacuation from Dunkirk: on 24 May Rundstedt ordered his armour to halt to allow the infantry to catch up. Basil Liddell Hart overstated the case when he attributed this controversial order entirely to the effect of Arras: Brian Bond strikes the right balance when he declares that it played a 'significant but not dominating part' in making up Rundstedt's mind. The panzer divisions had lost up to fifty per cent of their tanks – from breakdowns as well as hostile action – and were expected to establish bridge-heads over the Somme, take the Channel ports, then swing north to attack into Flanders. Kleist, whose panzer group included both Reinhardt's and Guderian's panzer corps, was frankly concerned at the state of his units and Rundstedt duly ordered the halt.

Although 150th Brigade of 50th Division raided towards Feuchy across the Scarpe and 5th Division seized a bridge-head over the river near Fampoux, it was obvious to Franklyn that to remain in the Arras area was to invite disaster as Rommel spun on to Béthune, 5th Panzer Division swung up through Mont St-Eloi to Souchez and the Lorette spur, and motorised infantry moved against Arras. On the evening of the 22nd the Germans crossed the Scarpe north of Monchy. They made no impression on the defences of Arras and as they nudged up on to the Lorette spur they were delayed by the few surviving tanks of the French 3rd Light Mechanised Division, then checked by 7th Royal Tanks, firing from positions around the cemetery. Late on the 23rd Gort ordered Petreforce to hold out to the last, but the order was mercifully countermanded and the Arras garrison retired northwards. About five kilometres outside the town a German post blocked one of the withdrawal routes, but was attacked by Lieutenant the Honourable Christopher Furness and the Bren gun Carrier Platoon of 1st Welsh Guards in a spirited action, which won Furness a posthumous Victoria Cross.

While the Arras battle was in progress, Weygand, the new French Commander-in-Chief, flew north to co-ordinate plans for a large-scale counter-attack. His trip was bedevilled by misunderstandings and instead of seeing all senior commanders in the north simultaneously, as he had hoped, he had a number of separate meetings. Their venue was, by another of the ironies of the 1940 campaign, in the mayor's office of Ypres town hall – not then in its present location, but in the north-east corner of the square. The first of the day's three meetings was with King Leopold of the Belgians, commanding his army in the field. Weygand hoped that the Belgians would fall back on to the line of the Yser, as they had in 1914, but Leopold professed, not without reason, that the army was too exhausted for such a move. Billotte then arrived, and Weygand outlined his plan for an offensive southwards around Cambrai and northwards from the Somme, its claws meeting in the area of Bapaume. Weygand thought that the events of the past fortnight had left their mark on Billotte, but felt that he understood the need for urgency. Unfortunately, although Billotte emphasised the tattered state of his army group, he suggested that the BEF still constituted a useful offensive force.

Gort was not in a position to state his own case. On the night of the 22nd the BEF had fallen back on to its old positions along the Belgian frontier between Halluin and Maulde, and Gort had moved his headquarters from Wahangies to Premesques – just north-west of Lille – where he was dutifully awaiting Weygand. Weygand did not know this, and remained at Ypres till 7 p.m. that evening, when he was visited by Admiral Abrial, Commander-in-Chief of French naval forces in the north. Abrial warned that German air power now made it impossible for Weygand to fly back to Paris and put the motor torpedo boat *Flore* at his disposal.

Flore left Dunkirk in an air raid, leaving the harbour at full speed amongst gouts of water hurled up by bombs. Weygand, still nettled by Gort's non-appearance, travelled to Cherbourg via Dover and eventually reached Paris on the morning of the 22nd.

There were two unkind twists still to come. Gort eventually got word of the meeting and reached Ypres after Weygand had left. His action was to fuel French speculation that he had stayed away deliberately, having already become convinced that the BEF would have to be evacuated. When Billotte told him of the Weygand plan, Gort was unimpressed: he was now fully committed at Arras and thought the projected operation had little chance of success. The Belgians eventually agreed to fall back from the Escaut to the Lys, slightly shortening their line and freeing some British troops for the offensive. A glance at the map shows the evident danger of such a policy. The Yser line was straight and reasonably defensible: but the Lys had less natural strength and invited an attack in the area Menin–Halluin. At the end of these wearing meetings Billotte, the only real link between Weygand and Gort, was mortally injured when his car crashed into a lorry. Such was the state of the French chain of command that it took three days to replace him with General Blanchard of the 1st Army and to move Prioux up to relieve Blanchard.

On 22 May the British Prime Minister, Winston Churchill, visited the French Premier, Paul Reynaud, and Weygand in Paris. Impressed by Weygand's unrealistic assurances that the great counter-attack would indeed take place, he sent a telegram asking Gort to attack towards Bapaume and Cambrai at the earliest moment – 'certainly tomorrow'. This flight of fancy did nothing for morale at Gort's headquarters. Relations with the French were worsened by the decision to withdraw from Arras, which left the 1st Army exposed around Douai and Valenciennes. The news rippled back to Reynaud, who told Churchill that the withdrawal had imperilled the entire Weygand Plan. It certainly served as a pretext for the cancellation of an offensive by the 6th and 7th Armies south of the Somme, and Weygand ordered General Besson, their commander, to establish himself in depth on the river. Despite this, there were Anglo-French discussions on an offensive astride the Canal du Nord, to begin on the 26th, but already the French government, influenced by Weygand's assertions that 'we have gone to war with a 1918 army against a German army of 1939' was beginning to consider the possibility of making a separate peace.

General Dill, Vice-Chief of the Imperial General Staff, visited Gort on 25 May. Neither Gort nor Blanchard had yet heard that Weygand had abandoned the attack northwards from the Somme and Dill believed that it should go ahead as planned. That evening Gort took the decisive step of cancelling his contribution to it. He had heard that the Germans had broken the Belgian front on the Lys, and were across the river between Wijk and Courtrai: the Belgians were being folded back northwards. Gort told General René Altmayer of V Corps, whose collaboration in the attack had already been reduced to a single division, of his decision, but was unable to speak to Blanchard and the episode put still more strain on the alliance. When Blanchard heard the news he ordered a withdrawal on to a bridgehead formed by the Aa Canal, the Lys and the Canal de Dérivation, and Weygand approved it, though not without more moonshine about the imminence of the attack across the Somme.

The situation on 26 May in many respects resembled that which might have been produced had Ludendorff's March 1918 offensive reached the sea. The Allied armies were split. The French held the line of the Somme, while the northern Allied armies had their backs to the sea. Weygand seems to have believed that this bridgehead could be held indefinitely and Blanchard understood that it was his mission to do so. Gort, however, had been informed that day that the 'safety of the BEF will be a predominant consideration. In such conditions the only course open to you may be to fight your way back to coast where all beaches and ports east of Gravelines will be used for embarkation.' The plight of the Channel ports was such that estimates of further evacuations were depressingly low. On 24 May Ironside gloomily opined that 'We shall have lost practically all our trained men by the next few days' and on the 25th Gort warned that 'the greatest part of the BEF and its equipment will inevitably be lost even in the best circumstances'.

The need to keep the BEF supplied, now that it was cut off from its bases in Normandy, had already encouraged the British to take an interest in the defence of the Channel ports. 20th Guards Brigade, consisting of two newly raised battalions of Irish and Welsh Guards, was sent to Boulogne on 23 May. They found a hotchpotch garrison consisting of assorted Frenchmen, most of whom had become separated from their units, some RAF ground crew and the Rear Headquarters of the BEF, based at nearby Wimereux, under the Adjutant-General, Lieutenant-General Brownrigg. There was also a surprisingly useful force of about 1,500 men of the Auxiliary Military Pioneer Corps under Colonel D. C. Dean. Dean's men, only one in four initially armed, had been working in the Doullens area when the Germans attacked, and he had managed to get them back to Boulogne, collecting some survivors of the Buffs and Royal West Kents as he did so. Dean himself had won a VC in the First World War and was to show himself a more redoubtable leader than might have been suspected from someone holding the unmartial appointment of Commander 5th Group of Pioneer Companies.

Boulogne was not easy to defend with such a small and heterogeneous force. The Guards' Brigadier, Fox-Pitt, walked round the perimeter with his two commanding officers and decided to deploy the Irish Guards south of the River Liane and the Welsh north of it. The Germans soon appeared and gained some ground off the Irish Guards near Outreau, and early on 4 May Boulogne was encircled. The Fort de la Crèche, towards Wimereux in front of the Welsh left flank, fell early on. General Brownrigg had departed for England in the small hours of the morning and as soon as Fox-Pitt realised that he was in undisputed command he deployed Dean's Pioneers – given extra rifles and Bren guns from those abandoned in the town – to strengthen the northern perimeter.

As the day wore on the defenders were attacked from air and land, and the Germans established themselves on the high ground around the town. The navy had been evacuating soldiers during the day under heavy air attack and in the evening the guns of the Fort de la Crèche joined in. Early in the afternoon Fox-Pitt was ordered to evacuate, but the message did not reach him until 6.30 and as it did so the heaviest air raid of the day damaged the destroyer *Vimy*. Destroyers came close inshore to engage German infantry and guns as the last of the Guards were snatched from the quays. The Pioneers crossed the bridges over the Liane minutes before they were blown and fell back on the *Gare Maritime* with little hope of escape, but in the early hours of the 24th HMS *Vimiera* slid in and picked them up. The remnants of a Welsh Guards company, with some stragglers from the Irish Guards and the Pioneers, held the seaward end of the mole until early on the 25th when, short of food and ammunition, they surrendered.

The fate of the French garrison was a microcosm of much of what went on during those desperate days. General Lanquetot, commander of 21st Infantry Division but cut off from it, had established himself in the *Ville Haut*, behind Philippe le Hurepel's stout walls. There had been no co-ordination between British and French defence of the town, and the French received no word that the British were proposing to evacuate. This would probably have made little difference, for the French were under orders to stand and fight, but the lack of communication added to French suspicions of betrayal. Lanquetot's men gave a good account of themselves when the Germans attacked the *Ville Haut*, but eventually an 88mm was hauled up to breach the walls near the cathedral and German infantry clambered through. By the 25th the French held only the Calais Gate and, warned that the town would be destroyed if they did not surrender, duly did so.

If a degree of optimism had led to the garrisoning of Boulogne, worse was to follow at Calais. With the BEF cut off from ports and bases in Normandy, Calais was the best port for the supply of food and ammunition. In London it was believed that the real problem was not holding Calais, but in ensuring safe transit of stores between the port and the BEF through a countryside allegedly infested by German light armour. The Calais garrison, therefore, consisted not of conventional infantry – though some were available if required – but of the specialist 30th Infantry Brigade. This comprised two motor battalions – 1st Kings Royal Rifles and 1st Rifle Brigade – with a Territorial motorcycle reconnaissance battalion, Queen Victoria's Rifles, with 3rd Royal Tank Regiment added at the last moment.

The brigade commander, Brigadier Claude Nicholson, had initially been ordered to get supplies through to the BEF and was not aware of the strength of German forces in the area. The Royal Tanks and the QVR arrived on 22 May, still having no idea that they were under Nicholson's command, and the former were immediately sent off to St-Omer by a liaison officer from Gort's headquarters. It was difficult to unload tanks and equipment at docks that had already been bombed and whose British dockworkers were exhausted. Not till the afternoon of the 23rd were the Royal Tanks able to set off and just outside Coquelles, on the Guines road, they met the leading elements of 1st Panzer Division. Although the British had the advantage of the initial clash with the leading Panzer Mark IIs, they were forced to withdraw as heavier German tanks arrived on the scene. The QVR, who had arrived without their motorcycles or 3-inch mortars, had been sent off to block the main roads into the town and patrol the beaches to prevent a German landing. The remainder of the British garrison consisted of an anti-tank battery in Nicholson's brigade, some base details, two searchlight and two anti-aircraft batteries, while the French garrison included naval gunners for the coast-defence pieces, and an assortment of reservists and stragglers.

At this juncture Nicholson arrived with his two regular battalions. While still in Dover he had encountered General Brownrigg who had ordered him to move at once to Boulogne, but it was immediately obvious that this was out of the question. The men of a searchlight battery, fighting as infantry, had checked elements of 1st Panzer Division at the appropriately named Les Attaques, but were eventually overwhelmed, and the Royal Tanks had lost heavily in their attempt to reach St-Omer. Nicholson decided to use his regular battalions to hold the old fortifications, with the Rifle Brigade in the east and the King's Royal Rifles in the west: the searchlight batteries and the QVR were to form an outer screen.

Nicholson then received another counter-order, which was 'to override all other considerations': the War Office directed him to take 350,000 rations to the BEF. He duly withdrew part of his infantry from the perimeter to escort the convoy and, late on the night of the 23rd, sent a squadron of the Royal Tanks to reconnoitre the Dunkirk road. The squadron found its way blocked by the rearguard of 1st Panzer Division, whose main body was making for Gravelines: three tanks broke through but the rest were destroyed. No news of this action got back to Calais and a second squadron, with a company of the Rifle Brigade, repeated the exercise without success. The Royal Tanks were now down to twenty-one tanks: there was one panzer division on the Dunkirk road and another, the 10th, began to shell the western and south-western parts of the town prior to launching its own attack. Breaking out was evidently impossible and Nicholson once again disposed his troops for defence, pulling the QVR, searchlight and anti-aircraft detachments back into the perimeter.

The air of tragi-comedy persisted. Major-General McNaughton, commander of 1st Canadian Division, was asked to prepare a brigade group with the aim of reopening communications between Calais and the BEF. McNaughton, travelling with a handful of staff officers and a small escort, actually reached Calais, but was unable to meet Nicholson. He recommended that Calais should be reinforced, but the War Office signalled Nicholson, early on the morning of the 24th, that its evacuation had been decided in principle.

Heavy attacks, which followed shelling on the morning of the 24th, made little headway except at one point in the southern perimeter where a counter-attack soon restored the position. Fort Nieulay was surrendered by its French garrison in the afternoon and the French marines manning Fort Lapin, on the coast to its north, withdrew after disabling the guns. The Germans got into the southern part of Calais during the afternoon and thereafter the defenders were galled by snipers and parties of infantry in the houses. They were also impeded by the consequences of the evacuation order. When they received it the Royal Tanks began to destroy their tanks to prevent them from falling into German hands and by the time Nicholson intervened only nine were left. An ambulance train was at the docks and the movement control staff loaded the wounded aboard ships which already contained some of the equipment for troops fighting ashore. Despite the furious complaints of regimental officers, the ships then sailed for England.

A further message confirmed that evacuation would take place, though the fighting troops would not leave until early on the 25th. Nicholson's perimeter was already creaking, and this encouraged him to fall back on to the shorter line of the Marck Canal and the Boulevard Léon Gambetta, and later still to withdraw behind the canals encircling the old town. Unfortunately, the bridges crossing the canals had not been prepared for demolition by the French and Nicholson's men had neither time nor equipment for the task. Even more unfortunately, the War Office now informed Nicholson that he came under the orders of the French General Fagalde, who had forbidden evacuation: he was therefore ordered to fight on in the interests of inter-Allied solidarity, even though the harbour was of no use to the BEF. He would receive ammunition, but no reinforcements. The message added, inaccurately, that the BEF's 48th Division had already started marching to the relief of Calais.

On the evening of 24 May the Germans dropped leaflets stating that Boulogne had fallen and warning that the bombardment would be intensified unless Calais surrendered. On the following morning the Mayor of Calais, captured when Nicholson withdrew to the old town, appeared, with a German escort, bearing a message asking Nicholson to surrender. The Mayor was detained and his escort returned to German lines.

No formal response was made to these surrender demands and the ensuing bombardment turned much of the old town to smoking rubble. The last remaining anti-tank guns were knocked out and the Royal Tanks soon had only three tanks left. The Rifle Brigade and detachments of the QVR, on the east side of the perimeter, came under heavy ground attack, and an attempt to counter-attack failed when the Bren gun carriers got stuck in the sandhills. Soon the Germans were across the canals, and the defenders fought on around the *Gare Maritime* and the quays. On the western flank, held by the King's Royal Rifles and elements of the QVR, the fighting centred on the bridges, one of which was eventually taken. The Citadel – now Nicholson's headquarters – was firmly held by its Anglo-French garrison and another surrender demand, made on the afternoon of the 25th, received the reply: 'The answer is no as it is the British Army's duty to fight as well as it is the German's.'

At nightfall 10th Panzer Division suspended its attack. During the night the War Office exhorted Nicholson to fight on, adding, 'Have great admiration for your splendid stand.' When the struggle resumed next morning its result could not be in doubt. The defenders were forced back to the northern end of the old town. The Citadel fell in mid-afternoon and Nicholson was captured. Isolated parties of men held out until dark and a few were taken off when HMS *Gulzar* crept into the harbour early on the 27th.

The defence of Calais and the destruction of 30th Brigade remain a matter of controversy. While the defence of the Channel ports did inflict some delay on three panzer divisions and helped give Gort's III Corps time to form a southern flank protecting Dunkirk, it is difficult to agree that holding Calais to the end was a useful gesture, even given the parlous state of the Anglo-French alliance. The navy had continued landing ammunition, evacuating wounded and engaging shore targets until fighting ceased, and could have taken off a substantial part of Nicholson's force. It is impossible to resist the conclusion that Churchill was unable to keep abreast of events: on 27 May, after the town had fallen, he suggested that Gort should have sent a column to relieve it.

On the morning of 26 May, with the fighting for Calais still in progress, Gort had visited Blanchard to discuss lines of retirement and at midnight Operation *Dynamo*, the evacuation from Dunkirk, officially began. It was shrouded in misunderstanding. On the morning of 27 May there was an Anglo-French conference at Cassel to discuss the organisation of the Dunkirk bridgehead: it was decided that Admiral Abrial would have overall command, with General Adam of III Corps in command of the British sector and General Fagalde the French. Gort was surprised to discover that Weygand's representative knew nothing of any evacuation: the French still thought in terms of holding the bridgehead indefinitely.

Gort had good reason for haste. On 27 May the Germans were only six kilometres from Dunkirk and began to shell the town. Worse still, the Belgians, under savage attack on the Lys, were nearing the end of their tether. They had asked for a British counter-attack into the left flank of the German 4th Army – moving across the British front – to relieve the pressure and, although this was probably a more realistic option than British sources generally admit, it was not attempted. On the morning of the 25th the Belgians told Gort of their desperate plight, and made it clear that they could not fall back on to the Yser under air attack and along roads jammed with refugees. The most they could do was to try to hold the line Ypres–Roulers and flood the low-lying country between Ypres and the sea by opening the sluices. Gort sent two divisions to cover his left and hold the Yser towards Dixmude. On the 26th the Belgians warned Weygand that their situation was grave and on the 27th King Leopold requested a ceasefire.

The Belgian armistice sent ripples of fury pulsing through the alliance. Reynaud saw it as a betrayal without precedent in history, Churchill suggested that history would blame Belgium for having caught up Britain and France in her ruin, and an exasperated French population reacted by vilifying

Belgian refugees. The case for the armistice is far stronger than Anglo-French historiography acknowledges. The Belgians had given the British and French repeated warnings of their plight, their army had fought hard against heavy odds and Leopold himself, a controversial figure during and after the German occupation of his kingdom, showed great moral courage by electing to stay on rather than seek safety in England. Nevertheless, in the climate of late May 1940 the armistice was hailed as the triumph of perfidy and in practical terms it opened a huge gap on the BEF's left.

While the BEF fell back across Flanders, fighting over ground familiar to many of its officers from their experiences twenty years before, steps were taken to organise the defence of Dunkirk and prepare its harbour for a large-scale evacuation. Post-war building has not obscured the distinctive character of the Flanders coastline around Dunkirk, despite port developments west of the town. A gently shelving sandy beach ends in sea walls at Dunkirk itself and the little resorts of Malo-les-Bains, Bray-Dunes and De Panne. Undulating dunes, held together by marram grass and sea thistle, run inland and a thin belt of scrubby ground connects the dunes to flat meadowland, intersected by canals. Of these many canals, two are important to our story. One runs parallel with the coast to Veurne (Furnes) and beyond, and another runs inland to the pretty fortified town of Bergues, where it joins the Colme Canal, which curls back to Veurne in the north and down to join the Aa between Gravelines and St-Omer in the south. The land between the canals is flat and easily flooded, and off-road movement by armour was difficult in 1940.

The Battle of the Dunes was fought north of the town in 1658 and after it the French honoured the terms of their treaty with the Commonwealth, and handed it over to the English. It remained in English hands until Charles II, perennially short of cash, sold it to the French in 1662. Dunkirk was the birthplace of Jean Bart (1650–1702), a successful commerce raider who overcame his origins as a member of a local privateering dynasty to become first a captain in Louis XIV's navy and then a nobleman. His rough manners and plain speaking made him few friends at court. When he was taking the Prince of Conti to Poland he narrowly escaped from a much superior force. The Prince remarked that they had been close to capture, but Bart briskly replied that there had been no danger of that: his son had been ready to fire the ship's magazine on his order. He advised Vauban over the building of Dunkirk's fortifications (most of his handiwork has now disappeared), and defended the town against the English in 1694 and 1696. Jean Bart is commemorated by a statue in the square that bears his name and is buried in St-Eloi's church, 200 metres to its north.

It is perhaps appropriate that Jean Bart, who had done so much damage to English merchant shipping, was Dunkirk's local hero, for as the BEF beat its path back towards Dunkirk the fissures in the alliance deepened. The Allies were fortunate that their own disarray was mirrored by serious differences of opinion within the German high command. The co-ordination of Army Groups A and B presented a problem as their area of operations contracted, and there was much debate before the canal line Gravelines–St-Omer–Béthune–Douai was established as the boundary between the army groups.

Rundstedt, as we have seen, had ordered a pause on 24 May to allow his infantry to close up with the armour, but that day Hitler met Rundstedt at Charleville and declared that the armour would soon be needed for operations south of the Somme, and in any case tightening the ring around Dunkirk would constrain the *Luftwaffe*. Further debate resulted in Rundstedt being allowed to decide when the advance should be resumed and he was concerned at the difficulties of operating with armour along the Flanders littoral. It was not until the 26th that Hitler was persuaded that Rundstedt should be told to get on and even then it took sixteen hours for the order to be implemented. Momentum had been lost and time wasted.

Lieutenant-General Sir Ronald Adam was appointed to command the British sector of the bridgehead, which lay in the quadrant defined by the coast, the Colme Canal and the Bergues Canal: the French were to hold the corresponding quadrant west of the Bergues Canal. The bridgehead was divided up into three corps sectors, each with a collecting area outside the perimeter, a section of perimeter and a sweep of evacuation beach: II Corps was allocated the eastern sector, I Corps the centre, and III Corps the western. Preparations were well in hand when, on 26 May, Ramsay, in his headquarters deep below Dover castle, sent the signal announcing that Operation *Dynamo* was to commence.

The possibility of an evacuation of some sort had been considered as early as 14 May, when a BBC announcement asked owners of 'self-propelled pleasure craft between thirty and one hundred feet in length' to send particulars to the Admiralty. A Small Vessels Pool was established and a list of other suitable shipping was drawn up. Forty small Dutch coasters, known as *schuyts*, which had sailed to England when Holland fell, were requisitioned and manned by the Royal Naval Volunteer Reserve. Ramsay held his first planning meeting on 20 May and preparations were well advanced by the time he received orders to commence the operation. The loss of Calais meant that the shortest sea route to Dunkirk, which involved crossing the Channel towards Calais and following the coast, was now covered by shore-based guns. A northerly diversion, which kept ships out of range of the guns around Calais, was duly swept for mines, but it meant that evacuation craft were more fully exposed to air and surface attack.

The air battle was to assume great importance over the days that followed. On the 16th the British Cabinet had agreed to send four extra fighter squadrons to France and Churchill promised Reynaud another six when they met that day. However, Air Chief Marshal Sir Hugh Dowding warned that this would reduce the fighters available for the air defence of Britain to an unacceptable level and there were too few airfields available for the new squadrons as the advancing Germans overran RAF bases. The Advanced Air Striking Force, its Blenheim and Battle squadrons badly knocked about, had pulled back southwards on the 16th. On 19 May it had been decided to send the surviving aircraft of the BEF's Air Component back to England: seventy-five of its 261 Hurricanes had been destroyed and another 120 damaged machines were abandoned in France.

A compromise between Churchill's promise of the fresh squadrons and Dowding's well-founded objections was achieved by basing six additional Hurricane squadrons in southern England and using them over France. This was too small a force to wrest air superiority from the *Luftwaffe* and the Hurricane's limited range meant that it could not operate far inland. Nor did using bombers based in Britain against tactical targets in France have much effect: the time lag between reconnaissance and attack meant that the bombers found it hard to strike targets in the panzer corridor. However, the Channel ports were so close to the airfields of southern England as to give Hurricanes reasonable loiter time, and at both Boulogne and Calais they intervened with effect against the German aircraft attacking troops and shipping. On the first day of *Dynamo* Fighter Command provided sixteen squadrons for the battle over Dunkirk and throughout the evacuation the RAF did its best to impede the *Luftwaffe*'s attacks.

The land battle gave cause for concern. On the 27th, on the western flank, the Germans broke out of their bridgeheads on the Aa and by nightfall the French were back on the line of canals running Mardyk–Spyker–Bergues. In the centre, the British held Cassel and Hazebrouck in the face of heavy attacks, but St-Venant – taken by the Germans on the 24th and recaptured on the 25th – was lost. By nightfall most of the BEF was across the Lys, but the greater part of the French 1st Army had been cut off around Lille. The Belgian surrender on 27 May imperilled the British left flank around Ypres as German forces that had been fighting the Belgians pressed in on the British. Gort's headquarters left Premesques on the 27th, spent a night at Houtkerque and set up at De Panne, in the Dunkirk bridgehead, on the 28th.

The BEF pulled back into the Dunkirk perimeter like a collapsing balloon, with Brooke's II Corps bearing the brunt of the rearguard actions. Abandoned stores and burning vehicles heightened the air of confusion and refugees and

stragglers – French and British – helped jam the roads.
Although repeated fighter sorties helped keep the *Luftwaffe*
off British shipping, the passenger ship *Queen of the
Channel* was sunk and several other vessels were damaged,
persuading the navy to restrict daylight evacuations to naval
vessels and small craft.

The withdrawal of II Corps went on. The garrison of
Cassel – 4th Oxfordshire and Buckinghamshire, 2nd
Gloucesters and 1st East Riding Yeomanry – held out until
it received the order to withdraw on the 29th. Its stand, on
this knuckle of commanding ground, had been immensely
valuable to the BEF, but most of the garrison was killed or
taken as it tried to fight its way back. By midday on the 30th
almost all the surviving BEF was back within the perimeter
and the evacuation was in full swing, with nearly 59,000
men taken off that day. The issue of evacuation placed still
more strain on the alliance. As late as 31 May Admiral
Abrial, nominally in command of the beachhead, did not
know that the evacuation was in process. Weygand, for his
part, demanded that British troops evacuated from Dunkirk
should be landed south of the Somme to continue the fight
and went on to complain that the French were being left
behind. On 31 May Churchill assured the French that the
evacuation would proceed on equal terms – '*Bras dessus,
bras dessous*' was how he put it in his robust French.

Over the next few days the perimeter was held by the
surviving units of II Corps, on the eastern flank, and I Corps,
to the west: most of III Corps had moved to the beaches or
already been evacuated. The Germans managed to cross the
canal near Furnes on the 30th, but were pushed back by a
counter-attack by 1st Coldstream Guards. On the following
day the German high command made operations around
Dunkirk the responsibility of Army Group B, freeing
Rundstedt's armour for operations further south. The
British, too, rejigged their chain of command: Gort was
formally ordered to hand over to a corps commander – his
capture would be 'a needless triumph to the enemy'. He
placed Major-General Alexander in command of I Corps,
whose commander had felt the strain of battle badly, left

him instructions to co-operate with Abrial and to ensure that French troops were evacuated, and left for England on the 31st.

On 1 June some 39,000 members of the BEF held a much-reduced perimeter, with the De Panne area now abandoned. The French sector, to the west, had also contracted and Abrial, in discussion with Alexander that day, envisaged a further withdrawal to an intermediate line running Bergues–Uxem–Ghyvelde. Alexander doubted if this would hold for long and stood forward on the canal for another day, pulling back, on the night of 1–2 June, behind the intermediate line, now held by the French. In the early hours of the 3rd, Alexander and Captain W. G. Tennant, who had been Senior Naval Officer at Dunkirk for much of the operation, were evacuated and Abrial followed suit at dawn on the 4th. The French defenders of Dunkirk, some 40,000 in number, surrendered that morning.

A total of 338,226 Allied troops were evacuated during *Dynamo*, nearly 124,000 of them French. Some of the scenes in the perimeter, on the beaches and in the harbour itself led to further inter-Allied recriminations, but it must be emphasised that the French played a key role at Dunkirk by holding the eastern flank of the perimeter throughout, and then by securing the intermediate line while the remainder of 1 Corps departed. Nor should we forget the French 1st Army in the Lille pocket. Prioux, who had taken command from Blanchard when the latter succeeded Billotte, eventually surrendered on 29 May and the garrison of Lille, under General Molinié, fought on until the 31st when it was accorded the honours of war. If there were panic-stricken French stragglers in Flanders that summer, so too were there determined Frenchmen who held Lille with courage worthy of their ancestors who had defended it against Marlborough.

Although the popular image of the Dunkirk evacuation is one of troops being taken off the beaches by the 'little ships', and the contribution made by the shoal of small craft was indeed memorable, in fact the majority of those evacuated were taken off the Eastern Mole in destroyers. Much of the mole was remodelled after the war, but its seaward end still consists of the criss-cross pattern of ferro-concrete that stands out so clearly in many of the photographs taken at the time. The German pillbox at the base of the mole post-dates the occupation, but stands in the area where the Canadian-born Commander James Campbell Clouston RN, the big, rugger-playing pier-master, established the post controlling entry to the mole. After 'doing noble service' there he returned to Dover for final instructions and the motor launch taking him back was crippled by German aircraft. He waved away another launch in case it was attacked trying to pick them up and tried to swim to a ship a couple of miles away. Realising that he would never reach it, he turned back to the wreckage of the launch, but was never seen again. The official dispatch on the operation, published in 1946, paid him the tribute of declaring: 'Commander Clouston had been of the utmost service in helping the escape of nearly two hundred thousand men under frightful conditions of strain and danger. It was a grief to many that he did not live to see the lifting brought to an end.'

The withdrawal from Dunkirk was certainly not the end of the war in the west, even as far as the British were concerned. The main British administrative area had lain south of the Somme and the German armoured drive had left substantial British forces there. They included two divisions: Major-General Fortune's 51st Highland Division, which had been up in the Maginot Line when the Germans attacked, and Major-General Evans's recently arrived 1st Armoured Division. Evans flew to Amiens on 16 May and, after a number of changes of plan, his division landed at Cherbourg. He was told that the French 7th Army would shortly be attacking Amiens and ordered to co-operate. Accurately briefed on neither the state of French forces on the Somme nor the condition of German formations to its north, and urged on from London, he determined to do his best, and duly sent part of his division to the Somme bridges at Pont-Remy, Picquigny, Ailly-sur-Somme and Dreuil-les-Amiens. His tanks were largely light tanks armed with heavy machine-guns, stiffened by some A13 cruisers, equipped with a 2-pounder gun but thinly armoured and with a poor record for reliability. On 24 May a troop of The Bays and a company of 4th Borders attacked each of the four bridges mentioned above. Only at Ailly was there even slight success: the Borders managed to cross under cover of the tanks' guns and seized the village of St-Sauveur, but could not hold it.

After the fiasco of the 24th, Evans found himself subordinated to General Robert Altmayer, commander of *Groupement* A – soon to be the 10th Army – and asked to assist the French in pinching out the German bridgeheads over the Somme south and west of Abbeville. There were two objectives. On the right, 2nd Armoured Brigade was to assist in the attack on the high ground around Huppy, south of Abbeville on the N28 Blagny road, while 3rd Armoured Brigade was to advance on St-Valéry. The attack was dogged by poor communications. The 10th Hussars did not get word that the attack on Huppy had been postponed and went forward on their own, reaching the village but losing two-thirds of their tanks before falling back. 3rd Armoured Brigade made better progress, getting as far as the outskirts of St-Valéry and pushing a patrol to Boismont, close to the spot where the British had crossed the Somme on their way to Crécy. The casualties sustained in the attack on the Abbeville bridgehead, together with the losses of the 24th and a dreary toll of mechanical breakdowns, effectively left 1st Armoured Division non-operational.

A fresh attempt was made on the bridgehead on 28 May by the French 4th Armoured Division under General Charles de Gaulle. He had jabbed into the flank of the panzer corridor at Montcornet on 17 May, and although his division had suffered some losses there and others on the march, it had been reinforced to bring it up to useful fighting strength, and included two weak battalions of powerful *Char* B heavy tanks, some excellent SOMUA S35 medium tanks, lorried infantry and a strong artillery group. Not only did he have a concentrated armoured punch, which 2nd Armoured Brigade had lacked, but his heavy tanks were proof against most German anti-tank guns. Attacking from the direction

of Hallencourt, de Gaulle overran Huppy on the left and
Bray-les-Mareuil on the right on the first day, and on the
second he took Moyenneville and Bienfay, and swung round
to the line Cambron–Yonval–Caubert. This was the high-
water mark of his progress. A visit to the calvary on the
Monts de Caubert, on the little road connecting the D925 Eu
road with the N28 to Blagny just west of Abbeville shows
the magnitude of his task. The hill itself was well defended
and German guns reached out from across the river. On the
31st de Gaulle withdrew with only eight of his *Chars* B in
battleworthy condition.

While the fighting was going on around Abbeville,
Weygand recognised that his attempt to cut through the
flanks of the panzer corridor was doomed to failure. He
therefore changed his plan, and ordered the last-ditch
defence of the lines of the Somme and the Aisne, with
defensive 'hedgehogs' around farms and villages. By the time
this policy was adopted both the British 1st Armoured and
the French 4th Armoured Divisions had wasted their
strength in battering at the German bridgeheads.

With the bulk of his army now holding the line of the
Somme, Altmayer made another attempt to take Abbeville.
This time its nucleus was the 51st Highland Division, a
good Territorial formation with extra engineers, artillery,
infantry and reconnaissance. The partly reconstituted
French 2nd Armoured Division, an Alpine division and the
remnants of the two light cavalry divisions, which had
accompanied 1st Armoured Division on the 24th, operated
under the command of Major-General Fortune of 51st
Highland. The attack, launched on 4 June, was aimed at the
Monts de Caubert and Yonval from the south, but the by
now familiar misunderstandings and communications
problems robbed it of any real chance of success, although
1st Gordon Highlanders and 4th Black Watch took Cambron
Wood and reached the river opposite Gouy.

Behind the wafer-thin line on the Somme, Major-General
de Fonblanque's lines of communications troops were being
patched together into extemporised brigades. On 23 May
Lieutenant-General Sir Henry Karslake was appointed to

command the lines of communication. Ironside decided to make him corps commander of all troops south of the Somme, but he was succeeded as CIGS by Sir John Dill on 27 May and the appointment was not confirmed. Dill did, however, send Lieutenant-General Sir James Marshall-Cornwall to head a liaison team at Altmayer's headquarters, though he did not tell him that Karslake had also been sent out.

Despite these bizarre command arrangements, the Allied forces holding the line of the Somme might have given a better account of themselves had it not been for a combination of German luck and skill. The German plan for the breaching of the Somme defences and the advance southwards, *Case Red*, gave XV Panzer Corps the task of crossing the Somme downstream of Amiens. The Germans noticed that the two railway bridges between Flixecourt on the north bank and Condé-Folie on the south had not been destroyed, and they pushed infantry across them on the night of 4–5 June. By a cruel mischance, when the French formation holding the sector, a horsed brigade of 3rd Light Cavalry Division, was relieved by 5th Colonial Division, the incoming Senegalese failed to picket the bridges. German infantry secured their southern exits, engineers removed the rails so that tanks could cross on the sleepers and Rommel's tanks rolled over at dawn, cutting the Senegalese to pieces despite their brave resistance and reaching the Amiens–Rouen road by nightfall. The scene of the French misfortune remains evocative. The minor road between Condé-Folie and Hangest is crossed by two railway bridges, although only one now carries track. There is a single bridge across the Somme, but the location of the other, destroyed in 1945, can be clearly seen.

Rommel's action was a stunning blow to the Allied chain of command, leaving Weygand, who visited Altmayer on the 6th, close to hysteria and so alarming Altmayer that he pulled his headquarters back to the outskirts of Paris, losing touch with his troops. Marshall-Cornwall signalled to London that he had lost confidence in French commanders and the events that followed proved him right. Evans had prepared a robust plan attacking the German right flank, but was ordered instead to hold the line of the River Andelle north-west of Rouen. This forlorn attempt to hold a long frontage too thinly met with predictable results and the Andelle line was soon in tatters.

On the Allied left, meanwhile, the Germans broke out of the bridgeheads at Abbeville and St-Valéry, inflicting heavy losses on 51st Highland and 31st Alpine Divisions. General Fortune asked Marshall-Cornwall – who had no executive authority – for permission to fall back on the line of the River Bresle. Marshall-Cornwall received Altmayer's reluctant approval, though the new line was 'to be held at all costs'. The scale of the German breakthrough to the south meant that such a policy could have only one result. The Highlanders held well on the forward edge of the Forest d'Eu, south of the Bresle, and by the time that they were incorporated in the French IX Corps and ordered to fall back by stages their rear was threatened by German armour hooking up towards the coast. A plan to evacuate the corps from Le Havre foundered when the Germans reached the coast on the 10th. The remnants of IX Corps fell back on to the coast at the little harbour of St-Valery-en-Caux and preparations – bedevilled as ever by misunderstandings between the French and British authorities – were made for evacuation.

The last hours of IX Corps were a nightmare. Troops were packed into St-Valéry, blazing after air attacks and shelling. The vehicles which had carried them in were destroyed where they stood and men waited, in mist and heavy rain, for the ships. A substantial flotilla was lying off the coast but few vessels had radio and the risks of coming inshore in bad visibility were too great. The artillery had already been destroyed and the western cliffs, which commanded the town, evacuated, but Fortune considered attacking to recover the high ground with the support of naval gunfire. By this time, early on the morning of the 12th, French troops were surrendering independently and although Fortune declared that his division would fight on whatever the French did, he received a note from the corps commander saying that fire would cease at eight o'clock that morning. A German tank arrived at his headquarters shortly afterwards. Some 8,000 British troops were captured at St-Valéry. A few managed to make their way to the Seine and some were evacuated from Veules-les-Roses. Arkforce, a detachment of the division formed at Arques-la-Bataille, near Dieppe, had managed to reach Le Havre before the Germans encircled IX Corps, and reached England safely.

The last act of the campaign, as far as the British were concerned, occurred so far beyond the borders of Flanders, Artois and Picardy as to lie properly outside the scope of this chapter. However, it is a loose end that deserves tying off. Over a week before the Highland Division had been swamped at St-Valéry, General Brooke had been ordered back to France to command a new II Corps consisting of 1st Armoured and 51st Divisions and the composite Beauman Division, made up of line-of-communication troops. He would be reinforced by 52nd Lowland Division, and the 1st Canadian Division and the reconstituted 3rd Division would follow soon afterwards. Brooke was strongly against the venture, arguing that it could achieve nothing from a military point of view and might even produce a second Dunkirk. He reached Cherbourg on 12 July and arrived at Le Mans, where Headquarters, Lines of Communication Troops was still sited, on the 13th.

Brooke ordered de Fonblanque to set about the evacuation of the 100,000 lines of communication troops, retaining only sufficient personnel to deal with his four divisions. He sent Karslake home, ending that unfortunate officer's brief sojourn in France. When he met Weygand on the 14th, Brooke heard that the French army was no longer able to offer organised resistance and was fast disintegrating. Weygand added that Churchill and Reynaud had agreed that a last position should be taken up across the neck of the Brittany peninsula. Although Brooke had heard about the plan for a 'Breton Redoubt' before he left England, he was deeply disturbed to get the news from Weygand, all the more so because the French Commander-in-Chief clearly had so little confidence in the project.

The politicians agreed on 13 June that the Brittany scheme was no longer a practical proposition, although on the following day Weygand urged Brooke to sign a joint declaration stating that they would press on with the venture as ordered, though they believed it to be unsound. It was not until he phoned Dill that Brooke discovered that the idea was already dead, and was given permission to evacuate all British troops except those under the command of the French 10th Army. This order was soon varied to include all British troops, and Brooke gave Marshall-Cornwall command of those with the 10th Army and passed on London's ambiguous instruction that he should co-operate with the French withdrawal but steer his own course so as to be able to embark for England.

On 15 June the French government, now at Bordeaux, decided to ask the Germans what armistice terms they would offer. Reynaud, who had favoured fighting on in Algeria, tried to buy time by requesting Churchill to release France from her undertaking not to conclude a separate peace and again asked the Americans for help. Churchill agreed on the condition that the French fleet was sailed to British harbours immediately and authorised Brooke to increase the pace of his evacuation. On the 16th Reynaud resigned and his successor, Marshal Pétain, immediately requested an armistice. There was a scrambled evacuation from Cherbourg, Brest and St-Nazaire, and much equipment, which would have been useful to a Britain pitifully short of up-to-date weaponry, was destroyed to prevent capture. It was an ignominious end to an enterprise impelled more by desire to show inter-Allied solidarity than by military logic.

The armistice was duly signed on 20 June, ending a campaign that had cost the Germans just over 156,000 killed, wounded and missing. French losses cannot be computed with certainty, but were in excess of two million, the overwhelming majority of them captured. The British lost just over 69,000, about half of them in the area south of Calais, in the defence of Channel ports and the often forgotten fighting on and south of the Somme.

The armistice brought peace of a kind to the north. The flood of abandoned equipment was mopped up, often to good effect: many of the thousands of 75mm field guns which fell into German hands saw further service, sometimes in fortifications intended to keep the British and Americans out of France. It soon became clear that the *départements* of Nord and Pas de Calais would suffer a more rigorous occupation than the remainder of occupied France. They were the responsibility, through the *Oberfeldkommandantur* in Lille, of the Military Governor of Brussels, while the rest of the occupied zone was answerable to the Military Governor of Paris. The north-west was initially a 'forbidden zone' separated from the rest of France by an internal frontier on the Somme. This was abolished in December 1941, and from March 1942 legislation passed by Pétain's Vichy regime was deemed applicable to the Nord and Pas de Calais.

The proximity of England made the north-east vital to the Germans throughout the war and Dr Fritz Todt's construction agency was soon busy. The triangle St-Omer–Boulogne–Calais contained several *Luftwaffe* bases and the Channel ports were havens for E-boats. In 1941 work began on the Atlantic wall. This consisted of coastal batteries, observation posts and fire control centres, later supplemented with anti-landing devices on the beaches. Although a number of bunkers have given way to industrial developments a good selection of the Todt Organisation's handiwork still remains.

Just north of Dunkirk, with a large blue-and-white water tower about 400 metres to its east serving as a useful landmark, is the Fort des Dunes, used, like so many old forts, as an assault course area until the French army abandoned conscription. Nearby stands the Zuydcoote battery, built on top of a pre-war French gun position. There are four concrete gun positions, with a two-storey observation post. A French military cemetery includes the grave of Général de Division Janssen, killed commanding the 12th Infantry Division in 1940. Twelve resistance workers, one of them a woman, were shot in the fort's ditch in September 1944. Further along the coast there is a battery position west of Grand

Fort-Philippe and an observation post, heavily disguised as a church tower, at Les Huttes d'Oye. North of the D119 coast road between Le Fort Vert and the Hoverport is a track which leads to the Oldenburg Battery, originally mounting two 24cm Russian naval guns. Further east, past the Leisure Centre – which started life as a Todt Organisation camp – is the Waldam Battery, with two gun positions, a rotating armoured turret, a fire control centre and troop shelters.

Some German guns were sited to take on shipping passing close to, or attacking, the coast. Heavier pieces were intended to interdict the Straits of Dover or to hit southern England. Many of these were mounted on rails and protected by what the Germans called a *Dombunker*, a 'Cathedral Bunker' with a sloped roof designed to deflect bombs. There is a surviving *Dombunker* on the north-west edge of Calais, on land now owned by Peugeot. Cap Griz Nez was heavily fortified. Between Audinghen and Audreselles on the D490, near the hamlet of Inglevert is the Todt Battery, which mounted four 38cm guns and remains in good condition. One of its bunkers now contains the *Musée Mur d'Atlantique*, with some 2,000 items of equipment, some post-dating the Second World War and many in need of conservation. They include a Krupp 28cm K5 railway gun, built in 1941, which could reach out to eighty-six kilometres with a special projectile and could fire up to fifteen rounds an hour. To its north are a radar bunker and two gun positions, while at the tip of Griz Nez are the remains of the *Grosser Kurfüst* Battery, which mounted four 28cm guns. South of Griz Nez, at the junction of the main N1 and the D242, is the Friedrich August Battery, whose bunkers housed three 30.5cm guns.

Wimereux, north of Boulogne, contains a bunker intended to serve as Hitler's headquarters during Operation *Sealion*, the invasion of England, and subsequently used as part of the Atlantic Wall defences. There are more German bunkers round the French Fort de la Crèche, and amongst the defences of Boulogne itself is a German battery built on a French fort overlooking the Hoverport. Further down the coast, on the Pointe du Touquet, near the resort town of

Le Touquet-Paris-Plage, is another battery and at the nearby Stella-Plage there are a fire control centre and three gun positions.

More sinister military architecture lurks behind the coastline. The Germans launched their V-weapons against England from sites in the Pas de Calais. The V-1 flying bomb was launched from a sloping ramp forty-five metres long and delivered 850 kilos of explosive to a range of 250–300 kilometres. Of the 22,384 launched, just under half were aimed at England and the remainder, in the last months of the war, at Antwerp, Liège and Brussels. The V-2, in contrast, was a rocket, launched vertically, carrying up to a ton of explosive between 270–350 kilometres. The V-3 was different yet again. This was a launching tube a hundred metres long, with boosters fitted at intervals to accelerate the projectile.

A common problem faced German permanent missile-launching installations in northern France. They were impossible to conceal and, once detected, were subjected to pulverising aerial bombardment. The Germans eventually changed their policy and obtained better results from mobile launchers than from purpose-built installations – however solid. At least two of the V-weapon installations survive in reasonable condition. In the Forest of Eperlecques, just off the D221 near Watten, thirty kilometres from Dunkirk, is the first V-2 launch bunker to be built in France. A massive rectancular concrete bunker was begun in March 1943, but on 27 August 185 Flying Fortresses dropped 366 2,000-pound bombs on the site, doing much damage and killing many of the Russian slave labourers who worked there. Successive raids persuaded the Todt Organisation that the site could no longer be used for its original purpose and it was developed instead as a liquid oxygen plant. Further raids, including two which dropped 12,000 'Tallboy' bombs, wreaked even greater havoc. Interestingly, even the Tallboys failed to pierce the bunker's roof: damage made by the one bomb to hit it was repaired, though another, which fell nearby, left a huge crater, which is now a pond. The site, taken by Canadian troops on 6 September 1944, is open to the public. One is impressed not by the scale of the damage but by the relatively untouched interior: the 216-ton doors slide smoothly shut at the touch of a button.

A different technique was used at the V-2 site at Wizernes, near St-Omer. Here connecting tunnels beneath the chalk led to a huge octagonal chamber covered by a protective concrete dome (the cupola from which the site takes its name) seventy-one metres in diameter and five metres thick, with a weight of 55,000 tons. The work was carried out by conscripted French labour and by Russian prisoners of war, the latter treated appallingly. Work started in late 1943 and the Allies began to bomb the site in March 1944. Although initial attacks were ineffective, on 22 June a raid by USAAF B-17s did serious damage to the roads and railways supplying the dome, and on 17 June RAF Lancasters dropped 6-ton 'Earthquake' Tallboy bombs, which undermined part of the dome's support. Hitler ordered the abandonment of the place at the end of the month. The La Coupole Museum was opened in 1997 after ten years of work on the site. The visitor guided by headphones, walks down a long tunnel to lifts which lead up to the chamber beneath the dome. Part of the exhibition is devoted to space flight and part to the Second World War history of the site, with some especially harrowing details of the plight of the slave labourers who worked there. The only V-3 site is on the D249 between the villages of Leubringhen and Landrethun-le-Nord, just east of the N1 near Cap Griz Nez. This originally consisted of ten groups of five launching tubes beneath a concrete carapace. It was heavily bombed – and still bears the scars of a Tallboy hit – and was largely demolished by the British when they took the area in 1944.

The heavy hand of occupation lay on the industries of the north-west. The mining and chemical industries came under German control and from February 1944 most of the large industrial concerns in the area were compelled to give at least three-quarters of their output to Germany. Outright opposition by the workforce was crushed mercilessly. In May–June 1941 a miners' strike led to hundreds of miners being deported. As the war went on, growing numbers of Frenchmen were sent to Germany for forced labour – *Service du travail obligatoire*, or STO. Many went underground and the Germans mounted large cordon-and-search operations to catch reluctant workers. There were worse fates than forced labour. Anti-Semitism had been well-developed in France long before the political polarisation of the inter-war years and French Jews suffered terribly during the occupation. In September 1941, for example, the little community of Polish Jews living in Lens was rounded up by French and German police: most of them perished in Auschwitz.

Before the war the Moscow-dominated Communist Party had been strong in the industrial towns of the north, but because of the Ribbentrop–Molotov Pact it was unable to adopt a frankly anti-German stance. The German invasion of Russia in June 1941 changed all that and the Communists soon began to play an active role in the Resistance. It is an over-simplification to speak of the Resistance as a single, unified organisation, because it began by being, and to some extent remained, diverse. It started as an 'Underground Railway' intended to help Allied aircrew escape into the Unoccupied Zone of France and thence into Spain. A measure of control, together with some weapons and specialists, was provided by the British Special Operations Executive.

Nevertheless, Frenchmen who wished to play their part in resisting the occupation joined a specific organisation rather than the Resistance in general. The OCM – *Organisation Civile et Militaire* – was strong along the coast and in Arras, but in 1943 it was crippled by the arrest of its leaders. The Gaullist *Voix du Nord* was well supported in Arras, while the socialist *Libération Nord* appeared in 1943. The Communists became increasingly influential. Their *Organisation spéciale de combat* took the field in 1941 and in July 1942 it took on its better-known title *Francs-tireurs partisans* – FTP.

The Resistance was strengthened by escaped Russian workers and by men on the run from STO, and assassinations

and acts of sabotage were increasingly frequent: when the
Allies invaded Normandy in 1944 there were widespread
attacks on telephones and railways. Yet even at this late
stage the Germans were a dangerous enemy: on 11 June three
companies of the FTP were wiped out at Bourlon. The price
paid by the Resistance during the occupation is nowhere
better demonstrated than at Arras, where memorials weave
together history, ancient and more modern. Just to the north
of Vauban's citadel, whose main gate with the emblem of the
Sun King glares out on to the *Avenue du Général de Gaulle*,
is a tiny road which passes the CWGC's *Faubourg d'Amiens*
Cemetery and leads to the *mur des fusillés*. A symbolic post
in the centre of the ditch marks the spot where 206
Resistance workers were shot by firing squad and the brick
walls of the ditch bear plaques commemorating those killed.
Each gives the individual's name and home town, together
with the Resistance organisation to which he belonged. It is
a noble fellowship of death. Colonels, teachers and mayors
are remembered alongside painters and blacksmiths, and
there is one very English name – James Tozer – probably the
son of an English soldier who had settled in France after
the First World War. Above all, this chilling spot testifies
to the role played by miners in the local Resistance. With
their frequent strikes and strong cohesion the miners of the
north had caused successive French governments more than
a little concern. But, as the firing wall at Arras shows, they
were even more of a thorn in the flesh of the Germans.

The *Faubourg d'Amiens* Cemetery contains both the
Arras Memorial to the Missing, with some 35,000 names of
those missing in the sector between 1916 and August 1918,
and the Flying Services Memorial, commemorating the
missing of the Royal Naval Air Service, the Royal Flying
Corps and the Royal Air Force: it includes the names of the
RFC VCs Mick Mannock and Lanoe Hawker, as well as
eleven other holders of the Victoria Cross. Amongst those
commemorated on the cloister walls of the main Memorial
to the Missing is Second-Lieutenant Walter Tull of the
Middlesex Regiment (Bay 7). He was the grandson of a slave
and his father had arrived in England from Barbados in 1888.

After their mother and father died, Walter and his brother
Edward were brought up in an orphanage in the East End of
London. Walter began his apprenticeship to a printer, but
after playing for the amateur club Clapton he was signed by
Tottenham Hotspur, becoming only the second black
professional footballer in Britain. He went on to play
for Northampton Town, though in October 1909 the
Northampton Echo reported that 'A section of the spec-
tators made a cowardly attack upon him in language lower
than Billingsgate . . . Let me tell these Bristol hooligans that
Tull is so clean in mind and method as to be a model for all
white men who play football . . . Tull was the best forward
on the field.' In 1914 he enlisted in 17th Middlesex (1st
Football Battalion). Soon promoted to sergeant, after serving
on the Somme he was commissioned in May 1917 despite
official prohibition against 'any Negro or person of colour'
being made an officer. He was killed in action near Favreuil
while serving with 23rd Middlesex on 25 March 1918, but
his body could not be recovered. He was the first black
officer in the British army and the panel bearing his name
has become the focal point of pilgrimage for those who
recognise his remarkable achievement.

Let us return to the Second World War. There was a
growing belief, amongst Frenchmen and Germans alike,
that the Allies would launch their invasion across the
beaches of the Pas de Calais. They were, after all, a logical
choice, only a short sea-crossing from England, with V-
weapon sites within easy reach of the coast, and the German
industrial heartland of the Ruhr was closer than to any other
likely invasion area. The pre-invasion deception plan
heightened the air of expectation, and radio intercepts
suggested that 'Army Group Patton' was concentrated in
Kent and Sussex: Rundstedt, Commander-in-Chief West,
remarked that 'if I was Montgomery I would attack the Pas
de Calais'.

The German forces preparing to meet the invasion formed
Field-Marshal Rommel's Army Group B, its 7th Army in
Normandy and its 15th covering the Channel coast. The
Allied descent on Normandy caught the Germans at a

disadvantage, but it was not until late August that the Allies broke out. There were bridgeheads over the Seine by 20 August and Paris was liberated on the 25th. The Germans had hoped to hold the Kitzinger Line, running from Abbeville to Amiens along the Somme, and thence to Soissons, Châlons, St-Dizier, Chaumont, Langres, Besançon and the Swiss border, but defences had only been prepared on the Abbeville–Amiens section, and the speed of the Allied breakout and shortage of German troops in the north – so many had been sucked into the Normandy fighting – enabled the Allies to cross the Kitzinger Line without drawing breath. The British took Amiens on 31 August, surprising the headquarters of the German 7th Army and capturing its commander, General Hans Eberbach.

In early September Montgomery's 21st Army Group liberated the north-west. The Canadian 1st Army advanced up the Channel coast, with the British 2nd Army moving across Picardy on its right. Near Bapaume 2nd Army joined the US 19th Corps, the left-flank formation of the US 1st Army, part of General Omar Bradley's 12th Army Group. This broad front advance was not what Montgomery had sought. He had argued in favour of a concentration of 'a solid mass of some forty divisions' on the Allied left, driving onwards to the Ruhr by the most direct route, but was overruled by Dwight D. Eisenhower, the Supreme Allied Commander.

Debates over Allied strategy were the last thing in the minds of the inhabitants of the war-worn towns of Flanders, Artois and Picardy as their liberators arrived in September 1944. Only on the coast, where the rump of the 15th Army still held the Channel ports, was there serious fighting. Lieutenant-General Ferdinand Heim, unemployed since his dismissal as a corps commander in Russia in 1943, was appointed military commandant of Boulogne shortly before the Canadians attacked. Although Heim and all his officers had sworn to hold out to the last man, Boulogne surrendered on 23 September: Heim was to argue that his situation was hopeless and there was no purpose in sacrificing lives. Lieutenant-Colonel Schroeder surrendered Calais on 1

October after five days' siege. The batteries on Cap Griz Nez were taken by two Canadian battalions on 28 September: Dover was freed from the danger of bombardment for the first time in four years. Dunkirk, whose garrison was believed to be more enthusiastic, and which lacked the strategic importance of Boulogne and Calais, was simply 'masked' rather than attacked directly. It was still holding out, watched by 21st Army Group's Czech Armoured Brigade, when German forces in the west surrendered.

Considerable destruction was inflicted on the north-west during the Second World War. The coastal strip suffered most heavily, although bridges, railway centres and V-weapon sites further inland had been heavily bombed, and some factories had been hamstrung by the removal of plant. The business of post-war reconstruction was made no easier by continued industrial unrest. There were major miners' strikes in 1947, 1949 and 1963, and the number of working mines shrank steadily.

In the 1990s politicians and population alike hoped that industrial diversification, together with the Channel tunnel and its high-speed rail link would restore prosperity to the north-west. At the time of writing, however, the cancer of recession still gnaws its way into the blighted industrial suburbs of Lens, Lille, Roubaix and Tourcoing. Unemployment runs at well above the national average, encouraging emigration, which makes the Pas de Calais 'a sort of demographic reservoir for the rest of France'. Graffiti urge voters to support the PCF (the French Communist Party); but a dangerous whiff of hard-right anti-immigration ideology seeps from the gaunt terraced houses and grey tenements.

But all is not gloom. The 'Live in the Pas de Calais' programme has had a visible impact on the environment, with the establishment of regional national parks and long-distance paths. Tourism has increased and now generates more money than local agriculture. While the majority of English visitors move quickly across the north, perhaps spending a restorative night in Arras or stopping off at a hypermarket in Calais to scorch through their remaining francs, some spend longer, for the north has much to offer once the relics of King Coal are left behind. Arras itself is an attractive town, with its two adjacent squares, the *Place de Héros* and the *Grande Place* attractively restored after the damage they sustained in the First World War. The town hall gives access to tunnels, known locally as *boves*, and part of the tunnel complex leading eastwards to the 1917 front line is due to open in 2008.

It is hard for even the most transitory visitor not to catch the extraordinary poignancy of the area, sprinkled as it is with the white-on-green Commonwealth War Graves Commission signposts and place names that read like the battle honours on regimental colours. Perhaps future generations will cast only a bemused glance at the wedge of Vimy Ridge, the fingers of chalkland on the Albert–Bapaume road, the meandering Somme, and the acres of white headstones incised with half-forgotten cap badges and symbolising so many personal tragedies. There are no certainties in human affairs, but we will probably see no more British soldiers slogging their way up to Arras with rifle and pack, or calling for a *vin blanc* and an omelette as a welcome antidote to the monotony of bully beef or Maconochie's tinned stew. And the world is none the worse for that.

V

Lorraine

LORRAINE LIES LIKE the trace of a mighty fortress between France and Germany. From the Sarre, its eastern border, an invader must cross the glacis of the Lorraine plateau before encountering the Moselle, with the Oolitic limestone of the Moselle heights rising like ramparts behind it. Then comes a second glacis, the Woëvre plateau, with the Corallian limestone of the Meuse heights at its back. There are bastions on both flanks. To the north, the Ardennes rise sharply from the Meuse, while to the south the Vosges frown out across the Alsace plain and the Rhine valley.

Forests once thickened the obstacles as a palisade defends a ditch. Today's extensive woodlands between Nancy and St-Mihiel and on up to Spincourt are remnants of the great forest that once covered the Woëvre. On the Meuse mixed woodlands give way to conifers, planted in a deliberate attempt to hide the ravages of war. The border between Lorraine and Champagne is marked by the forested hills of the Argonne, crossed by narrow defiles.

Roads and railways followed creases on the ground. A north-south route, important when Spain held the Netherlands, ran up the Meuse to Stenay and Sedan. The east-west routes made Lorraine a military corridor, stamped flat by successive armies. One Roman road ran from Paris to Rheims, through the Clermont gap in the Argonne, to Verdun and on to Metz, while another went from Rheims and Bar-le-Duc to Commercy, Toul and Nancy. The railway entered Lorraine through Sedan and Montmédy, and went on to Thionville, where one line swung up to Luxembourg

LORRAINE

✿ Seré de Rivière or older forts

⬡ Maginot Line artillery forts

⋯⋯ Frontier 1871-1919

while another took the Moselle down to Frouard – junction with the Châlons–Paris line – and then followed the Meurthe to Lunéville. A frontier line connected Metz with Alsace.

It is surprising that a central axis, linking the military centre of Châlons-sur-Marne with the fortress of Metz, was not completed before 1870. Here the rise and fall of the land provides a clue: the Meuse heights were one obstacle and the Woëvre, with its heavy, poorly drained soil, presented its own challenges. The lack of a central railway line cost France dear in the Franco-Prussian War and it is yet more evidence of the way that the ground helps shape what happens on it.

What lies beneath the ground had an importance all of its own. The triangle Longwy–Briey–Metz was once a thriving centre of the iron industry, although its towns now resemble those of the mining belt of the north-west. There is a rich coalfield around Carling and St-Avold, and between Dieuze and Château-Salins on the Lorraine plateau are salt mines that once produced two-thirds of France's salt supply.

The name Lorraine results from the Treaty of Verdun, signed in 843, when Charlemagne's empire was divided between his three grandsons. Lothar received the central portion – *Lotharii regnum* in Latin, Lothringen in German and Lorraine in French. The 'reunion' of Lorraine to the French crown came about gradually. For much of the Hundred Years War the Counts, later Dukes, of Bar, whose territory, the Barrois, lies on the limestone massif between Toul and St-Dizier, supported the Valois. Duke Edward, his brother John and his nephew Robert fell at Agincourt. In the confusion accompanying the conflict between the houses of Burgundy and Armagnac, Lorraine was torn by brigandage. English and Burgundian war bands, German and Breton soldiers of fortune, dispossessed peasants and local noblemen with an eye for the main chance, looted and pillaged across the whole of the region.

On 6 January 1412, Joan of Arc was born in Domrémy, on the Meuse between Neufchâteau and Vaucouleurs. Her birthplace, the thick-walled house of a well-to-do peasant family, is open to the public and there is now an extensive visitors complex. The church, though much modernised, dates from her time and the local war memorial capitalises on her presence. Just outside Domrémy, on the Coussey road, is the Basilica of Bois-Chenu, on one of the spots where Joan heard the 'voices' of St Catherine, St Margaret and St Michael. The basilica was built between 1881 and 1926, and contains frescos telling the story of Joan's life. It is an appropriate stop for a visitor to the battlefields of Lorraine, for it is a place of prayer for peace and for soldiers, living and dead.

In 1429 Joan rode out of the French enclave of Vaucouleurs, whose governor, Robert de Baudricourt, initially sceptical of her claims to be sent by God to lead the French to victory, gave her a small escort for the dangerous journey to Chinon, where she was to meet the Dauphin, the future Charles VII. She rode out through the Porte de France on 23 February. Traces of the original gate remain between the Rue de l'Observatoire and the Avenue Jeanne d'Arc, and in the nearby ruins of the castle where Joan met Baudricourt is a lime tree, which is reputed to date from her time. Several of the towers of the old town wall remain. Driving north out of the town along the Avenue André Maginot, the visitor passes between the Tour du Roi, on his left, and the Tour des Anglais, on his right. The equestrian statue of Joan, which stands outside the town hall, has a chequered history. It was erected in Algiers in 1951, badly damaged shortly before French withdrawal in 1961 and re-erected in Vaucouleurs five years later.

In the second half of the fifteenth century Lorraine was torn between the power centres of France and Burgundy. In 1477 Charles the Bold was defeated and killed outside Nancy, cut down by a Swiss halberd, which clove him through helmet and skull. There was scarcely more peace in the first half of the sixteenth century. In 1544 an Imperial army marched from Metz to Commercy, looting and burning as it went, until checked by the resistance of St-Dizier, a feat which earned it the Legion of Honour, belatedly awarded in 1905. In 1552 Henry II came to an agreement with the Protestant princes of Germany and invaded Lorraine with an army of 40,000 men, occupying the 'three bishoprics' of Metz, Toul and Verdun. The Emperor Charles V responded with an attack on Metz, but the city was energetically defended by Francis, Duke of Guise.

During the Wars of Religion Lorraine was a corridor for the transit of German troops who supported the Huguenot cause. Its own nobility was divided between the rival camps. In 1562 a party of Huguenots, commanded by François de Béthune, father of Henry IV's chief minister Sully, tried to take Verdun by surprise to permit the passage of a column of German heavy cavalry on its way to relieve Gaspard de Coligny, under siege in Orléans. The Bishop of Verdun got wind of the enterprise and the Huguenots were repulsed. Between 1584 and 1589 the Protestant garrison of Jametz, on the Loison south of Montmédy, raided into the countryside towards Verdun. In January 1588 a column from Verdun took Stenay, cutting Jametz off from the Protestant stronghold of Sedan, and Jametz fell the following year. It had proved difficult to take because of the efforts of Jean Errard de Bar-le-Duc, an engineer in the service of the Protestant Lamarck family, Dukes of Bouillon and Princes of Sedan. Errard was a pioneer of bastioned fortification and was responsible for transforming many of the fortresses of the north from their traditional high-walled aspect to the low, geometrical trace that was to become standard in the seventeenth century.

In the seventeenth century Lorraine echoed to the tread of French, Spanish, Austrian and Swedish armies, using it for transit between theatres of war. Jacques Callot was born in Nancy in 1594 and his engravings, *The Miseries of War*, provide an insight into what went on in Lorraine during the Thirty Years War. Here trees groan beneath the weight of hanged soldiers, dangling like so much sinister fruit: there a village is ravaged, its inhabitants butchered and their homes burnt. In 1637 Louis XIII sent the Marshal de Châtillon to 'clean the Meuse' of its Spanish garrisons. Châtillon marched north along the river to Stenay, held by the French since 1632, and a column under Henri de Bellefonds worked its way between Meuse and Chiers, taking a score of little garrisons – Vilosnes, Sivry, Louppy-sur-Loison, Inor, Chauvency, Brouennes and La Ferté. These were all medieval works and had little hope of holding out once Bellefonds's cannon came up. They were demolished to prevent the Spaniards from re-establishing themselves, but there are substantial remains of the tower at Louppy and old walls are hidden behind a hedge of *Leylandii*. Another of Châtillon's columns, under the Marquis of Feuquières, set off for the more modern fortress of Damvillers. His gunners made slow progress but his sappers did rather better, undermining first a ravelin, then one of the bastions: the place capitulated on 24 October. No sign of the walls now remains, but in the town square there is a statue of Marshal Etienne Maurice Gérard, a local boy who joined a volunteer battalion in 1791 and was a colonel in 1800. Ennobled by Napoleon, he was a corps commander at Ligny in 1815. A supporter of the 1830 revolution, Gérard became a marshal that year and commanded the besieging force at Antwerp in 1832. He lived just long enough to bask in the favour of Napoleon III and died in 1852.

Between 1650 and 1654 there was fighting between the French and the followers of Charles, Duke of Lorraine, who supported Condé and Turenne during the *Frondes* and allied himself with Spain. The last act of this drama was the lengthy siege of Montmédy. Perched high on a rocky outcrop above the Chiers, the castle of Montmédy occupied a position of enormous natural strength. The site was first fortified by Arnold, Count of Chiny in 1221, and passed to the Duke of Luxembourg in 1395 and on to the Duke of Burgundy in 1462. When Charles the Bold's daughter Mary married Maximilian of Hapsburg, it became Austrian and on the abdication of the Emperor Charles V it was amongst the northern provinces given to Philip II of Spain.

Montmédy had been strengthened by new bastioned fortifications when the French attacked in 1657. Its governor, Jean d'Allamont, came from a Lorraine family, which had governed Montmédy for the Spaniards for generations, and he made a resolute defence against a much-superior French force commanded by Marshal Henri de la Ferté. Louis XIV was present at the siege and Vauban, not long in his service – he had been fighting for the *frondeurs* when captured by Ferté in 1653 – reckoned that the siege was one of the hardest of his entire career. D'Allamont was mortally wounded by a roundshot while directing operations on the St Andrew Bastion and Montmédy duly fell.

Modern Montmédy consists of two distinct towns, the high town, inside the fortress on its knoll, and the low town sprawling out along the Chiers. The fortifications of the high town, two distinct layers with a deep ditch between them, are in excellent order and the tourist office issues a short guide, which enables the visitor to walk the ramparts. There is a staid eighteenth-century church within the fortress and a good museum, which tells the history of fortification and celebrates the local artist Jules Bastien-Lepage, who is buried (and commemorated by a Rodin statue) at Damvillers.

The extant fortifications of Montmédy are a mixture of styles, with some of the medieval work still visible beneath St Andrew's Bastion. Much of what we now see is the work of Vauban, who fortified the place when it became French by the terms of the Treaty of the Pyrenees in 1659. The high town was besieged in 1814 and in 1870 German bombardment caused extensive damage to the buildings within the enceinte: the town hall and sub-Prefecture were rebuilt in the lower town. Montmédy was tied into Seré de Rivière's frontier defences, but both its location – away from anticipated German thrusts – and exposed position meant that little new work was carried out. In August 1914 its garrison attempted to break out to reach French lines, but was caught in the open and lost 600 men killed and 8–900 captured. With Verdun only forty kilometres away, the town was an important German supply base during the First World War. It resumed some of its former military importance after 1936, when the so-called 'Montmédy bridgehead' was built around it, extending the Maginot Line to the north. The sight of Montmédy citadel against a bright autumn sky is one of life's little pleasures, but the town itself seems to be dying on its feet: in the spring of 2007 a beautiful building dated 1635 appeared to be falling down.

It was not until 1766, with the death of Stanislas Lesczynski, Louis XV's father-in-law, that Lorraine became definitively French. It was natural that the Lorraine frontier should be threatened in 1791, when the stiff armies of old Europe moved to the rescue of Louis XIV. In the summer of 1792 the Duke of Brunswick crossed the frontier between Sedan and Metz, and passed across the barriers of Lorraine, taking Longwy and Verdun, and marching into the Argonne. The decisive encounter of that campaign took place at Valmy and is examined in a later chapter.

After the Napoleonic wars the fortresses of the eastern frontier were improved in an effort to keep pace with new technology. There were two major lines of defence, originally laid out as part of Vauban's *pré carré*. The first, lying in the Moselle and Meurthe valleys, comprised the fortresses of Thionville, Metz and Nancy. The second, on the Meuse and the upper Moselle, included Montmédy, Verdun and Toul. In the 1830s and 1840s existing works were strengthened and refurbished, and in the 1860s engineers responded to the threat posed by rifled artillery by planning outlying forts, which would keep a besieger out of range of the main body of the fortress.

Metz, corner of the Lorraine frontier, was at the forefront of the new fortification. Its position, at the confluence of the Moselle and Seille, and the junction of three major Roman roads, had encouraged the Gallo-Romans to fortify it as a bulwark against Germanic invasions and its defences had been improved over the centuries. In 1552 it was besieged by Charles V with an army of 80,000 men and 114 guns, but the garrison, under Francis of Guise, held out against repeated assaults. The city's resolve gave it the nickname *Metz la Pucelle* – Metz the Virgin. Parts of the old city walls remain, notably the *Porte des Allemands*, a fortified bridge with defences on either side of the Moselle.

The Cathedral of St Etienne dominates Metz. Built of yellow Jaumont stone, it has the third highest nave in

France and a wonderful selection of stained glass, including a fourteenth-century rose window. To its south-west, where the parkland around the Esplanade runs down to the Moselle, is the eighteenth-century Palais de Justice. The nearby arsenal, once surrounded by Vauban's citadel, houses both a centre for the performing arts and a new hotel, sympathetically integrated into this fine structure. On the other side of the cathedral is the Museum of Art and History, which includes the military collection of the painter JOB (Jacques Onfroy de Bréville).

With the advent of rifled artillery, Metz's position was a disadvantage, for the city could be engaged by the guns on the high ground around it. Engineers sited four new forts, St-Julien to the north-east, Queuleu to the south-east, and St-Quentin and Plappeville to the west, to form an outer carapace. Metz was well worth defending. Not only was it an important route centre, but it contained the Artillery and Engineer *Ecole d'Application*, where young officers did their practical training (formerly a convent, and now the mess for the officers and senior NCOs of the garrison), and its arsenal was full of war material. The forts were not quite complete when war broke out in 1870. Bazaine, ill-starred commander of the Army of the Rhine, declared that they were 'unfinished . . . incomplete . . . in no state to render serious or prolonged service'. The commandant of Plappeville found contractors at work when he arrived and reported that the place was like a builder's yard.

The Metz forts had a varied history. Taken by the Germans in 1870, they were developed and strengthened by new works when the Germans turned Metz and Thionville into the *Moselstellung*, on which Schlieffen's wheeling march was to pivot. The forts were taken back into French service in 1918 and given suitably Gallic names – Fort *Kronprinz* became Fort Driant, named after a hero of Verdun. In 1944 the Germans held them against Patton's 3rd Army, which had the unenviable experience of fighting its way through a defensive system laid out by the French, improved by the Germans, restored by the French and finally garrisoned by the Germans.

It is not hard to find forts around Metz – though it is more difficult to visit them legally and safely. Here a cautionary tale is in order. When I was old enough to know better I found myself wandering around Fort Driant with a companion, fondly imagining that the 'Keep Out' signs were addressed to someone else. We slid down the ditch and climbed up through an entrance to the rear of the fort. I had no torch, but knew that the long vaulted barrack room, lit only by shafts of light striking down from embrasures covering the ditch, must lead to a central corridor and stairs, which would take us up to the top of the fort.

Halfway along I stepped into nothingness. I had fallen into a fissure – the result of structural damage inflicted by American 1,000-pound bombs in 1944. Only the fact that my left leg jammed in a crack in the concrete stopped me from going on down to the next storey. I heaved myself out, lay on the floor in agony and spent the next few minutes attempting to preserve what was left of my dignity, trying not to cry out in pain and announcing, not altogether convincingly, that I was perfectly all right. Though there was a good deal of blood, nothing was broken, and I made my way painfully across the ditch and down the hill to our car, on the little road that hairpins up from Ars-sur-Moselle. The story had a happy ending, but its moral was clear enough: we should not have been there in the first place and it was sheer folly to explore underground workings without proper equipment.

The forts cover the line of the Moselle, dominating the hills to its west and providing an extra line of defence on the east bank. Fort Driant holds the vital ground above Ars-sur-Moselle, where the Mance ravine tumbles down into the Meuse. To the north, the *ouvrages* of Vaux, Bois la Dame and St-Hubert line the saddle between the Mance ravine and the Verdun road. The forward slopes of the Montvaux ravine, on to the north, are covered by the fortified groups of Jeanne d'Arc and Guise. The flat hilltop above Scy is crowned by St-Quentin on its east end and Girardin on its west. Plappeville secures the next ridge to the north and the Lorraine fortified group dominates the upper end of the

Saulny valley, above the village of Plesnois. More forts go around the eastern edge of the city, from Fort St-Privat, north of the Metz-Frascaty airfield, through Queuleu, Lauvallière and Champagne, with St-Julien, in the woodland north-west of St-Julien-les-Metz, securing the river approaches.

The forts of the 1860s had a geometrical trace of which Vauban would have approved, with bastions covering ditch and glacis. The German works were markedly different. In plan they were shaped like decapitated triangles facing the direction of enemy threat. Their guns were originally mounted on the *terreplein* and their ditches were defended by caponiers, standing like massive pillboxes in the ditches. The armoured cupolas developed at the end of the nineteenth century might be placed on the body of a fort, or positioned in detached batteries nearby to make a cluster of defensive positions, or *feste*. Fort Driant has four such detached batteries, two mounting three 100mm guns and the other two three 150mm guns. Just outside the perimeter is the Moselle Battery, its three 100mm guns sited to cover the valley. Underground galleries linked detached batteries with the fort and the garrison lived in barracks in the dead ground behind it. Many of the forts around Metz retain their metal railings, intended to keep attacking infantry out of the ditch, and there is also much surviving barbed wire, strung from angle-iron pickets whose cruelly sharpened tops were themselves no mean obstacle.

Metz was at the centre of the first phase of the Franco-Prussian War. French plans envisaged the formation of three armies, the Army of the Moselle, based on Metz, the Army of Alsace, based on Strasbourg, and a reserve army at Châlons. Last-minute changes, which helped turn the mobilisation into a nightmare etched indelibly on French military history, resulted instead in the deployment of the Imperial Guard and six corps with no intermediate headquarters between the corps commanders and Napoleon himself. To make matters worse the Emperor was in agony from a kidney stone and could not furnish the leadership his corps commanders demanded.

In late July the northern wing of French armies concentrated east of Metz, with 2nd and 3rd Corps in the valley
between St-Avold and Forbach, 4th Corps on the left flank
around Boulay and 5th on the right at Sarreguemines. This
concentration was accompanied by a compendium of
incompetence, which resulted in columns meeting on the
march and hapless infantry waiting for hours in the
unseasonable rain while staff officers vainly tried to sort
things out.

On 2 August, 2nd Corps took the heights of the
Wintersberg and the Reppertsberg above Saarbrücken. The
tiny German garrison inflicted a few casualties before
slipping away, but the French made no attempt to exploit
their victory. Four days later the Germans struck back.
Moltke had concentrated his armies in two wings, with
Prince Frederick Charles's 2nd Army and General von
Steinmetz's 1st on the right, moving on Lorraine, and the
Prussian Crown Prince's 3rd threatening Alsace. Moltke had
intended that the advance should not begin until all
arrangements were complete, and hoped to catch the French
with 1st and 2nd Armies between Metz and the Saar. The
headstrong Steinmetz swung his army further south than
Moltke had intended, heading for Saarbrücken instead of
Saarlouis. By doing so he failed to envelop the French left
wing as Moltke had intended, cramped the march space
allocated to 2nd Army and crashed straight on into the
French concentration.

On the morning of 6 August Steinmetz's cavalry found
that the French had pulled back off the high ground
dominating Saarbrücken and reported that they were in
general retreat. This encouraged Steinmetz and Frederick
Charles to pursue, and led to an encounter battle which a
more astute French command might have turned to its
advantage. Far from retreating, General Frossard of 2nd
Corps had established himself in a formidable position from
which his *Chassepot*-armed infantry could dominate the
surrounding countryside.

The tangle of road, railway and A4 autoroute in the
Forbach–St-Avold valley, and the industrial excrescences

between Forbach and Stiring-Wendel, make the battlefield
of Spicheren an unlovely place. Indeed, it was not notably
attractive in 1870, and the fighting amongst the factories
and goods yards of Stiring-Wendel was an appropriate
overture to western Europe's first experience of industrialised war. Even now there can be no mistaking the
commanding character of Frossard's position. It can best be
appreciated from the sharp-sided Rotherberg, reached by the
little D32c, which runs north out of Spicheren, and well
signed from the village as *Site des Hauteurs*. A tall cross,
erected to commemorate the French soldiers who fell in the
battle, and now embellished with panels giving details of the
French units engaged, stands right on the chin of II Corps's
position, and dominates the Forbach valley to the west and
the Saar valley to the north. The trees in the area obscure
much of the view, but one can see just how strong the main
position was. An enemy who took the Rotherberg would
have to cross the narrow neck of ground between it and
Spicheren. Even if he accomplished this, and cleared the
houses of Spicheren into the bargain, his task had only just
begun, for behind the village the Pfaffenberg rises to block
exits from it. On both flanks the ground, now more wooded
than it was in 1870, rises steeply, offering excellent fields
of fire to the *Chassepot* and an exhausting climb for an
assailant.

There were risks inherent in this superb position. It could
be bypassed by an enemy who crossed the Saar at Volkingen
and struck down the Rosselle valley to Rosbruck; the loss of
the Rotherberg would deny the defender observation into
the Saar valley, and the defender on the forward slopes of
the high ground at the mercy of weapons which outranged
his own.

Frossard was not popular amongst peers or subordinates,
but his dispositions that morning were sound. He put
Laveaucoupet's division on the high ground, Vergé's in the
valley to its left and his third division, Bataille's, in reserve
at Oetingen. Outpost work was poor – the French, with
years of campaigning in Algeria behind them, were reluctant
to risk small patrols – but Leboeuf, Napoleon's chief of staff,

had telegraphed to warn Frossard and Bazaine of 3rd Corps of an imminent attack, and in any case German horsemen were clearly visible on the high ground before Saarbücken on the morning of the 6th.

The four divisions of Bazaine's corps lay between St-Avold and Sarreguemines – within an easy march of Frossard. Steinmetz's leading division, Kameke's of VII Corps, attacked just before midday after an artillery duel which established the superior quality of German guns and their percussion-fused shells. The right flank of the attack was brought to a stop by Vergé's men at Stiring-Wendel, and Laveaucoupet blunted the attack on Spicheren and out-flanking movements through the woods of the Gifertswald and Stiftswald to its north-east. Although the Rotherberg was stormed, the French position remained intact elsewhere. Bataille sent his men up to reinforce the forward divisions and Laveaucoupet counter-attacked into the Gifertswald, re-establishing his right flank. While this was going on, Bazaine moved two of his divisions up to Bening and Thoding, just behind Frossard, and in the early after-noon, in response to Frossard's warning that the battle was developing, he ordered the division at Sarreguemines to march on Spicheren. By his own calculation he had put two divisions at Frossard's disposal and sent a third directly to his aid, keeping only one under his hand. This assessment was badly wrong. The divisional commanders at Bening and Thoding were not clear that they could be ordered up by Frossard, and the message to Sarreguemines had been sent by galloper rather than telegraph. In consequence, when Frossard urgently needed troops he found that none were readily available.

By mid-afternoon Frossard needed all the help he could get. Advancing German formations did not require elaborate orders but marched to the sound of the guns. Two divisions of 1st Army crossed the river at Volkingen and moved down the Rosselle valley, while 2nd Army's III Corps, under the competent Constantin von Alvensleben, attacked into the Gifertswald. These attacks were inconclusive. The Rotherberg was fully secured and guns were hauled up to dominate the col leading to Spicheren. Reinforced by Bataille, Vergé's men riposted through Stiring-Wendel, sending a shoal of Germans back to Saarbrücken. Laveaucoupet shrugged off repeated attacks and withdrew, battered but in good order, to the Pfaffenberg at about last light. Frossard had lost only his first line of defence and had exacted a heavy price for it: the slopes below the Rotherberg and the corpse-filled Gifertswald bore eloquent witness to French firepower.

Frossard had accomplished this only by committing his entire corps, confident that 3rd Corps was marching to his aid. And so it was, but perilously late. Metman, at Bening, laboriously checked with Bazaine when Frossard ordered him up, and Montaudon, at Sarreguemines, did not receive his orders till late afternoon. When Bazaine got a frantic message announcing that Frossard was 'gravely compromised' he sent forward all of his third division at Thoding and part of his fourth at St-Avold. By the time the first of these reinforcements neared the battlefield they found their path blocked by fugitives. The German 13th Division had pushed on down the Rosselle valley and reached Forbach. Bataille's division was committed forward and only a handful of reservists were available to defend Forbach. By 7.30 p.m. the Germans were in possession of the village. With its flank turned, Frossard's fine position was little better than a trap. He ordered a withdrawal to the south, towards Cadenbronn, but it was no easy task to extract his exhausted troops. The French had lost 2,000 men in the fighting and to these were added another 2,000 missing, most of them captured. German casualties, at 4,500, reflected the risk of launching frontal attacks on infantry armed with a breech-loader.

Spicheren set the pattern for what was to follow. If the French dominated the infantry battle, the Germans were superior when it came to artillery. Not only were their Krupp breech-loaders better than the French muzzle-loaders, but their artillery doctrine, which stressed the rapid commitment of all available guns, produced a volume of fire denied to the French, who retained most of their 12-pounder batteries in corps artillery reserves, available for use in concentrations but in practice often not committed until it was too late. Above all the battle showed the superiority of the German command. Not in the launching of a set-piece battle – for Moltke's plans had already been ruined by Steinmetz – but in its ability to take risks and generate purposeful action. When I stand on the Rotherberg I cannot help admiring Frossard's infantry: they deserved to win, but their commanders would not let them.

Spicheren was one of two blows which jarred the French command that day. Away in Alsace, the same story was repeated. MacMahon's 1st Corps was beaten by the 3rd Army at Froeschwiller, and dragged 5th Corps back in its retreat. MacMahon made his way, by road and rail, to Châlons, where he was to command the Army of Châlons, whose cautionary tale will be told in another chapter. For the moment, though, his defeat meant that the Vosges, right-flank bastion of Lorraine, were ungarrisoned, enabling 3rd Army to march north-westwards to throw its weight into the balance around Metz.

The French response to the twin defeats of 6 August was characteristic of a command lacking cohesion and self-confidence. Despite the concentration of French troops east of Metz the sick and demoralised Napoleon ordered a general retreat on Châlons. When his horrified prime minister warned of the disastrous political consequences of such a move, Napoleon changed his mind: the army would concentrate on Metz. Retirement on Metz was shot through by the same lamentable staff work that had bedevilled the advance. Worse still, there was doubt as to who was actually in command of the army. Bazaine had been given authority over 2nd and 4th Corps in addition to his own 3rd 'for military operations only' and on 9 August this power was widened to make him in effect commander of a little army of three corps. That day saw his force back on the French branch of the River Nied east of Metz, with the Guard, still not under his command, alongside. Moltke, meanwhile, had directed the 1st Army back on to its original northern route through Saarlouis to Boulay, while 2nd Army used the main

St-Avold–Metz road. The Crown Prince was to march via Saar-Union on Dieuze, swinging south of Metz.

We must not let the order that military history strives to impose upon events that participants see as chaos persuade us that the French were uniformly incompetent while the Germans were universally capable. The German cavalry, still comparatively cautious, lost contact with both MacMahon and Frossard. The same rain that pelted down upon the French soaked their enemies and an observer in the Vosges passes, or on the Metz road out of St-Avold, might have been forgiven for noticing little difference between wet and exhausted Frenchmen and wet and exhausted Germans. The quintessential distinction between the two armies lay in command. The German armies were animated by the directing will of Moltke, transmitted through his general staff: the French, under the nominal command of a sick man with no general staff worthy of the name, were afflicted by a progressive sclerosis whose consequences were ultimately fatal.

On 11 August the French abandoned the Nied position and fell back under the guns of Metz. 6th Corps, which had concentrated at Châlons, was sent forward to Metz by railway. Because the direct Châlons–Verdun–Metz line was unfinished, it had to make the perilous journey up the Moselle valley from Frouard, where the line was already within reach of German cavalry. Most of the infantry got through safely, but the line was cut before the cavalry, artillery and engineers could join them. Other reinforcements, unable to get forward, concentrated at Châlons, where they were to be joined by the retreating 1st and 5th Corps, and by 7th Corps, isolated at Belfort by the first defeats.

By mid-August the French had two respectable armies, one five corps – 180,000 men – strong at Metz, and another coalescing at Châlons. Napoleon, warned by his Empress that the regime would never survive his return, defeated, to Paris, decided to go to Châlons to oversee matters there. His Chief of Staff, Marshal Leboeuf, had been discredited by his assertion as Minister of War that the army was ready 'to the last gaiter button' and was dismissed by the Council of Regency. Canrobert of 6th Corps was the senior marshal, but knew his limitations and would not accept command of the army, and MacMahon's reputation was in eclipse after Froeschwiller. It was therefore to Bazaine that Napoleon turned, and on 12 August that plump but distinguished veteran of Algeria, the Crimea and Mexico became commander-in-chief of the cruelly misnamed Army of the Rhine. Ironically, Bazaine himself was a Lorrainer: his family farm was at Scy, at the foot of Mont St-Quentin. He was unsure just how far his authority extended, at least as long as the Emperor remained with the army, and he kept his own Chief of Staff, Jarras, at a distance, further weakening an already fragile chain of command.

Over the days that followed Metz exercised a magnetic attraction on Bazaine. It was full of stores and ammunition, and its governor, by an unwise piece of double-hatting also the commander of the army's engineers, protested that it would fall if the army abandoned it. On 13 August Napoleon told Bazaine to do exactly that and on the 14th the withdrawal began. Napoleon had had the foresight to order his engineers to throw pontoon bridges over the Moselle and Seille, but they were damaged by a sudden rise in the river caused by the unusually heavy rain. Four were eventually repaired and used to supplement the three permanent bridges in Metz, although fudged staff work resulted in one of these being forgotten.

No sooner had the army begun to work its way into the constricting funnel of Metz than the Germans attacked. Steinmetz, his knuckles rapped for the bull-at-a-gate business of Spicheren, had been proceeding with marked caution, in contrast to the Crown Prince, whose advance guards were already over the Meuse at Dieulouard and Pont-à-Mousson. However, once again the leading German formation commanders got the bit between their teeth and on the 14th they attacked the French position between Queuleu and St-Julien. Bazaine's old corps was holding the line while the others thinned out, and it gave a good account of itself, checking the Germans in the Lauvallières and

Notre-Dame ravines in front of Borny. Although Bazaine interrupted the withdrawal long enough to support 3rd Corps, he failed to launch the counter-attack that might have given the overextended Germans a bloody nose. There was little co-ordination on the German side and at nightfall Steinmetz attempted to order his two leading corps commanders back to the Nied. Both conveniently declined to obey and before Steinmetz could press the matter Royal Headquarters forbade withdrawal. The Battle of Borny cost the Germans 5,000 men and the French 3,500, including General Decaen, Bazaine's successor at the head of 3rd Corps. Bazaine had lost twelve hours in retiring on Verdun and it was to be enough to ruin him.

Getting across the river was not the end of Bazaine's difficulties, for the Moselle heights were another natural choke point. The army used the single road that winds up through Rozerieulles. This forked at Gravelotte, one road leading to Verdun by way of Conflans and Etain, the other going directly there through Mars-la-Tour. There were some practicable routes north of the Rozerieulles road, which would have enabled part of the army to reach the Gravelotte plateau without waiting in an endless traffic jam, but while Bazaine's staff had excellent maps of Germany they had few of Lorraine. General Ladmirault of 4th Corps asked if he could use the Woippy–Briey road, but Bazaine, worried by false reports of a German force to his north, ordered him to stay closed up.

Dawn on 16 August found the French halted on the Gravelotte plateau, with a cavalry division at Vionville, 2nd and 6th Corps at Rezonville, the Guard at Gravelotte and 3rd Corps, now under Leboeuf, at Verneville. Ladmirault was still filtering his way out of Metz, whose defence had been entrusted to Laveaucoupet's division of 2nd Corps, badly knocked about at Spicheren. Bazaine had agreed to postpone the army's march until Ladmirault arrived, and the day began with a classic juxtaposition of order and counter-order as units who were ready to march off were told to stand down and wait. The Emperor left for Verdun that morning: Bazaine saw him off at the crossroads in

Gravelotte, receiving the haunting valedictory: 'I entrust France's last army to you. Think of the Prince Imperial.' With that he set off in his carriage, with the Guards light cavalry brigade jingling along as escort, down the Briey road. He will re-emerge in our story, pale and haunted, when we consider the unhappy fate of the Army of Châlons.

It is not surprising that the Germans assumed that the French retreat had gone better than it had. On the 15th Moltke, scenting the possibility of catching the French on the Woëvre, allowed 2nd Army to push on across the Moselle south of Metz, while Steinmetz masked Metz with a single corps and sent the rest of his army across the river on Frederick Charles's right. Frederick Charles's cavalry had found Forton's division at Vionville, but the Prince assumed that this was Bazaine's rearguard and ordered a general pursuit westwards. Only his III Corps, with IX Corps far behind it, was to hook up on to the Gravelotte plateau. The scene was set for an encounter battle with the concentrated French army being assailed by a much inferior force.

The battle of 16 August has several titles, but perhaps Rezonville is the most appropriate, for that village was at the centre of Bazaine's position and its name subsequently enabled French soldiers, retreating when they believed themselves undefeated, to quip about the battle of *Rentr'en ville*. The battlefield is well preserved, seldom visited, and quite extraordinarily evocative. As the Verdun road (now, commemorating the route of Patton's 1944 advance, the *Voie de la Libération*, with distinctive kilometre markers) climbs the Moselle heights the industrial haze of Metz is left behind and the traveller swings across the crest line with bare slopes opening up to his right and goups of white crosses marking mass graves of men from both contending armies. Solidly built farms dot the open fields: Moscou and Leipzig, built by First Empire veterans, on the slopes to the right, St-Hubert by the road itself, Malmaison and Chantrenne further on. The Mance ravine slashes out between St-Hubert and Gravelotte, and the road rolls across downland to Rezonville, Vionville and Mars-la-Tour. The crests are big, with capacious folds of dead ground between them. There are thick woods between Rezonville and the Meuse, and more north of the main road – the Bois Pierrot and the Bois de St-Marcel, fronting the Roman road that strides parallel with the modern road. Further on is the cluster of woodland due north of Tronville.

The villages are one-street affairs with closed housefronts sending the noise of the traffic echoing back and faded signboards advertising hotels in Verdun. The battlefield is rich in monuments. On the western side of Mars-la-Tour is the French memorial to the 1870 battles, with a French soldier falling into the arms of an opulent female embodiment of France, who crowns him with victor's laurels as he dies. She stares defiantly across the post-1871 Franco-German frontier into the lost provinces, towards German memorials which commemorate the prowess of long-forgotten regiments: Ziethen Hussars, 35th Brandenburg Fusiliers, 1st Foot Guards. There used to be a little museum in the main street of Gravelotte, but it is now closed indefinitely.

Just before nine on the morning of the 16th the first German shells, fired by the horse artillery of Rheinbaben's 5th Cavalry Division, burst amongst the unsuspecting French cavalry in Vionville, causing a panic. Before long III Corps had come up on Rheinbaben's right, and Alvensleben sent one division straight against the French at Rezonville and another out towards Mars-la-Tour to cut the Verdun road. He was hoping to snap up the French rearguard, but soon found that he had caught a tartar: his infantry were in serious trouble and only the burgeoning firepower of his artillery, on the crest line above Flavigny, kept them propped up. Alvensleben quickly realised what had happened and concluded that his only hope of avoiding disaster was to persuade Bazaine that the entire German army was there. He redirected the division making for Mars-la-Tour to Vionville, which fell to an attack supported by the full weight of his artillery. But his men then ran into the concentrated fire of the French gun line, thickening up between the Bois Pierrot and the north-west corner of Rezonville, and could get no further forward: III Corps was played out.

The French position was formidable. 2nd Corps was deployed in thick lines before Rezonville, with 6th Corps on its right. The Guard was making its stately way forward from Gravelotte, 3rd Corps had begun to edge down from Verneville and Ladmirault's men were stepping out from Metz. Bazaine never glimpsed the opportunity and was obsessed with the threat of an attack boiling up out of the Gorze ravines to cut him off from Metz. His gaze was fixed on his left, growing hourly stronger, rather than his right, where he could have cleared the Verdun road and jabbed southwards to do crippling damage to Frederick Charles.

Frossard was bearing the heat of the day. He started the battle without Laveaucoupet's division, Bataille was hit soon after the fighting started and his infantry were taking heavy punishment from the German gunfire. Frossard asked for help and Bazaine sent the Cuirassiers of the Guard to support the 3rd Lancers, already under Frossard's command. The lancers charged first and as they eddied back the cuirassiers went in. The regiment advanced in three echelons, its path impeded by 2nd Corps' campsite and the hedges and ditches between Rezonville and Flavigny. German fire tore down most of the first echelon, but it is to the credit of this fine regiment that some of its members actually reached the German line north-east of Flavigny. The gigantic Quartermaster-Sergeant Fuchs was seen in the ranks of the German 52nd Regiment, sword gone, swinging his helmet by its chin-strap. The charge cost the cuirassiers twenty-two officers and 244 men, and at the best it earned Frossard a brief respite.

Nobody could fault Bazaine's courage. He had been badly bruised at Borny by a shell splinter, which had spent its force in breaking his epaulette, but spent the day of Rezonville in the saddle, white neck cover fluttering under the brazen sky. First he was here, siting two guns, then there, leading a battalion forward with the cheerful '*Allons, mes enfants, suivez votre maréchal*'. When Prussian hussars shepherded the cuirassiers back to the French lines he was caught up in a swirl of cavalry, but found time to tell an excited subaltern that it was nothing.

While Bazaine was allowing activity to count for achievement, his opponents were taking desperate measures to
shore up their position. It was midday. Alvensleben was
concerned at the plight of his left flank, battered by the
guns of 6th Corps and under imminent threat of attack
by Canrobert's infantry. He ordered von Bredow's cavalry
brigade, drawn up in a re-entrant just north of Tronville, to
charge Canrobert's guns so as to take off some of the
pressure. Bredow went about his task methodically and was
not ready to attack until about 2 p.m. He dropped off a
squadron from each of his two regiments to screen his left
flank and moved his brigade up in column, through the
shallow valley running north from the road junction on the
western end of Vionville, to a natural depression between
the Tronville copses and the track that runs from the
western edge of the Bois de St-Marcel to the Verdun road.
Here he shook out into line, 7th Cuirassiers on his left, 16th
Uhlans on his right, and charged.

Canrobert's guns were drawn up in the open between
Rezonville and the wood line. Most of the infantry was well
behind them and Forton's cavalry division – victim of the
early-morning panic at Vionville – was also to the rear,
where the Rezonville–Villers-aux-Bois road passes the end
of the Bois Pierrot. A walk along the southern edge of the
Bois Pierrot to the nick in the wood line which separates it
from the Bois de St-Marcel takes the traveller to the right-
hand end of Canrobert's gun line. The batteries ran across
the field to the Verdun road, and were firing steadily at the
German guns and infantry around Vionville. When Bredow's
troopers came out of the valley in front of them the gunners
had little time to swing their pieces on to the cavalry and
replace shell with canister for this close-range target. We
should not believe stories that French gunners were
forbidden to change targets without asking the permission
of higher authority: the fall of the ground in front of the gun
line makes it clear that perhaps a minute would have
elapsed between the first sighting of the horsemen and their
arrival on the position. The Germans were over the gun line
in a welter of lance thrusts, sabre cuts and pistol shots. As

they approached the Rezonville–Villers-aux-Bois road, Forton's division crashed into their left flank and they were rolled back to Vionville. The charge cost the brigade almost half its strength of 800 men, but it brought Alvensleben a respite.

The respite was all too brief. Ladmirault's corps was now in action on the French right. It cleared most of the Tronville copses and reached the Verdun road, while du Barail's cavalry division entered Mars-la-Tour. Ladmirault hesitated before going on into Tronville and while he paused X Corps arrived at last. Its leading division, 20th, went straight to the south-eastern corner of the Tronville copses and the 19th was sent off to fall on the French flank. This attack went awry and ran straight into two of Ladmirault's divisions. The Germans advanced into the Fond de la Cuve, north-west of Mars-la-Tour, and were brought to a dead stop by the rifle fire of Grenier's division. The ensuing counter-attack caused a panic on the German left flank, and in Tronville Colonel Caprivi, X Corps' Chief of Staff, ordered the confidential documents burnt. But Voigts-Rhetz had no intention of giving ground: he sent the Guard Dragoons and a regiment of Cuirassiers to check the French pursuit, and Cissey's division was halted just north of the Verdun road.

Shortly before nightfall there was serious fighting on the extreme western flank, between the Mars-la-Tour–Jarny road and the valley of the Yron. Rheinbaben's cavalry division was hovering there with a view to attacking Ladmirault's right and was itself assailed by 4th Corps cavalry division, whose commander, General Legrand, shouted to one of his brigadiers not to bother about carbines but to attack with the sword. Legrand was run through the chest in the mêlée that followed (he may be the last divisional commander in European warfare to be killed by the sword), and the battle soon attracted all the French and German cavalry in the vicinity. Although weight of numbers should have told in favour of the French, the Germans seem to have had the best of the encounter, which ended with the French falling back on Bruville and the Germans retiring on Mars-la-Tour.

The last clashes of the day took place south of Rezonville. Steinmetz had pushed VIII Corps across the bridge at Corny late in the afternoon, and it debouched through the Gorze ravine with IX Corps hard on its heels. Bazaine had been worried about his left flank all day and the forces which he had piled up around Rezonville were enough to check the German advance. Frederick Charles also made a final effort at nightfall and his attack caused some confusion, with an ugly panic on the Verdun road sending fugitives reeling through the streets of Rezonville.

Both sides had lost around 15,000 men. About one-third of the French casualties were missing, most of them wounded, captured when the French vacated the battlefield. Bazaine recognised that, despite its many encouraging tactical successes on the 16th, his army's position was in reality very poor. The main Verdun road was cut, and such was the confusion in French ranks that it would take at least a day to get the army sorted out and restocked with ammunition from Metz before it could set out for Verdun by the Etain road. At midnight he ordered his corps commanders to fall back on to the Moselle heights and the withdrawal began – with more than a little exasperation amongst soldiers who thought that they had won the battle – at dawn on the 17th. The French spent the day getting into position. In the south, 2nd Corps held the bend of the Verdun road in front of Rozerieulles, with the outpost of St-Hubert stoutly garrisoned. The central slopes, with the farms of Moscou, Leipzig and La Folie, were the responsibility of 3rd Corps and 4th Corps held the line on to Amanvillers.

So far, so good. The French had excellent positions and time to dig in: Frossard was a sapper and Leboeuf a gunner, and needed no reminding of what would happen when German guns wheeled into line on the slopes above Gravelotte. The real problem was in the north. 6th Corps had been incomplete before it fired a shot in anger, and its already weak artillery had been mauled by Bredow's horsemen and Alvensleben's gunners. Bazaine had intended to post it forward at Verneville, like a breakwater in front of his position, but Canrobert protested. The position was exposed and its fields of fire were poor: could he not fall back on to St-Privat and continue the army's line up to the north? Bazaine gave way at once and Canrobert's men duly arrived at St-Privat late on the 17th. They had little time to dig, and in any event digging equipment had been one of the casualties of the corps' hasty concentration. Bazaine positioned himself behind his army's left rear at Fort Plappeville, with the Guard and the artillery reserve close by, and sent messages to Châlons saying that he hoped to resume the retirement on Verdun soon, if he could do so 'without compromising the army'.

We cannot be certain what Bazaine really thought. He later spoke both of wearing down the Germans and then moving off again, and of waiting for the Army of Châlons to relieve him. It is more likely that no lofty conceptions actually entered his mind. In Michael Howard's magisterial words: 'He was living from day to day, confining himself to the routine details of administration which he understood, and trusting to his good luck to pull him through.'

The Germans were positive in their approach but wrong in their assessment. Moltke left Steinmetz as a pivot on his right, and ordered Frederick Charles to fan out to the north and fight Bazaine where he found him. This was to result in Frederick Charles – who thought that the main body of the French had escaped, leaving only a rearguard on the heights – marching across Bazaine's front, exposed to a bruising counter-attack. There was little chance of that: at nine on the morning of the 18th, when Leboeuf reported that great masses of German troops were sliding across his front, Bazaine told him to sit tight. By mid-morning Frederick Charles believed that he had found the French left, and swung the Hessians of IX Corps eastwards towards Amanvillers. By the time Moltke ordered him to push further to the north, the battle had already started.

No sooner had the guns of IX Corps opened fire than Ladmirault's men deluged them with rifle fire and came roaring down the slope to capture four guns. The battle in the centre then became an artillery duel as the Hessians waited for the attack to develop elsewhere. Moltke had been so afraid of Steinmetz's propensity for counter-productive action that he had made it clear that he would not commit his infantry against the positions opposite and had with-drawn VIII Corps from his control. It was a wasted precaution: when the Hessians joined battle, VIII Corps attacked anyway. Successive assaults on the southern section of the French line did little but add to German casualties. St-Hubert was eventually taken, but, try as they might, the Germans could not get beyond it. However, a gun line was established north of Gravelotte and was extended to the south as VII Corps came up: soon there were 150 guns firing at a target they could scarcely miss, and the French positions around Moscou and Leipzig were pounded into rubble. The mass graves by the roadside testify to the casualties caused by this fire and to the damage done by the *Chassepot* to the German attackers. A regimental memorial by the convenient loop of disused road above St-Hubert shows what it could do to a single regiment, whose list of officer casualties is headed – as so many are – by its colonel, commanding the brigade that day.

The fall of St-Hubert persuaded Steinmetz, sure that he was facing a rearguard, to throw all available troops into the battle. They could only advance across the Mance ravine and made as little impression on the French as previous attackers had. By late afternoon the ravine was choked with dead and wounded, with panic-stricken men and riderless horses rushing back towards Gravelotte. A counter-attack at this moment could have had the gravest consequences for the Germans, dividing 1st and 2nd Armies, and leaving Frederick Charles marooned, with the French across his communications. Bazaine had no thought of such a stroke. He spent the day at Plappeville, where freak acoustics prevented him from hearing just how fierce the battle was, until he rode out to check his left at Mont St-Quentin.

The real danger lay on his right. Once the Germans had passed Verneville they could see how far the French line really extended. Frederick Charles left the Guard facing 6th Corps and waited for the Saxons of XII Corps to begin their outflanking movement. Ste-Marie-aux-Chênes, held by two battalions of the 94th Regiment, fell just after 3 p.m., but the defenders scampered back up the slope to the main position, having imposed a delay at trifling cost to themselves. The Saxons then went on round the northern flank of the French position before swinging back in towards Roncourt to begin rolling it up. Had the Germans waited for this promising move to gain momentum, they might have won a cheap victory in the north. They had built up a massive gun line level with the St-Ail–Habonville Road, and the defenders of St-Privat were subjected to the same merciless hammering as those of Moscou and Leipzig. Canrobert had only cavalry behind him – a regiment of *Chasseurs d'Afrique* and a brigade of Forton's division – and at 2.30 p.m., watching the attack develop, he had asked Bazaine to honour an earlier promise and send him the Guard. The French Guard had not appeared when the infantry of the Prussian Guard intervened. The decision to commit the Guard was made by its commander, Prince Augustus of Württemberg, with the approval of Frederick Charles. Perhaps Augustus was deluded, by the slackening of French fire, into believing that Canrobert was on the move; perhaps he believed that the Saxon attack was about to bite home; or perhaps he simply wanted the honours of the day to be carried off by Prussians, not Saxons.

The Guard attacked in echelon from the right. At about 4.45 p.m. 3rd Guards Brigade advanced through the Bois de la Cusse, only to be stopped by rifle fire in the cutting of the unfinished Verdun–Metz railway east of Habonville. At 5 p.m. 4th Guards Brigade set off from St-Ail towards Jerusalem, on the southern edge of St-Privat. It has been estimated that its opponents, Lafont de Villiers's division, were firing 40,000 rounds a minute. So intense was the fire that a flock of sheep which dashed between the two armies was killed in the twinkling of an eye. The guardsmen were scarcely more fortunate: 4th Brigade was halted over 500 metres from the French line, with a major as its senior surviving officer. At 5.45 p.m. 1st Guards Brigade attacked north of the main road and shared the same fate

The Guard suffered over 8,000 casualties that day, mostly

between 5 and 6 p.m. The scene of its misfortune can be visited easily. The little *rue de la tour*, leading off the St-Privat–Ste.-Marie-aux-Chênes road, runs along the crest line. It is named for the Guard corps memorial, a crenellated tower, looking like a large castle chessman, with a spiral staircase inside leading to a viewing platform, inaccessible in the spring of 2007. The memorials along the crest line read like the Almanach de Gotha: the roll of honour of the 4th Queen Augusta Guard Regiment bears the names of two Princes of Salm-Salm, Félix and Florintin. As the road, by this stage unmetalled, approaches the bridge over the autoroute the view opens up to the west, with Ste-Marie and St-Ail both clearly visible. One can imagine the Guard coming up the slope, dense company columns behind a skirmish line, field officers mounted and drummers rattling out the step, only to run into musketry whose sheer intensity rivalled British infantry fire at Mons or even German machine-guns on the first day of the Somme.

On this part of the front the French were formed two lines deep. The 25th and 26th Regiments, six battalions in all, were drawn up on the slope where the track now crosses the autoroute, with the 28th and 70th Regiments 400 metres behind them. Military historians are right to snipe at the staff work of the army of the Second Empire and to point to the mercurial morale of its soldiers, but there can be no gainsaying the resolution of Canrobert's infantry, who had endured hours of shelling on these slopes, innocent of any cover. Charismatic leadership helped: old Canrobert, with his big moustache and wispy hair, rode up to the 70th Regiment under a blistering fire and greeted its brigade commander cheerfully: 'Good day, Chanaleilles, I am pleased to see you: this is indeed the place for a gentleman and a soldier.' We must also give credit to French company officers, many of them old and ill-educated ex-rankers, but with enough moral authority to keep their men closed up as the iron tempest blew through their ranks.

Until six that evening the French had every reason for self-congratulation. They had repulsed the German infantry wherever it had attacked and had stood fast under a concentration of fire without precedent in history. All this ended when the Saxon flanking attack at last bit home. Canrobert had been able to spare few troops to cover his right and by 7 p.m. the Saxons were in Roncourt. There was no sudden collapse and Canrobert's men fought doggedly in the blazing ruins of St-Privat, killing the leading Saxon brigade commander; but their position was fatally compromised, and at 7.30 p.m. the village fell to the combined efforts of the Saxons and the Guard.

French military doctrine emphasised that the Guard was a reserve, which should be used intact at the crucial moment, and although Bazaine had promised to support Canrobert with the Guard he had done little to direct its distinguished and flamboyant commander, General Bourbaki. It was early evening when two of Ladmirault's staff officers approached Bourbaki, in the dead ground behind the French line, to ask for support: Ladmirault had been deftly counter-attacking to take some of the pressure off 6th Corps and believed that the Guard could strike a telling blow at the Germans opposite him. Bourbaki agreed to help and set off along the Amanvillers road at the head of his *Voltigeur* division, only to discover, as he reached the high ground, that 6th Corps had already given way. Even if recapturing St-Privat was beyond him, Bourbaki might have checked the Germans and given 6th Corps a chance to rally. As it was, he chose the moment to indulge in the unsuitably dramatic. He vilified the officer who had led him forward, turned his men about and cantered off. The sight of the Guard departing was the last straw for 6th Corps and part of 4th, and even the Guard itself became infected with panic. As darkness brought fighting to a halt, there was no doubt that the entire French right wing had collapsed.

At the same time that Canrobert's men were giving way in the north, the Germans were putting in one last effort in the south. Steinmetz had continued to carpet the slopes of the Mance ravine with Prussian dead and, arguing that he had actually carried the heights and simply needed to clinch the victory, asked for permission to make one last effort. The King of Prussia, who had just reached Gravelotte, allowed him to use the newly arrived II Corps. The French saw II Corps crossing the ridge and were awaiting Steinmetz's final piece of butchery. Their line crashed into life and the Germans broke with horrifying suddenness in a panic worsened by the fact that II Corps had accidentally fired on the exhausted German defenders of St-Hubert. Demented soldiers ran through the streets of Gravelotte under the horrified gaze of Royal Headquarters and the staff failed – even with the assistance of the King himself – to beat men back with the flat of their swords. II Corps held the line to ward off the counter-attack which never came, but to Royal Headquarters, now in the crammed village of Rezonville, it seemed that the attack had failed all along the line and that the French had won. Only when word of Frederick Charles's success came in after midnight did Moltke realise that he was the victor. It was evident to all that it was an expensive victory. The Germans had lost just over 20,000 men and the French perhaps two-thirds as many.

Bazaine fell back on to Metz that night. He told Napoleon that he proposed to set off again once his troops had recovered and by 22 August his army was reprovisioned with ammunition. Moltke quickly reorganised his armies to take account of the changed situation. He divided the 2nd Army into two, giving the Crown Prince of Saxony command of three of its corps, now called the Army of the Meuse, and combining the remaining four corps with 1st Army to invest Metz. Frederick Charles was given command of the investing force: when Steinmetz showed resentment at his new subordination he was shunted off to be governor of Posen and disappears from our story, leaving only those well-filled grave pits above the Mance ravine to mark his passing.

Over the weeks that followed Frederick Charles tightened the ring around Metz, throwing up field fortifications which would make it difficult for Bazaine to break out. Bazaine, indecisive in open field, was even more prey to uncertainties shut up in Metz, deprived of news about the Army of Châlons and doubtful as to when and in what direction to break out. A breakout to the north-east was scheduled for 26 August, but the troops were already formed up for the venture when a council of war agreed that the project was doomed to failure and abandoned it. At the end of the month Bazaine tried again, this time after receiving a smuggled message saying that the Army of Châlons was on its way to meet him. It is a tribute to the quality of the weapon under his hand that the attack, out towards Noisseville and Borny to the east of Metz, cut deeply into the German lines, but hopeless staff work and the lack of killer instinct at the top turned the episode into yet another fiasco.

On 6 September Frederick Charles arranged an exchange of prisoners to ensure that Bazaine heard of the defeat of the Army of Châlons at Sedan and German outposts thoughtfully provided Parisian newspapers announcing that the Empire had fallen. The Army of the Rhine shrivelled with hunger and inanition. On 6 October Bazaine considered yet another breakout, only to abandon it in favour of a large-scale food-gathering expedition, which crumbled under the fire of German guns.

If Bazaine found the military situation taxing, his political plight was even more alarming. Napoleon's fall left him as the only Imperial authority on French soil, commanding an army that included an Imperial Guard and most of whose officers were of Imperialist sympathies. Bazaine's attitude to the Government of National Defence was equivocal and he may have believed that his army could play some sort of political role. A shadowy character called Edmond Regnier hoped – without a shred of diplomatic experience or any credential save a picture postcard of Hastings with a few words from the Prince Imperial on the back – to persuade the army at Metz and the garrison of Strasbourg to capitulate in the name of the Emperor and form the nucleus of a new Imperial government, which would overthrow the Government of National Defence. The project teetered on the sharp edge between tragedy and farce, and eventually collapsed when a reluctant Bourbaki, allowed to leave Metz, travelled to England only to find that the Empress would have nothing to do with the scheme. Bazaine was on his own.

As hunger strengthened its grip there was friction between soldiers and townspeople – and also between officers and men, for the former could afford rocketing food prices while the latter could not. The city was packed with troops. Although the comfortable routine upon which beleaguered armies often fall back absorbed much of their time, it could not stop angry talk, a lot of it directed at Bazaine. Desvaux, who had succeeded Bourbaki at the head of the Guard, was a real fire-eater, and amongst the more junior ranks an engineer captain named Louis Rossel made a name for himself by demanding a breakout at all costs. Rossel came to a sad end, for his extreme opinions led him to support the Commune in 1871, and brought him before a court martial and a firing squad at Satory outside Paris. The object of this wrath was rarely seen. Bazaine was living at the Villa Herbin in the suburb of Ban St-Martin on the left bank of the Moselle and, so his detractors claimed, spent his time playing billiards, with the snick of ivory on ivory soon competing with the drumming of rain on the roof.

October started fine, but the weather broke in the middle of the month and it became one of those Lorraine autumns when the wind drives clouds low across the Moselle heights and tugs at the neatly ranked poplars. The number of sick grew alarmingly, and row upon row of railway wagons were packed into the Place Royale (now the Place de la République) to increase hospital space. Horsemeat was issued to the army as other rations dwindled, and soon the horse lines witnessed a sad race between the farrier and starvation.

A final attempt to find a political solution, this time through the efforts of General Boyer, Bazaine's senior aide-de-camp, failed. On 25 October Bazaine sent old General Changarnier, who had left the army in 1851 but valiantly appeared at Metz in a borrowed uniform in August, to see what terms Frederick Charles would offer. There were none save unconditional surrender and Cissey, who made the same depressing journey that afternoon, got no further. A council of war in the Villa Herbin on the 26th agreed that Jarras should be sent to negotiate a capitulation. The bleakness of the terms was brightened only by an offer that the French could march out with the honours of war, but Bazaine accepted these only on the condition that they were not carried out: he made a number of excuses about the weather and the state of the troops, but it is hard to resist the conclusion that he feared to hazard his person in the presence of the men who had expected so much but been given so little.

One of the last acts of the Army of the Rhine was a mournful microcosm of so much else that had gone wrong. Bazaine ordered that all the regimental colours should be handed in at the arsenal. Desvaux wisely saw the colours of the Guard burnt and a few steadfast commanding officers had theirs destroyed or hidden. Through mischance, malice, or both, the fifty-three colours in the arsenal were surrendered to the Germans along with nearly 180,000 men, seventy-two *mitrailleuses*, 500 field and 900 fortress guns. Even on the brightest day I cannot walk along the Esplanade at Metz, looking across to Fort St-Quentin on its bluff over the Moselle, without thinking of the captured eagles ranged outside Frederick Charles's headquarters while Mutte, the great bell of Metz, tolled out the city's misery from her tower above the cathedral.

The Treaty of Frankfurt ceded Alsace and much of Lorraine to Germany: Metz became a German city. It was a heavy blow. The question of national sovereignty had not been raised even after the defeat of Napoleon, and the loss of large areas of territory that were French by history, language and culture caused much grief. The deputies from Alsace and Lorraine made stirring speeches in the National Assembly of 1871 and Alphonse Daudet's tear-jerking story, *The Last Lesson*, described an old schoolmaster teaching his last lesson in French. The new provinces were assimilated into the German Empire with some difficulty, at least in the early stages, and as late as 1914 there was an ugly incident at Saverne – or Zabern, as it then was – when a German officer sabred a civilian who had insulted him. However, by 1914 the inhabitants of Alsace and Lorraine were in the main resigned to being German, and their political leaders were concentrating on securing as much autonomy as was possible within the fabric of the German Empire.

The French responded to defeat in the Franco-Prussian War by elaborating plans for a renewal of the conflict. These eventually became offensive in character, but long before the French felt strong enough to contemplate attacking to recover the lost provinces, they had set about fortifying the new frontier. With Metz and the Moselle heights lost, the Meuse was the next natural barrier between the Germans and the French heartland. Seré de Rivière stretched his defensive curtain along the Meuse, from the entrenched camp of Verdun, through a series of smaller work, down to Bourlemont. Another network of forts secured the confluence of the Meurthe and Moselle around the entrenched camp of Toul. Then came the Charmes gap with the fortifications of the upper Moselle, from Epinal to Motbeliard – beyond it.

Many of Seré de Rivière's forts have survived the ravages of time and high explosive, though the traveller who seeks to visit those not formally open to the public should bear in mind that they are private property and, as my experience at Fort Driant shows, dangerous places. They were originally designed to mount heavy guns (120mm, 138mm and 155mm breech-loaders using black powder) on the *terreplein*, with masonry shelters (*traverses-abri*) between each gun to give some cover to the piece and to protect gunners awaiting fresh orders. The guns pointed out across a polygonal trace defended by a ditch, usually six metres deep and eight wide, both scarp and counterscarp revetted with masonry. Low caponiers, armed with the Hotchkiss *canon-revolver* and an 1858 pattern 12-pounder converted to breech-loading, formed the ditch defence. Integral barracks, with bakery, kitchens and hospital, housed the fort's garrison.

In their original form, the forts were built of stone about 1.5 metres thick, covered with three to five metres of earth. Many were modernised, but there are some good examples of forts which remain much as Rivière designed them: both Génicourt on the Meuse, and Villey-le-Sec on the Moselle retain their original features. As a response to the development of melinite and lyddite in the mid-1880s, some forts were given a shell of concrete between 1.5 and 2.5 metres thick, with a burster layer of sand, to absorb the concussion of shells, between the original masonry and the new concrete. A layer of earth up to four metres thick covered the concrete. At the same time the original caponiers were replaced with counterscarp defences – *coffres* – almost like pillboxes built into the counterscarp. There was usually a double *coffre* at the fort's chin and single *coffres* at the flanking angles.

The new explosive posed a dreadful threat to unprotected guns. As early as 1874 attempts had been made to shelter guns in armoured casemates and with the development of armoured, revolving cupolas it was possible to make at least some guns safe from all but a direct hit by a heavy weapon. Yet these cupolas were no panacea. They were very expensive and their addition to existing forts required extensive work. Because they retracted almost flush with the ground when not in action, the guns they housed had to be short-barrelled, which reduced their range. The only long-barrelled pieces that could be tied into the modernised forts were 95mms (soon replaced by 75mms) mounted in *casemates de Bourges*, designed in 1899. These were added as 'earpieces' to existing forts, able to cover the interval between the fort and its neighbour but not at risk from frontal fire.

We can trace the effect of modifications on the mighty Douaumont, largest of the forts in the Verdun complex. Completed in 1885, it originally mounted some sixty guns on its *terreplein*. In 1887 work began on adding concrete to its masonry defences and in 1889 the *coffres* were installed to defend the ditch. A Bourges casemate was added in 1901–3, its two 75mm guns covering the nearby *ouvrage* at Thiaumont. At the same time, two *cloches Digoin*, armoured observation cupolas, were added, as were two *cloches Pamard*, armoured, non-retracting machine-gun turrets. The big turrets arrived surprisingly late: a turret for a 155mm was added in 1907–9 and another for two 75mms in 1911–13. Each of these turrets had its own observation cupola and was connected to the main underground galleries of the fort. Work was in hand on a 75mm turret to the east of the fort and another for two 155mms to its south when war broke out.

The forts on France's eastern frontier seem undergunned when one considers their size. Even allowing for the fact that protection and ammunition supply accorded to a single turreted piece were deemed to make it the equivalent of a battery of similar guns in open field, it is remarkable to consider that Douaumont – 300 metres from the counter-scarp gallery at its chin to the ditch at its rear and over 400 metres from ditch to ditch at its widest point – mounted four 75mm guns and a single 155mm. A large part of the answer to the problem of undergunning lies in the fact that, despite all the expensive improvements carried out on the major forts, it was recognised that they would remain vulnerable to high explosive. Guns were moved out into smaller batteries between the main forts and extra turrets – like those under construction at Douaumont – were built close to the forts to turn them into centres of resistance rather than individual strongpoints. The forts would provide secure observation posts and shelter infantry. Douaumont could house some 800 men in its two-storey barracks, though its wartime garrison was only intended to be 500 men.

When war broke out in 1914 the value of fortresses was questioned by the fall of the Liège forts under the bombardment of German heavy howitzers. The field armies had an insatiable appetite for heavy guns and the fortresses of the east were denuded of many of their pieces. The guns that departed were mainly early breech-loaders, designed to fire from the *terreplein*, which lacked the recoil-absorbing mechanism found on more modern guns. Turreted pieces could not be usefully removed, then or subsequently – a large number of 75mms still rust quietly amongst the undergrowth.

It was not until the war had been in progress for nineteen months that the Verdun forts assumed great importance. However, fighting had broken out in Lorraine at the very start of the war. The French government had ordered its armies to pull back ten kilometres from the frontier to make it clear that the Germans were the aggressors, and this withdrawal was followed by small-scale actions around Pillon, Arrancy and St-Pierrevillers, north-east of Verdun, as German patrols crossed the frontier. Over the next few days the French prepared for their own offensive. Joffre had modified Plan XVII to take some cognisance of the German interest in Belgium and envisaged two main thrusts from the eastern frontier. One, made by the 3rd and 4th Armies, would stab up into Belgium north of Montmédy: the other, involving the 1st and 2nd Armies, would lunge hard across the frontier south of Metz towards Morhange and Sarrebourg.

Both thrusts splintered against a technologically superior defence. In the south, the French pressed on into the lost provinces, infantry sweating under their long blue great-coats and the flat-trajectory 75mms singularly ill-suited for the hilly, wooded countryside. The attack reached its culminating point on 20 August, when it became clear – even to a French army brought up on the doctrines of *l'offensive à l'outrance* – that courage and dash were no match for machine-guns and howitzers. The story was the same in the north, where vicious battles around Virton and Neufchâteau ended, on the 23rd, with the French falling back across the Chiers and the Meuse north of Verdun.

Victory in the battles of Morhange-Sarrebourg encouraged Crown Prince Rupprecht of Bavaria, commander of the German 6th and 7th Armies, to ask Moltke for permission to counter-attack his shaken enemy. Moltke gave way, although German interests were best served by keeping the French overextended in Lorraine while the decisive battle was fought north of Paris. His general directive of 27 August abandoned Schlieffen's single-minded concentration on the wheeling flank and ordered a general offensive in which the 6th and 7th Armies would advance between Toul and Epinal – through the Charmes gap. The isolated fort of Manon-viller, built to command the Paris–Saverne railway line, was hit by 17,000 rounds and surrendered: the fumes produced by the shells made the place untenable. The attractive town of Lunéville lay in the German path. Its château – 'the little Versailles' – had been the favourite residence of King Stanislas, who had died there in 1766. It was also a town with long-standing military associations, for a cavalry division had been based there both before and after the Franco-Prussian War. An equestrian statue of Antoine Charles Louis, Comte de Lasalle, darling of Napoleon's hussars, stands in the château's *cour d'honneur*. He announced that any hussar still alive at thirty was a blackguard and once, on a lightning visit to Paris, stated that he had only sufficient time to order a new pair of boots and to make his wife pregnant. He was killed at Wagram in 1809 at the age of thirty-four. A section of the museum in the château tells the story of the town's cavalry garrison. The building was badly damaged by fire in 2003 and repairs to the roof were completed in 2007, though it is not certain when it will reopen. All its military heritage could not save Lunéville in 1914: the Germans took it and pushed on, to be halted just short of the Moselle. Rupprecht had been less successful to the north, where Foch, commanding a corps in front of Nancy, fought with a tenacity that marked him out for greater things and the French retained their grip on the high ground of the Grand Couronné.

The German crossing of the Meuse north of Verdun and the success of their attacks on the right bank east of the city enabled them to develop a pincer movement, which threatened to encircle it. In September the German 4th and 5th Armies rolled down the left bank south-west of Verdun, reaching the line Andernaye–Vassincourt–Rembercourt–Heippes. On 10 September only ten kilometres separated the Germans east of the Meuse from the 5th Army's main offensive around Heippes, and there was every risk that attacks on the line of the Meuse would break through this narrow corridor and cut off Verdun altogether.

Fort Troyon, an unmodernised Seré de Rivière work on the Meuse, stood in the way. It was held by Captain Heym's company of the 166th Infantry and its guns – four 120mms and twelve 90mms, with a *canon-revolver* and an old 12-pounder in each of its caponiers – were manned by a battery of the 32nd Artillery. The Germans began to shell the fort on 8 September and on the 10th the fort's guns – and those of the nearby works at Génicourt and Paroches – inflicted heavy losses on infantry moving up to the river line. The fort was relieved on 13 September as part of a counter-offensive, which pushed the Germans back as far as Montfaucon on the left bank, and the threat to Verdun had, for the moment, passed. Fort Troyon lies just off the D964, between Troyon and Lacroix-sur-Meuse: a monument stands where the track to the fort leaves the road. The fort is open on summer weekends and public holidays. The two forts that supported it in September 1914 are in rather better order. Génicourt lies downstream, on the right bank of the Meuse, between Dieue and Génicourt itself, overgrown but a good example of an unmodernised fort. Paroches is upstream, on the left bank, in the woods north-west of the village.

The failure of the Germans' first attempt to encircle Verdun brought only a brief breathing space. On 20 September they renewed their efforts, attacking on both banks of the Meuse. On the left bank they took Varennes and drove deep into the forest of the Argonne, only to be fought to a standstill on the wooded *butte* of Vauquois, Le-Four-de-Paris and La Harazée, a line running parallel with the main routes to Verdun, the Rheims road (the modern N3) and the railway line. Both were now covered by German artillery fire, a fact of no little importance to what follows.

On the right bank the Germans bit into French lines between Nancy and Verdun, taking the abbey town of St-Mihiel and establishing a bridgehead across the Meuse. The fort of Camp des Romains, another unmodernised Seré de Rivière work, dominated the river line. The Germans had already neutralised the guns in Fort Paroches, so Camp des Romains was unable to profit from supporting fire. Early on the morning of 25 September two battalions of the 11th Von der Tann Infantry Regiment assaulted after a bombardment, crossed the ditch in the north and seized a section of the ramparts. After some bitter underground fighting the fort's commander agreed to surrender. The French had lost fifty killed and between sixty to seventy wounded, the Germans twenty-three killed and seventy-three wounded. The loss of St-Mihiel left the Germans in control of the Meuse valley north and south of Verdun. The main routes into the city were either in German hands or blocked by their artillery fire. This left a narrow-gauge railway from Bar-le-Duc and the minor road beside it as the only routes into the largest fortress of France's eastern frontier.

French recognition of the importance of Vauquois led to a series of costly attacks in February and March 1915. Thereafter both sides resorted to mine warfare, making Vauquois the most heavily mined spot on the earth's surface. In all, the French exploded some 300 mines and the Germans 200. The majority of these were camouflets, small mines designed to blow in the enemy's mine galleries. The largest mine was German: its sixty tonnes of explosive killed over a hundred men of the 46th Infantry Regiment on 14 May 1916. No less than forty Allied divisions were rotated through the Vauquois sector, two of them Italian and one American. It was this latter, the 35th Infantry, which at last took Vauquois on 26 September 1918.

The *butte de Vauquois* is heavily scarred by mining and tangles of wire on rusting pickets sprawl across the grass around the craters. The summit is dominated by a memorial to the men who fought there: there is an orientation table, and a fine view which show why the *butte* was so important. At its foot, where the road swings left to climb the feature, a small pillar commemorates Henri Collignon, a former prefect and member of the Council of State who volunteered for service at the age of fifty-eight and was killed at Vauquois as a private in the 46th Regiment. Little remains of Vauquois village except occasional shards from rooftiles. A good deal of work has been carried out at Vauquois over the past ten years. There is now a little museum at the car park below the *butte* and some of the tunnels beneath the feature are open to guided tours on the first Sunday of every month. I have not always been fortunate in finding guides on hand when I expected them, but the visit is a rewarding one.

The small town of Varennes-en-Argonne lies north-west

of the *butte*. It was here that Louis XVI and his family were detained on 21 June 1791 as they attempted to flee the country: a monument marks the location of the house to which they were taken. The Pennsylvania State Memorial, commemorating the sons of Pennsylvania killed there in 1918 stands where the D946 enters the town from the south. Alongside it is the *Musée de l'Argonne*, which contains material concerning the King's arrest, a gallery devoted to local arts and crafts – particularly the brightly coloured tin-glazed earthenware, *faience*, produced in the district – and a First World War gallery that deals with mine warfare and the American involvement here in 1918.

The area is rich in German bunkers and tunnel entrances. An especially historic bunker complex, the *Abri du Kronprinz*, lies down a track off the D38 Varennes–Les Islettes road. This was used by the Crown Prince Wilhelm, commander of the German 5th Army, in 1915, and the entrance that remains visible led on to a network of underground chambers. On the evocative *Route de la haute Chevauchée*, the little D38c, is the *Kaisertunnel*, some 350 metres long, with an assortment of side tunnels, usually open on summer Sundays.

The fighting for the Argonne spurs was costly but inconclusive. Yet there was more to it than the simple desire of each side to wrest more territory from its opponent. While the French held the spurs, the Germans could not envelop Verdun down the left bank of the Meuse, but as long as the Germans dominated the main road into Verdun, it was difficult for the French to guarantee Verdun's supply. From 5 August 1914 Verdun ceased to be a fortress in its own right and became the centre of the *Région Fortifiée de Verdun*, which stretched from Avocourt on the edge of the Argonne, through Fresnes-en-Woëvre and the long ridge of Les Eparges, down to St-Mihiel. Over the next three months forty-three heavy and eleven foot artillery batteries were withdrawn from the area: the Bourges casemates lost their 75mms, and most of the counterscarp machine-guns and cannon were sent off to the field armies.

Fortress or not, Verdun remained immensely strong. The

city itself, Roman *Virodunum*, lies inside a ring of hills. Its position, where the Roman road from Metz to Paris crossed the Meuse, gave it long-standing military importance. It was sacked by Attila the Hun in 455 and thereafter frequently besieged. Its defences lie round it like the rings of an onion, with medieval survivals – like the splendid double-fronted *Tour Chaussée* where the old Metz road crosses the Meuse– surrounded by the ashlar and earthworks of artillery fortification. At its centre is the citadel, a six-bastioned work built on the site of the Abbey of St-Vanne: Vauban retained one of the abbey's towers in his works. The town's defences were upgraded in the 1840s, when Vauban's bastions and ditches were refurbished.

After the Franco-Prussian War Verdun succeeded Metz as the cornerstone of the eastern frontier. It was encircled by rings of large forts and smaller *ouvrages*. On the right bank, the forts of Belleville, St-Michel and Belrupt secured the high ground close to the city. An intermediate ring, the *Ouvrage de Froideterre* and Forts Souville, Tavennes, Moulainville and Rozelier, curled round the city's eastern flank, with Douaumont and Vaux lying like ravelins in front of them. The left bank, too, was well secured, with the imposing Bois Borrus ridge, pointing down to the Meuse at Charny, held by Forts Bois Borrus, Marre and Vacherauville. There was no approach to the city that was not covered by the interlocking arcs of fire from the forts and *ouvrages*, supported by detached posts and batteries.

The removal of so many of the guns from Verdun's defences, together with GQG's insistence that the city should not be held if the Germans attempted to invest it, had led to its military governor, General Herr, preparing defences on the left bank behind Verdun. Herr warned GQG that he was concerned at the state of Verdun's defences and Emile Driant, a local parliamentary deputy now serving as a lieutenant-colonel on the Verdun front, repeated these warnings in the most forceful tones. Joffre was infuriated and remained convinced that the sector would remain what it had become in 1915: a quiet backwater.

All that changed in early 1916. In December 1915

Falkenhayn wrote a lengthy memorandum arguing that the war might be won if France was persuaded that she had nothing more to hope for. This could be accomplished by selecting an objective whose retention would force the French to throw in every man they had. 'If they do so,' he argued, 'the forces of France will bleed to death – there can be no question of voluntary withdrawal – whether we reach our goal or not.' The spot he chose was Verdun. The Kaiser, at his headquarters at Charleville-Mézières, approved the decision, and on 23 December Falkenhayn met General Schmidt von Knobelsdorf, Chief of Staff to the Crown Prince's 5th Army, at Montmédy station and told him of the plan. It was to be code-named Operation *Gericht*: Operation *Scaffold* might be the most idiomatic – as well as the most appropriate – translation.

Preparations went on with astonishing speed. Troops and guns were amassed north of Verdun: the forest of Spincourt was laced by railway spurs leading to gun positions. The Germans concentrated over 1,200 guns for an attack on a frontage of some 12 kilometres. Big concrete-busting 420mms and smaller 305mms stood ready to cope with the forts; 380mm naval guns prepared to reach deep behind the front to destroy communications; 210mm and 150mm guns could deal with front lines and gun positions alike, and the great mass of 77mms, workhorse of the field batteries, would support the infantry. The Germans stockpiled two million shells – the first six days' ammunition – turning their back areas into one huge munitions dump. They dug *Stollen* – massive dugouts, some of which could hold several companies at a time – to protect the assaulting troops and took good care to screen all these preparations from the French. But even though the Germans had air superiority over the Verdun sector, they could not conceal the fact that something was afoot.

The attack was scheduled to take place on 12 February, but bad weather caused its postponement. This took some of the edge off the German assaulting troops, cooped up in their *Stollen*, although things were scarcely better for the French, out in trenches below the windswept ridges. The storm broke on the morning of 21 February. It was ironic that the first shell of the German barrage – aimed at a bridge over the Meuse – burst just behind its target, damaging the bishop's palace. Between the city – target of naval guns sniping away from the Forest of Spincourt – and the front line, the French position was swept by a savage and concentrated bombardment, which turned forward trenches to piles of ploughed earth and drew curtains of fire between front-line troops and their supports.

It was not until 4 p.m. that the infantry scrambled from their *Stollen* on the narrow front between the Bois d'Haumont and Herbebois to draw the cheque made out by their gunners. On the extreme right of their attack, the Bois d'Haumont fell to VII Reserve Corps, whose flame-throwers had a numbing effect on the defence. The Bois de Caures, to its east, was held against XVIII Corps by the two *Chasseur* battalions commanded by Emile Driant. He held the wood in depth, with a network of outposts, strongpoints and main redoubts, and while the wood was smashed up as if some diabolical axeman had run amok in it, the *Chasseurs* were still holding out by nightfall. Although the Germans had done less well than they had expected on the first day, on the second they exploited their success: VII Reserve Corps took the Bois de Consenvoye and the village of Haumont, shrugged off counter-attacks and threatened Samogneux. The loss of the Bois d'Haumont exposed Driant's flank and the Bois de Ville, on its north-eastern quarter, had also fallen. That afternoon the Germans renewed their attack on the Bois de Caures after more shelling. This time the weight of numbers was too heavy for the skill and courage of Driant's men. They were driven back on the colonel's command post, and at about 5 p.m. he divided up the survivors into three groups, telling them to make for Beaumont, to the rear. As they left the wood they were caught by flanking fire and Driant was killed. The battle had its first hero.

On the 23rd the crack widened ominously. General Bapst, commanding the 72nd Division from his headquarters in Vacherauville, believed that Brabant, up on the Meuse in the left front of his sector, was untenable. He asked his superior, General Chrétien of XXX Corps, in Fort Souville, for permission to withdraw. Chrétien granted this, only to withdraw it soon afterwards and demand an immediate counter-attack. Bapst was in no position to comply and the Germans entered Bras at midday. Herbebois fell that afternoon and Beaumont was taken, after an epic defence, on the 24th. Although the French had now lost their entire front line in the attack sector, they were able to buy extra time on the Castelnau Line, an intermediate position between the old front line and the forts on the crest, established after a visit by General de Castelnau, Joffre's Chief of Staff, in January.

The Castelnau Line was breached when, believing that Samogneux had fallen, Herr ordered his own heavy guns to shell it, blowing the heart out of the defence. Two of Chrétien's divisions, 51st and 72nd, had effectively ceased to exist, and when he threw in his remaining division, 37th African, its soldiers, fighting in the freezing cold under the lash of paralysing shelling, fought with less than their usual ardour. By the time Chrétien's worn-out corps was replaced by Balfourier's XX Corps on the night of the 23rd, the Germans had gobbled up the whole of the Castelnau Line and were hacking their way up the slopes towards the forts.

On 25 February the 24th Brandenburg Infantry Regiment of III Corps attacked the Bois Hassoule, north-east of Fort Douaumont. The 2nd Battalion, on its right, had taken its objective with little difficulty and Sergeant Kunze, commanding the pioneer section on the battalion's left, decided to push on and look at Fort Douaumont itself. He managed to get through the wire protecting its ditch, then walked along the iron railings on the north-east face of the fort until a shell obligingly blew him into the ditch at a spot where the railings had been torn down. Most of his men joined him – shells were bursting on the fort's superstructure, so there was some merit in being in the ditch.

Kunze gained access to the fort by ordering his men to make a human pyramid, which he climbed to push his way in through an embrasure in the north-east counterscarp gallery. All but two of his men declined to follow the intrepid sergeant, who set off down a long corridor that eventually took him into the 155mm turret. The gun was banging away at some distant target, but Kunze, now alone, captured the gunners. They escaped when he lost his way in the passages below and Kunze wandered about in the bowels of the fort for some time until he found a well-stocked mess-room where he sat down for a late lunch.

Some of Kunze's comrades-in-arms had followed him into the fort. Lieutenant Radtke led his platoon into the ditch in the same place as Kunze had entered it and, using timber to improvise a ramp, climbed up on to the fort, crossing its glacis to enter the barracks, where he took some prisoners. Captain Haupt followed the same route with his company and the Germans soon met up, securing the garrison, including its unlucky commander, Warrant Officer Chenot, and taking steps to ward off counter-attack. Lieutenant von Brandis commanded the company on the extreme right of the 2nd Battalion and his men had suffered heavily from a French machine-gun in the spire of Douaumont church, causing them to lag behind the rest of the advancing battalion. He too took his men into the fort – without, by his own account, realising that any other Germans were there already – and had taken some prisoners by the time he encountered Haupt.

Brandis was eventually sent back to tell the battalion commander what had happened. He went on to spread the good news at regimental headquarters and his own part in the capture lost little in the telling. Brandis and Haupt received the *Pour le Mérite*. It was not until after the war, when Brandis, author of the best-selling *Stormers of Douaumont*, was a national hero, that Kunze, then a police constable, wrote to his old battalion commander asking him to set the record straight. Major von Klufer's well-researched account made it clear that Kunze and Radtke had indeed commanded the first two parties to enter the fort. It was by then a little late for the Blue Max, but Radtke was rewarded with a signed photograph of the Crown Prince and Kunze was promoted to police inspector: perhaps he came out of it best after all.

The loss of Douaumont rocked the defence back on its heels and there was something approaching a panic in the streets of Verdun. However, that day Castelnau himself was in Verdun and he reported to GQG that, Joffre's scepticism notwithstanding, the right bank could be held. He had already proposed that the 2nd Army, now in reserve, should be put into Verdun. Its commander was a big, rather stiff infantryman from the Pas de Calais – Philippe Pétain. He had been close to retirement, commanding the 33rd Regiment at St-Omer, when war broke out, and lightning promotions had taken him up to head the 2nd Army in the autumn of 1915. Pétain reported to GQG at Chantilly on the morning of 25 February and drove to Herr's headquarters at Dugny, on the Meuse south of Verdun. He went on to Souilly, on the Bar-le-Duc road, where he met Castelnau, who handed him a written order to defend the right bank and told him to take command at midnight that night. Pétain phoned both Balfourier and his opposite number on the left bank, General de Bazelaire, sketched out to his chief of staff the line he wanted holding and fell asleep in an armchair. He awoke with double pneumonia and for the next few days commanded by remote control, with his faithful staff shuttling between Souilly and the front.

Just as the Germans had used artillery to pulverise the defence, so Pétain looked to the power of the gun. While his infantry held the line of the forts, his gunners reached out to claw German troops moving across the raw-backed ridges north of Douaumont, and to interpose their own barrages between German attackers and French defenders. The voracious appetites of French batteries, and need to move up fresh troops and relieve exhausted ones, threw an enormous burden on to the Bar-le-Duc road, the only artery connecting Verdun with the outside world. It speaks volumes for the ingenuity of Major Richard, responsible for this tenuous lifeline, that it remained open even when a sudden thaw in late February turned it into a ribbon of mud. When the road was at its busiest, in June that year, a vehicle passed along it every fourteen seconds. By night it looked like some great pullulating caterpillar, its glowing coils stretching round corners and over crests.

The Germans paid ever more heavily for each success. On 4 March they took Douaumont village after a week's battle, which had decimated Pétain's old regiment: a young company commander, Charles de Gaulle, was among those captured. Many of the increasing casualties being suffered by the Germans were the result of the fire of guns from across the Mcuse: French batteries under the Bois Borrus ridge, and behind the parapets of the forts there, could fire down the re-entrants and on to the reverse slopes on the right bank.

The importance of the left bank had been recognised by both Seré de Rivière and pre-war German studies of Verdun, but it was not until early March that Falkenhayn agreed to put extra reserves at the Crown Prince's disposal to enable him to extend his offensive across the Meuse. He was also ordered to deal with Fort Vaux, which now threatened the left flank of the German attack. Although the new thrust came as no surprise, its early stages proved surprisingly successful: the Germans threw troops across the Meuse at Brabant, captured Forges and Regneville, established themselves on Goose Ridge and began to push up into the Bois des Corbeaux.

The slopes on both sides of the Meuse are so much more thickly wooded now than they were in 1916 that it is easy to miss the significance of the Bois des Corbeaux, now simply the eastern end of a pine forest that sprawls across to the Argonne. Modern woods cover the long ridge that dominates the country north of Bois Borrus. It has two peaks. One, with the sinister name of Mort-Homme, rises above the village of Chattancourt, and gives an uninterrupted view across the Meuse on the right flank and over the valley of the little Forges brook to its front. To its east stands the second peak, Hill 304, whose field of view stretches off to the Argonne. In French hands the Mort-Homme–Hill 304 feature enfiladed German gains on the right bank and cushioned the guns on the Bois Borrus ridge: in German hands it would imperil Bois Borrus and leave the forts on the left bank vulnerable to a turning movement through Esnes and Montzeville. Attack on the front of the ridge, across the Forges brook, would inevitably be costly, but infiltration along it, through the Bois des Corbeaux and the Bois de Cumières, offered better prospects.

The results of 6 March, the first day of the attack, confirmed the evidence of the ground. Frontal assault on the Mort-Homme failed, but French troops holding the Bois des Corbeaux were badly shaken. On the 7th the whole of the wood was in German hands, although a spirited counter-attack drove them out the following day. The wood changed hands again on the 10th, but by now the defensive crust had hardened and no cheap or easy success could be expected. The Germans were faring no better on the right bank. Although their initial attack took them into Vaux village and up to the edge of the fort, they could get no further forward.

With the Mort-Homme holding fast against attacks from front and right flank, the Germans shifted their gaze westwards to Hill 304. Again they recognised that the flanks offered better prospects than the front, and again early efforts were crowned with success: on 20 March the 11th Bavarian Division took the Bois d'Avocourt. This was as far as easy gains went, and thereafter the Germans took ground slowly and expensively.

On 9 April they stepped up their efforts, launching simultaneous attacks on both banks of the river, with General von Mudra in command on the right bank and General von Gallwitz on the left. This brought a slogging match of the most obdurate sort as the French fought for every yard. Gallwitz was an artillery expert and it was thanks to his efforts that Hill 304 was smothered by a two-day bombardment, with over 500 heavy guns pounding its slopes. It fell on 6 May, and with its capture the Germans were able to bring their full weight to bear on the Mort-Homme and it too was in their hands by the end of the month. Yet even though the Germans were at last across the ridge, there was no sudden breakthrough. Bois Borrus still snarled down on them, and the broad Esnes–Chattancourt valley lay between them and its forward slopes.

Like the land battle, the war in the air had also become attritional. The Germans enjoyed comfortable air superiority when they began Operation *Scaffold*, but the French quickly shifted six out of their fifteen fighter squadrons – including the famous *Cigognes*, the Storks – to the Verdun front, and were soon shooting down scores of observation balloons and reconnaissance aircraft. In March the Germans regained the initiative, partly through the efforts of their ace Oswald Boelcke, based at Sivry-sur-Meuse, a few kilometres downstream from the battle. The French hit back, using large groups of fighters to protect their observation aircraft. In May the *Escadrille Lafayette*, a squadron of American pilots serving in French uniform, arrived on the Verdun front and on 24 May William Thaw opened the scoring for the Americans in that sector by shooting down a Fokker over Fort Douaumont.

The pendulum might easily have swung the other way. Boelcke developed the *Jagdstaffel*, or hunting-pack, but the death of Max Immelmann made it too dangerous to risk the loss of another ace: Boelcke was taken off flying duties and was unable to put his new tactics into practice at Verdun. When he was allowed back into action in July he was sent to the Somme, where the *Jagdstaffel* concept worked murderously well. It was fortunate for the French that it was never properly used at Verdun, for they never again lost command of the air there. And with command of the air went free use of observation aircraft for artillery spotting: from May onwards German balloons and aircraft were constantly at risk, while the French were able to send well-escorted observation aircraft over German lines.

As the French gained the upper hand in the air, the war on the ground grumbled on. A new offensive on the right bank was delayed by bad weather and the effectiveness of French guns on the far side of the Meuse. Mudra, who had grown pessimistic, was relieved by the more sanguine General von Lochow, but not before his gloom had infected the Crown Prince, who had become convinced that German gains at Verdun could never be commensurate with the losses. The

Crown Prince told Falkenhayn that he had decided to cancel the attack, but Falkenhayn swept down upon his head-quarters at Stenay and, with the support of Knobelsdorf, emphasised that the battle would go on as planned.

Pétain, too, was in difficulties with his superiors. For all his success in re-establishing the front and organising a system of *roulement* which replaced exhausted divisions with fresh ones, Pétain lost the favour of Joffre. He argued against diverting resources to the Somme that summer and believed that Verdun should be accorded priority. In late April he was kicked upstairs, promoted to command the *Groupe des Armées du Centre*, with his headquarters at Bar-le-Duc, and his place at the head of 2nd Army was taken by General Robert Nivelle. Nivelle had been a gunner colonel when war broke out and rose rapidly to command III Corps.

One of Nivelle's divisional commanders was Charles Mangin, a tough forty-nine-year-old colonial infantryman. On 22 May he had put in a sharp attack on Fort Douaumont: some of his men managed to set up a machine-gun on the fort's Bourges casemate, but the gain could not be secured and on the 24th the Germans snuffed it out. This abortive attempt was the last French offensive for some time: the long-awaited German attack was about to begin. Three corps – I Bavarian, X Reserve and XV – were launched in Operation *May Cup* on the front running from Vaux to Thiaumont. Fort Vaux, commanded by Major Raynal, had lost its single 75mm turret when a German shell exploded a demolition charge left there. The fort was full of wounded and stragglers. Nevertheless, when the Germans broke into the fort's counterscarp galleries on 1 June, Raynal's men defended the tunnels leading into the body of the fort foot by foot. Vaux was encircled on 3 June and although French guns – including the 155mm from Fort Moulainville to the south – repeatedly scraped Germans off the superstructure, the weight of numbers and firepower eventually told against Raynal. It was not until the garrison had used up all its water and a relief attempt had failed that Raynal at last surrendered. He had held out for a week and inflicted over 2,500 casualties on the Germans, and was received by the

Crown Prince at Stenay with the courtesy his gallantry merited.

The day after the Germans took Vaux they entered the *Ouvrage de Thiaumont*, which commanded the northern end of the Froideterre ridge and the long slope leading south-east, past Fleury, to Fort Souville. The French put in an immediate counter-attack and the work changed hands several times over the weeks that followed. Nivelle's policy of counter-attacking lost ground sometimes produced useful results, but more often it added to the butcher's bill to no purpose and stimulated Verdun's appetite for flesh. Joffre's insistence on pressing ahead with preparations for the Somme made it difficult to rotate fresh divisions through Verdun and it was hard to sustain morale as the same troops ascended the calvary of the Bar-le-Duc road for the second or third time.

Pétain was caught between the hammer of Joffre and the anvil of Nivelle, and his determination became tingled with a deeper pessimism and the suspicion – echoed by many of his subordinates – that the British should be doing more to help. It was every bit as hard for the Germans, who kept divisions in the sector for longer and topped them up with replacements. Verdun had now gulped so much blood that the French could not abandon it – just as Falkenhayn had predicted. Yet the Germans were jammed equally tightly on to the conveyor belt leading to what their soldiers' slang called 'the Mill on the Meuse'. The Crown Prince would have given up the attack had he been allowed to, but Falkenhayn drove him on. Even Falkenhayn was not able to bend his will to Verdun alone. The offensive brewing astride the Somme would consume some of his reserves and Austrian requests for assistance – the Russian General Brusilov had just ripped a huge hole through their armies in Galicia – also tapped into his finite resources. It was not until 23 June that he was in a position to administer the killing blow at Verdun, aimed at the ridge topped by Fort Souville: with this taken the Germans would have an uninterrupted view over Verdun.

The attack was preceded by a lavish bombardment with

the new 'Green Cross' phosgene gas, against which French gas masks were not totally effective. On their right, the Germans captured Thiaumont, overran the *Ouvrage de Froideterre* and reached the command post known, from its four ventilation shafts, as the *Abri des Quatre Cheminées*. On the left, the Germans took Fleury and their tide lapped at the foot of Souville. The defence was parlous but it held and over the next few days Nivelle stoked up the inevitable counter-attacks. The climax of the battle had passed. On 1 July the Allied offensive began on the Somme and over the weeks that followed it was the chalklands of Picardy, not the charnel ridges above the Meuse, that would dominate German strategy.

In early July the Germans made their last throw. On the 7th they took the Damloup battery south of Vaux, and on the 12th a small group of Germans broke through to Souville. From its roof they glimpsed the towers of Verdun cathedral, but were swept away by the surviving garrison before they could be reinforced. The failure convinced all but the most obdurate German commanders that they could never take Verdun and their efforts should be devoted to consolidating the ground they held. Knobelsdorf was sent off to command a corps in Russia and in late August Falkenhayn himself was dismissed. His successors, Hindenburg and Ludendorff, at once cut off the flow of men to the Mill on the Meuse.

In October the French launched a series of attacks for once supported by adequate artillery, including some new 400mm heavies to deal with German-held forts. Douaumont was the focus of French interest. Its interior had been little damaged – although an accident in an explosive store in May had killed 650 Germans, who lie walled up within it. The Germans had dug communication trenches up to the fort and had begun work on a tunnel leading towards their front line. On 23 October the French began to shell the fort with their 400mms, whose shells spun down through the concrete to explode deep within, causing fires and filling the place with fumes. The garrison commander evacuated Douaumont and although a stray captain eventually mustered a handful of men to defend it, when the meticulously planned assault surged in behind a curtain of fire on 24 October, the fort was taken as easily as it had been lost eight months before. Vaux was recaptured on 2 November and on 15 December the French pushed their line out still further, taking Louvemont, north-west of Douaumont, and Bezonvaux to its north-east.

Verdun was one of the most costly battles in the world's history. Both sides lost roughly the same: the French admitted 377,000 and the Germans lost at least 337,000. One estimate puts the total casualties even higher, at 420,000 killed and 800,000 wounded. To the dismal list of men physically wounded at Verdun we must add the even longer toll of those who would never recover psychologically from the experience. Any First World War battle imposed its own mental strain, but the circumstances of Verdun were so horrific as to make its pressures especially heavy. The weather was often appalling: the wind cuts across the ridges like a razor, and I have been cold on top of Vaux even in March with a coat on and a hot meal inside me. The concentration of artillery fire was unprecedented and there was little opportunity to bury the dead – nor any guarantee that they would not be disinterred by the next shell. Soldiers in the front line were usually short of food and water, for each side could interdict its opponent's lines of supply with artillery, and the unsung heroes of the battle were the soldiers on ration parties, toiling across the moonscape with water bottles and ration loaves.

It is easy to trace the scars of Verdun into the inter-war years. The evidence of Verdun was to be used to justify the construction of the Maginot Line and, appropriately enough, André Maginot himself – Minister of War when work began on the Line – had served at Verdun. Although Under-Secretary of State for War, he had volunteered for service in 1914, won the coveted *médaille militaire* as a sergeant in the 44th Territorial Infantry and been badly wounded near Maucourt, north-west of Verdun, on 9 November 1914. Pétain believed that Verdun proved the folly of fighting machines with men, a view that encouraged him to support the construction of the Maginot Line and later to recognise that French infantry could not beat German tanks. Indeed, young Frenchmen of another generation were to blame the moral paralysis of 1940 on the 'Men of Fifty' who had fought at Verdun.

Although Verdun was badly damaged by shelling, the destruction was not as complete as in some hard-hit towns of the north-west. The cathedral had been destroyed by one fire in 1048 and damaged by another in 1755, and the shelling of 1916 actually exposed romanesque work once hidden by eighteenth-century baroque. Some of the columns supporting the crypt date from the twelfth century, but others post-date the battle and their capitals are decorated with Le Bourgeois's scenes from the fighting of 1916. The eighteenth-century bishop's palace lies between the cathedral and the citadel, and the medieval Porte Chatel stands nearby. The Hôtel de la Princerie, once the residence of the Princier, senior figure in the diocese after the bishop, houses the municipal museum, which includes a room full of indifferently displayed arms and armour.

The shadow of 1916 lies heavily on the city and its monuments. The underground portion of the citadel, approached along the Rue du 5^e RAP – 5th Artillery Regiment Street – contains a small museum, and a sound and light show which tells of events in these long vaulted corridors during and after the battle, and there is a miniature railway tour of the underground galleries. On 10 November 1920 the French Unknown Soldier was chosen from among eight coffins in one of the underground galleries by Private Auguste Thin of the 132nd Regiment. He added together the regimental number on his collar and placed the little bunch of red, white and blue flowers, to signify his choice, on the sixth coffin, which now lies beneath the Arc de Triomphe. The other coffins were interred in the military cemetery in the Faubourg Pavé, under Belleville Ridge. Just outside the entrance to the *citadelle souterraine* is an avenue flanked by statues of the notable marshals and generals of France.

The Victory Monument, a cloaked and helmeted figure leaning on its sword, stands, flanked by two guns (they are

Russian pieces, captured by the Germans and captured in turn by the French), looking out to the Meuse at the end of the Avenue de la Victoire, and in the crypt below it is the Livre d'Or, many volumes rather than a single book, containing the names of French and American combatants who fought at Verdun in the First World War, as well as those of American soldiers who liberated the city in 1944. The Porte St-Paul, a fine example of an eighteenth-century fortress gate, complete with drawbridge, stands at the northern end of the Rue St-Paul, with Rodin's monument to the defence of Verdun in front of it. A monument to the dead rests against the earthworks of Vauban's bridgehead at the east end of the Pont Chaussée and there is a plaque to General Mangin on the wall nearby. There are military statues at both ends of the Quai de la Republique. At its northern end, appropriately enough, stands the fiercely republican General Sarrail, commander of 3rd Army in the Argonne battles of 1915 and subsequently commander of the expeditionary force sent to Salonika. At the other end, in the square named after him, stands François de Chevert, Verdun's eighteenth-century local hero.

The Memorial-Museum on the right bank is the best place to begin a visit to the battlefields. It stands near the site of Fleury-devant-Douaumont, one of the many villages totally destroyed and never rebuilt, though its site is made all the more striking by signboards which mark where farriers and shoemakers once worked. Unlike so many battlefield museums it does not consist of tired artefacts jammed into seedy cases. At its centre is a life-size representation of part of the battlefield, littered with wire, shells, mortar bombs, helmets and sniper shields, with the wreckage of a German 77mm gun in the foreground. Around it, on two storeys, are well-presented exhibits, and an audio-visual display tells the story of the battle. I find the film in the museum's cinema harder and harder to cope with. There is something remorselessly terrible about its soundtrack, from Brahms's *German Requiem*, and the slides of stolid Lorraine peasants, like their fathers and grandfathers cultivating a landscape that was shortly to be transformed for ever, always makes me cry. A little way south of the museum is the *Massif fortifié de Souville*, with Fort Souville at its centre, recently opened up with easily-accessible paths and helpful orientation tables.

North-west of the museum stands Douaumont ossuary, its long vaulted roof surmounted by a tower and lantern. Work began in 1920 and the ossuary was inaugurated in 1932. There are eighteen alcoves in the long, dimly lit building, each containing two granite tombs, surmounted by the name of one of the battlefield's sectors. Directly below lie the remains of unidentified soldiers, French and German, found in that sector. There were simply too many bones from the most fought-over sectors to fit into the designated space, so new vaults were added at each end of the ossuary to take the extra remains. There is a cinema in the ossuary, and halfway up the tower is small war museum. At the top, orientation tables enable the visitor to pick out the salient points of the battlefield. Beneath the ossuary lie the bones of some 150,000 men. Windows set at ground level in the building's exterior give a heart-stopping view of chalk-white relics of humanity; it is impossible to glance in and not be moved. The ossuary stands on Thiaumont crest and on the slopes below it lies Douaumont National Cemetery, containing the graves of 15,000 soldiers.

Charles de Gaulle had been wounded and captured as an infantry captain on this very crest and his visit to Verdun with West German Chancellor Konrad Adenauer in 1962 marked an important step in Franco–German reconciliation. The meeting of President Mitterrand and Chancellor Kohl in 1984 was scarecely less symbolic. Jean de Lattre de Tassigny, who, like so many of his generation, had fought at Verdun, commanded a division in 1940, led the First French Army in Alsace-Lorraine and Germany in 1944–5, and represented France at the surrender ceremony in Berlin on 8 May 1945. He then was High Commissioner and Commander-in-Chief in Indo-China, where his only son was killed in action, but was forced to return to France because of ill health and died of cancer in 1952: he was posthumously created Marshal of France. He is commemorated not far from the ossuary, near the ground he fought on as a young man.

Fort Douaumont is a short distance from the ossuary, approached along a road that crosses a section of communication trench and passes isolated graves. Its exterior has been so ravaged that it is useful to have looked at the model of Douaumont in the museum to remember what it was once like. The interior, dripping and gloomy though it is, is surprisingly intact. The 155mm turret is worth a visit – with Kunze's exploits in mind – and the walled-up vault at the western end of the main corridor, with a plaque commemorating the German soldiers killed in the accidental explosion of May 1916 is, in its own way, as moving as the ossuary.

South-west of the ossuary, past a memorial to French Jews killed in the war, the road passes the *Abri des Quatres Cheminées* and the *Ouvrage de Froideterre*. The latter played an important part in the battle, for its fire interlocked with that of the *Ouvrage de Charny*, on the left bank, to secure the valley. It includes good examples of *casemates de Bourges* and a pair of turreted 75mms.

West of the ossuary, along the road that leads to Bras, is the *Tranchée des Baionnettes*. During the fighting of June this sector was held by 3 Company of the 137th Infantry, which was effectively wiped out by German shellfire. After the war a section of filled-in trench was found with rusty bayonets protruding from it. Beneath each bayonet was a rifle and beneath each rifle the body of its owner. It was concluded that the men had been ready to repel an attack when they were buried by shellfire. The story gained wide currency and an American benefactor provided the structure that now covers the spot. The British historian, Alistair Horne, whose *The Price of Glory* remains one of the best books on Verdun, is right to cast doubt on this story. It is unlikely that a trench thirty metres long was filled with earth by a single salvo and that all the men in it were killed at once, but more probable that the soldiers of 3 Company were given hasty trench burial by the Germans, with a rifle left to mark each soldier's resting place. Seventeen

unknown soldiers still lie there: another forty identifiable
bodies were buried at Douaumont. I find it a dreary place.
Souvenir hunters have made off with most of the bayonets
and the concrete roof gives an air of bunker-like gloom.

Continuing on down into Bras, turning right along the
D964, then forking right in Vacherauville, takes the
traveller between the long shoulder of the *Côte du Poivre*
(Pepper Ridge) and the spur crowned by Hill 344, both scenes
of heavy fighting in February 1916. After a few kilometres
the memorial to Colonel Driant appears to the left of the
road and, just past it, a sign points to his command post.
This lies inside the Bois de Caures, still pocked by shellfire.
The command post itself is a simple concrete shelter,
surrounded by a low fence whose pillars bear the bugle of the
French army's *chasseur* battalions. It was surprisingly close
to the front line. About 500 metres to its north the D125
crosses a line of German *Stollen*. Although these are well
tucked away in to the woods, the communication trenches
leading up to them betray their positions. One, just west of
the road, still has its name – *Nilpferd*, Hippo – inscribed on
the lintel above its entrance. The original occupants must
have had a sense of humour, because the bunker is full of
water, just as so many were in 1916.

Fort Vaux can be reached from the museum at Fleury,
along the D913, which passes the lion monument at the
high-water mark of the German advance. The road to Vaux
turns off a kilometre onwards and crosses the Tavannes
valley, where there stands a monument to the *fusillés de
Tavannes*, Resistance workers shot by the Germans in the
Second World War. Fort Vaux itself is externally in the same
battered state as Douaumont – visible across the valley to its
left. When he could not get through by telephone or helio-
graph, Raynal used pigeons to communicate with Souville,
and a plaque on the rear wall of the fort commemorates
the pigeon-fanciers who died for France and Major Raynal's
last pigeon.

The left bank of the Meuse is visited less often. The D38
follows the river through Charny and Marre, and swings
hard left to Chattancourt. Here the tiny D36b climbs up to

the Mort-Homme. Although storms and reforestation occasionally open up some vistas, the pines mask the hill's military significance. A half-shrouded stone figure clutching a flag stands on the summit: it is a monument that manages to combine stunning ugliness with effective symbolism. Hill 304 is reached by going back to Chattancourt, then turning right along the D38. Shortly after leaving Esnes-en-Argonne the road forks and the D18 climbs up on to the ridge. At its top the short, straight D18b goes on to the huge Hill 304 memorial, with a small private memorial to a lieutenant of zouaves striking, to my mind, a more poignant note. The visitor who has time and energy should drop down to the other side of the ridge, where the road passes through the destroyed villages of Haucourt, Bethincourt and Forges. From the crossroads on the site of old Bethincourt a track leads south-eastwards and goes up the Mort-Homme from the German side, passing the entrance to the Crown Prince's tunnel, a subway 3,000 metres long, which enabled some Germans to reach the front line in relative safety.

The forts on Bois Borrus ridge can be seen from Marre on the D38. Looking south from the village to the ridge, Bois Borrus itself is in the woodland on the right; Fort Marre is in more trees, across a re-entrant to its left: the *Poste de la Belle Epine* is in the next clump of trees to the left and Fort Vachereauville stands, at the left-hand end of the spur, in yet more trees.

The road to Bar-le-Duc leaves the city under the guns of Forts de la Chaume and Regret. A few kilometres along the N3 the N35 swings off southwards, near the village of Maison Brulée and a memorial to the *Train des Equipages Militaires*, the French army's service corps, without whose Herculean efforts the battle of Verdun could scarcely have been fought, let alone won. The milestones along the N35, each topped with a *poilu*'s helmet and decorated with a frond of palm, bear the title given the road by a grateful country: *La Voie Sacrée* (The Sacred Way). Souilly is ten kilometres from the junction. It was Pétain's headquarters during the battle, and there is a small collection of photographs and memorabilia in the town hall. Pétain often stood on the steps leading up to the building from the road as troops marched up to the Golgotha above the Meuse and he watched their tired remnants trudging back.

Verdun remains schizophrenic about Pétain. After the Second World War the marshal was tried for treason and sentenced to death. His sentence – like Bazaine's, two generations before – was commuted to life imprisonment. Unlike Bazaine, who escaped to die in Spain, Pétain died in captivity on the Ile de Yeu, where he is buried. He had always wanted to lie at Verdun alongside the soldiers he had commanded and the issue of whether Pétain's bones should moulder by the Atlantic or beside the Meuse still generates more heat than light. I suspect that generosity of spirit will eventually prevail. Whatever Pétain may have done in later life, in 1916 his was the brain that nerved the defence of Verdun, his the spirit that inspired it. And Souilly is no bad place for us to take our leave of him, for it was here, on 10 April 1916, that he signed his general order with the concluding sentence: *'Courage! On les aura! . . .'*

There was little movement on the Verdun front between the French offensives of late 1916 and the last few months of the war. In August 1918 Foch asked General Pershing, commander-in-chief of the American Expeditionary Force, to pinch out the St-Mihiel salient so as to secure the Allied right flank. Pershing, with his headquarters at Ligny-en-Barrois, had four corps at his disposal for the operation, the 1st, 4th and 5th US, and the French 2nd Colonial Corps. The main attack was entrusted to the 1st and 4th Corps, driving northwards from the southern flank of the salient, while the 5th Corps attacked its western flank. Once the attack was under way, two French divisions would advance across the nose of the salient to occupy St-Mihiel itself.

The French and Americans enjoyed considerable numerical superiority, outnumbering the defenders by about four to one, and had 3,000 guns, 270 Renault light tanks and no less than 1,500 aircraft to support the attack. The Germans recognised that the salient was ultimately untenable and had planned to withdraw from it. The American attack, launched early on 12 September, caught them wrong-footed. The salient had disappeared by 15 September, and the Germans lost 16,000 prisoners and 443 guns. American losses, although light in view of the ground gained and casualties inflicted, nevertheless numbered 7,000.

The best starting point for a visit to the St-Mihiel battlefield is the circular colonnade of the American Memorial on the Butte de Montsec, in the middle of the salient and overlooking the Lac de Madine, a massive reservoir in the Parc Régional de Lorraine. There is an excellent field of view from this commanding point and orientation tables help the visitor to make sense of it. Over 4,000 Americans are buried in the graceful St-Mihiel American Cemetery on the D67 just north of Thiaucourt-Regnieville. It is a cemetery very much in the American style, with marble, bronze and well-manicured avenues of trees. The new TGV line, which cuts like an arrow across the salient, passes within about a kilometre of the cememtery.

There are untidier sections of battlefield. The *butte* at Les Eparges lies on the D154 south of the Verdun–Metz road. It is atop a long finger of ridge pointing out into the Woëvre and was bitterly contested in 1914–15: the mining beneath it rivalled that at Vauquois. There are several memorials on the mine-riven hill, including one to the missing – *Ceux qui n'ont pas de tombe* – and another to 12th Infantry Division, bearing its badge, the Gallic cockerel. At the very tip of the salient, just south of the village of Ailly-sur-Meuse, south of St-Mihiel, a track runs through the Forêt d'Apremont, along the front line of 12 September 1918 in the sector attacked by the French 39th Infantry Division. An obelisk pays tribute to the French 8th Corps and behind it lies the extensive remains of a trench system called the *Tranchée du Soif* – Thirst Trench – with its numerous concrete dugouts and troops shelters. The whole *Bois d'Ailly–Tranchée du Soif* sector has now been sympathetically developed, with circular works and orientation tables in three languages. The sector is well worth a visit, not least because it contains good examples of the use of reinforced concrete in front-line position in what was the proverbial 'quiet sector' for much of the war. The fire-step, that essential element of a trench, mounted by its defenders so that they could fire across the parapet, is built into the external wall of these bunkers.

Allied victory in 1918 was followed by the enthusiastic re-assimilation of the lost provinces into France. The veteran commander, General de Maud'huy, had sworn never to enter a theatre until Metz was French again and after its liberation he duly went to the theatre in Metz, accompanied by *chasseur* buglers blasting out their regimental march, 'Sidi-Brahim'. Pétain received his marshal's baton on the Esplanade of Metz – as Philip Guedella wrote, 'the unhappy ghost of the Army of the Rhine was laid.' The recovery of Alsace-Lorraine was not all bugles and cheers. Part of the enthusiasm for French return had been provoked by harsh German behaviour during the war and Alsatian autonomists were active soon after the peace. It took longer for an autonomist movement to gain momentum in Lorraine, but there were tensions between France and both her newly recovered provinces in the 1920s and 1930s.

Construction of the Maginot Line helped increase tensions, for the inhabitants of the frontier provinces were well aware that there would have to be large-scale evacuations from the area around the Line in the event of war. Lorraine was shielded by defence works of the Metz *Région Fortifiée*, which initially ran from Téting on the River Nied to north of Thionville and was subsequently continued by the 'Montmédy bridgehead'. There were twenty-seven small infantry forts and sixteen larger artillery forts between Téting and La Ferté, north-west of Montmédy.

This main defensive line – the *position de résistance* – lay behind *avant postes*, bunkers armed with anti-tank and machine-guns, with *maisons fortes*, the fortified barracks of the armed frontier police, close to the frontier. The Line itself posed a continuous obstacle to an attacker. Between the forts themselves ran anti-tank obstacles made from lengths of rail, reinforced by barbed wire. Interval case-mates, bunkers mounting anti-tank and machine-guns, covered the obstacle. The forts, situated an average of 5 kilometres apart, consisted of a number of combat blocks connected by underground corridors, with an underground railway linking the fighting part of the fort to its barrack and 'factory' area. The number and style of blocks depended on the type of ground and the importance attached to its defence, and forts ran from the smaller infantry works with a garrison of 200 men commanded by a captain, to larger artillery forts with 1,200 men commanded by a lieutenant-colonel.

Several of the Maginot Line forts in Lorraine may be visited and at least two, Hackenberg, east of Thionville, and Fermont, north-east of Longuyon, are fully restored and open to visitors. We will take Fermont as our example of the Maginot Line fort, just as Le Quesnoy served as the model of eighteenth-century fortification. There are two separate entrance blocks at the rear of Fermont, one for personnel and the other (now the visitors' entrance) used for stores and ammunition: narrow-gauge railways led off to the rear so that matériel could be moved up by train. Ammunition handling was assisted by a monorail system, which enabled the metal cages containing shells to be swung up off the external railway, into the lift and then on to the internal railway. Each entrance block is topped by armoured observation cupolas and defended by twin 8mm Reibel machine-guns, with an ingenious mounting which enables the machine-gun to be swung away and replaced by a 47mm anti-tank gun on an overhead rail. The narrow ditch is crossed by a metal drawbridge, and grenade dischargers enabled the defenders to deal with infantry who might seek refuge in it.

Inside the armoured air-tight door the short entrance corridor, itself defended by a machine-gun chamber, leads to another armoured door and the lift, which takes men and stores the thirty metres down to the barrack and factory area. This includes a power station with diesel generators, kitchens, a bakery, hospital – with a separate entrance for gas casualties – and barrack rooms for its garrison of 600 men. The accommodation is certainly short of creature comforts. There is no dining room: meals were eaten at collapsible tables along the corridors. Men slept cheek-by-

jowl on steel beds with long leaf springs and even the
commander's room is scarcely luxurious. The ventilation
system kept the fort at a slight over-pressure to reduce the
danger of poison gas being introduced into it and the air
intake was protected by a battery of air filters. The main
magazine (M3), built to contain 3,000 rounds per gun, lies on
the same level as the barracks. A loop of railway line runs
round the magazine chambers. The magazines at Fermont
now hold a museum of the Maginot Line, with a
comprehensive collection of arms and equipment from
Fermont and other forts.

An electric railway leads from the rear of the fort to the
seven fighting blocks 1 kilometre to the north. Block 1
contains a retractable cupola for twin 75mm guns, and
Blocks 2 and 6 a retractable cupola for twin machine-guns.
Block 3, the fort's main observation post, has a fixed
observation dome with machine-guns. Block 4 is the most
impressive: it is a battery of three short-barrelled 75mm
guns firing from a concrete casemate. There are two breech-
loading 81mm mortars in a retractable cupola in Block 5,
and Block 7 is a casemate housing a 47mm anti-tank gun
and machine-guns. Each block resembles the turret of a
warship. At its base is the block magazine (M2) with 2,800
rounds per gun. A lift and stairway, with rest accommodation for gun detachment, lead to the gun position with
its ready-to-use magazine (M1) of 600 rounds per gun. The
naval comparison is heightened by the signal dials used for
the transmission of fire orders from the command post to
the fighting blocks.

Although the short-barrelled guns installed in Maginot
Line forts had the same range limitations as those in the
Verdun forts, the distances between forts were short enough
for this not to be a serious disadvantage: Fermont enjoyed
the flanking support from Chappy Farm 2.75 kilometres to
the west and Latiremont six kilometres to the east. The
optical equipment in the fort was checked by the use of two
distant aiming points. One was the *crassier* in nearby
Longwy – the other, by the most bizarre irony, the tower of
the ossuary at Douaumont, 35 kilometres to the south-west.

The forts were first occupied during the Rhineland crisis of 1936. Lighting and heating were inadequate, and there were complaints about the damp: some remedial work was done in the period 1936–9. The experience also revealed shortcomings in day-to-day procedures within the forts and these were improved by studying the way that warship crews operated. Thereafter garrisons worked on a three-watch system – duty, standby and rest.

The Line began to prepare for war on 21 August 1939, before hostilities had officially commenced. Reservists arrived to bring garrisons up to strength, interval troops deployed and the civilian population was evacuated from the area around the forts. Given that the German invasion of Poland was the *casus belli*, Gamelin felt a moral obligation to do what he could to help the Poles, and planned an advance as soon as the French army had been mobilised and fully equipped. In the meantime he mounted a modest offensive, with strictly limited objectives, into the Saarland. This, in its way, was a ghostly replaying of past history. Just as the French army in 1870 had begun with an advance on Saarbrücken, so in 1939 it moved as cautiously over the same ground – although on a rather wider front. Nine divisions advanced as far as the outposts of the *Westwall*, but there was no serious fighting. Gamelin, weighed down by the psychological baggage of the First World War, had already announced: 'I shall not begin the war by a Battle of Verdun.'

The winter of 1939–40 was the coldest for years, and the weather combined with boredom and German propaganda to reduce morale amongst the troops holding the Line, now nicknamed *le trou* – the hole. There was some patrolling activity to its front and the guns of the large fort at Hochwald, north of Strasbourg, occasionally fired into Germany. Most forts – Fermont amongst them – were out of range of Germany and had nothing to patrol against, so the garrisons lived out their troglodyte existence, their outlook confined by the narrow horizons of steel and concrete, with only leave to look forward to. Even the opening of the German offensive did not change things instantly. The forts were initially left well alone and it was not until the Meuse front had been pierced that the Line came under attack. Forts at the northern end of the Line, where the flank had been turned, were most vulnerable. The small fort at La Ferté, defended by only 240 men and commanded by a subaltern, fell on 18 May after an explosion caused an electrical fire, which asphyxiated its garrison.

Fermont, under the command of Captain Daniel Aubert, was attacked on 17 June. Elements of the German 161st Infantry Division, which had outflanked the line from the north, attacked Fermont from the west. It had not been sited to resist assault from this quarter and the Germans were able to position an 88mm gun on the western edge of the Bois de Reuville – the wood that conceals the entrance blocks. One of the first rounds knocked out an observation cupola above the personnel entrance, killing Private Florian Piton, on duty there. Because most of the turrets above the fighting blocks retracted, they offered poor targets, so the gun turned its attention to the vertical concrete wall of Block 4, the three-gun casemate. It made excellent practice, hitting the same spot time and time again. Eventually it had broken right through, but the cloud of concrete dust prevented the gunners from seeing the damage they had done. They ceased firing and after dark the garrison repaired the smashed wall with quick-drying cement. When dawn came it seemed to the Germans that their shelling had had little effect, so that particular attempt was not repeated.

On 21 June the Germans shelled the fort all morning with four 305mm Skoda howitzers, three 210mm howitzers and six batteries of 105mm field guns. This was precisely the sort of punishment that Maginot Line forts had been designed to take and Fermont duly shrugged it off. When a German battalion attacked that afternoon it was stopped in its tracks by fire from the retractable turrets, and the Germans asked for a truce to carry off their dead and wounded.

Similar stories were repeated elsewhere. When the German 1st Army launched an attack on the front Saaralbe–St-Avold on 14 June, the small forts at Kerfent and

Bambesch were taken, but only after heavy bombardment and a most creditable resistance. The great majority of forts were still in French hands when the armistice came into force on 25 June and a radio broadcast ordered them to cease firing. It was widely expected that since the garrisons had not surrendered they would be permitted to depart for the Unoccupied Zone of France. This proved a vain hope and the garrisons, who marched out with the honours of war, went into captivity. Captain Aubert duly surrendered Fermont, but soon escaped from prisoner-of-war camp and joined the Free French: he was killed in action in North Africa.

Outside the fort at Hackenberg is a memorial to the US Army, which liberated the area in 1944, and there is an American M-10 *Wolverine* tank destroyer on the approach road. There is also a French memorial to the British Expeditionary Force 1939–40, and to all the Maginot Line garrisons and the interval troops who covered the gaps between the forts. At the top of the memorial is a saltire, presumably a reference to the fact that 51st Highland Division was sent to this sector as a gesture of inter-Allied solidarity. Although it was not caught in the line when the Germans broke through, we have already seen how it reached the coast at St-Valéry, only to be captured.

The last had not been heard of the Maginot Line. The Germans removed some of its guns and turrets for use in the Atlantic Wall. In late 1944 some of the forts were used against the advancing Americans of Patton's 3rd Army. The forts that caused the most serious difficulties were the cluster around Metz and it is to the assault on these, the culminating point of Patton's Lorraine campaign, that we must now turn our attention.

Lorraine had few attractions as a route into Germany in 1944. Its terrain was as unappealing to an army moving westwards as it was to one advancing from the east and the industrial area of the Saar was a less important objective than the Ruhr. However, Eisenhower's concept of a broad front advance put Lorraine in the path of the Allied armies. If the campaign was given part of its distinctive character by the ground it was fought over, another pungent ingredient was the personality of Lieutenant-General George S. Patton, commander of the US 3rd Army. Patton had made his reputation by the bold use of armoured forces and in August 1944 he had advanced rapidly from the Normandy bridgehead until brought to a halt by shortage of fuel at the end of the month. The ensuing delay was to have important consequences for the campaign, for it allowed the Germans to reinforce Army Group G, responsible for the defence of Lorraine, with forces from Italy, southern France and the northern part of the front.

Patton's tanks entered Verdun easily on 31 August, to the surprise of some older officers who remembered the ground from the First World War and feared that the Germans would defend the passes through the Argonne. The Pont Beaurepaire was captured intact thanks to Fernand Legay, of the local resistance, who thwarted a German attempt to blow it up: it has now been renamed in his honour. However, it was not until 5 September that 3rd Army received enough petrol to continue and Patton immediately sent Major-General Manton S. Eddy's XII Corps to capture Nancy. This plan hinged on Eddy's 80th Division bouncing a crossing over the Moselle at Pont-à-Mousson, and it failed when 3rd Panzer Grenadier Division drove back the one battalion which had crossed, showing clearly enough that tactics that worked well during the pursuit were no use against an organised defence.

When XII Corps tried again it was in a deliberate operation, which sent elements of 35th Division across the Moselle just south of Nancy while the 80th Division, supported by Combat Command A (equivalent to an armoured brigade) of 4th Armoured Division, crossed at Dieulouard. The appearance of the armour proved decisive and Lieutenant-Colonel Creighton W. Abrams's 37th Tank Battalion led a rapid advance, which swung down behind Nancy to join 4th Armoured's Combat Command B, which had crossed south of Nancy, between Lunéville and Arracourt. Nancy fell on 15 September. Patton hoped to get XII Corps on the move towards the Rhine, but the arrival of the extemporised 5th Panzer Army under the redoubtable

Hasso von Manteuffel produced a week-long battle around Arracourt and took the edge off the advance.

There was also heavy fighting in the Forest of Gremecey, west of Château-Salins, as 1st Army's 559th *Volksgrenadier* Division strove to push the 35th Division back across the Seille. The village of Moncel, on the Nancy–Dieuze road, was the German initial objective and the attackers achieved some success, threatening Pettoncourt and infiltrating through the forest to take on the Americans in savage little close-range battles amongst the trees. Eddy decided to withdraw, only to have the order countermanded by Patton, who told him to use 6th Armoured Division, now XII Corps reserve, to counter-attack. The battle ended on 2 October with the forest securely in American hands, but XII Corps had, for the time being, run out of steam.

Patton's other corps, Major-General Walton Walker's XX, had Metz as its initial objective. This was a harder nut to crack than Nancy: as we have seen, the area was thick with forts, made all the more dangerous because XX Corps, operating off Michelin road maps, had no knowledge of them. Just as XII Corps had first tried a direct attack, which had been roughly handled, so too XX Corps embarked upon a frontal assault, with its three divisions in line. On 5 September 5th Infantry and 7th Armoured Divisions attacked, the former moving directly on Metz from the west and the latter feeling south for crossings over the Moselle. North of 5th Infantry, 90th Infantry Division made for Briey and the northern suburbs of Metz.

In the centre the US 2nd Infantry Regiment, attacking two battalions up, advanced up the slopes west of Gravelotte against 462nd Infantry Division's *Fahnenjunkerschule* Regiment. This officer-cadet unit was composed of NCOs under training for commissions, fighting under the orders of their instructors. They held the forts on the crest line (the *Jeanne d'Arc* and *Guise* fortified groups) and knew the ground well. When the attack faltered they counter-attacked: on 8 September 1st Battalion 2nd Infantry was roughly handled in a pre-dawn raid. By nightfall that day the 2nd Battalion had taken Verneville, but the 1st Battalion had been stopped on the edge of Amanvillers.

In 1870 the weakness of the Gravelotte–St-Privat position had been on the French right, where VI Corps was badly exposed. We cannot be sure whether the Germans in 1944 had a better eye for the ground, or had simply read their military history: in any event, they declined to be drawn forward and when the Americans tried to outflank the position on 9 September they got over the ridge at St-Privat, only to be stopped on its far side by fire from the Bois de Jaumont and Fort Kellermann. Things went no better on 2nd Infantry's right, where the 3rd Battalion was committed in an attack towards Moscou Farm, only to be galled by fire from what the Americans called Gravelotte draw – the Mance ravine of evil memory. On 10 September three squadrons of fighter-bombers attacked the German positions around Moscou and Amanvillers, but the 500-pound bombs used by the P-47s made little impression on the forts and when the infantry resumed their efforts they were halted as definitively as before.

On 11 September the Americans tried a repetition of the 18 August 1870 manoeuvre. A detachment of 7th Armoured moved round through Roncourt, groping for the German flank, while the infantry tried another frontal attack. This project fared as badly as previous ventures: the armour was brought to a halt short of Pierrevillers and Semecourt: fire from the Canrobert fortified group helped check the attack. The American infantry in the centre were hard pressed, and there was fierce fighting around Montigny-la-Grange, centre of 4th Corps' position on 18 August 1870. On 14 September, with the attack stalled in front of Amanvillers, 2nd Infantry heard that it was being relieved.

Before the 2nd Infantry repeated history below Moscou and Montigny-la-Grange, 7th Armoured had swung south of the Metz–Verdun road, seeking a crossing over the Moselle. The Germans had blown all the bridges, and the best that the Americans could do was to push through determined rearguards in defiles leading down to the river – up which German infantry had marched on 16 August 1870 – to take Dornot and Le Chêne on 7 September with a view to staging an assault crossing the following day.

Preparations were impeded by rain, which made the Gorze ravine slippery and it was not until mid-morning on the 8th that assault boats of the 11th Combat Engineer Battalions ferried two companies of 2nd Battalion 11th Infantry over the river just north of Dornot. The troops approached along the little road that leads down from Le Chêne, and crossed the railway line and the short stretch of open ground beyond it (now the local sports field) to reach the river. It was a hopeless venture from the start. The Germans still held much of the west bank and were able to fire into the crossing from the flanks, while on the far side they were strongly posted in Forts Sommy and St-Blaise, still visible in the woods on the crest line which dominates the river.

The two assault companies got to the wire outside St-Blaise, where a sniper killed the leading company commander and the attackers, faced with the fortress ditch, called for artillery support. Elements of 2nd Battalion 37th SS *Panzergrenadier* Regiment, which had infiltrated round the American flanks, lunged in and the American survivors fell back to the bridgehead, held by their battalion's other two companies and a handful of men of 23rd Armoured Infantry. Night found the Americans clinging to a narrow strip of ground amongst the trees on the far bank.

The Dornot bridgehead was too exposed to be usefully developed, and Major-General Irwin of 5th Infantry Division sent two battalions of the 10th Infantry to cross between Novéant and Arnaville in the small hours of 19 September. German attention was focused on Dornot and by first light both battalions were across in strength. They pushed on to the crest to secure much of the high wooded ground between Arry and Corny, and were soon fighting off elements of 17th SS *Panzergrenadier* Division. Attacks on the bridgehead gathered momentum, but by now there were thirteen artillery battalions – soon firing 20,000 rounds a day – supporting the crossing, and 9th Air Force had agreed to release fighter-bombers from Brest, their primary target. The crossing site was still desperately vulnerable, overlooked by observers in the Moselle Battery at Fort Driant and gunners on the far bank. It was screened by the 84th Smoke Generator Company, but on 10 September the wind changed, enabling the Germans to bring down accurate fire. The generators were repositioned and by nightfall the valley was once more filled with a haze of oil smoke.

General Irwin had hoped to use troops from Dornot to reinforce those at Arnaville, but although the Dornot bridgehead was evacuated on the night of 10–11 September, the commander of the 11th Infantry reported that men who had fought there were no use for the time being. On the 11th the Americans managed to throw a bridge across the Moselle at Arnaville and 3rd Battalion 11th Infantry was sent across to assist the 10th by attacking towards Corny to take some of the pressure off the left flank. A better bridge was built on the night of 11–12 September and 31st Tank Battalion from 7th Armoured Division crossed at noon on the 12th. Its arrival made the bridgehead secure, and on the following day tanks and infantry expanded it as far as Lorry-Mardigny, just short of the modern autoroute.

The expansion of the bridgehead encouraged General Blaskowitz of Army Group G to ask his superiors for a decision on the future of Metz. The matter was referred to Hitler, who ruled that the shoulders of the salient must be reinforced to prevent the town's encirclement. In consequence, the Americans ran into fierce resistance when they tried to break out and there was hard fighting at Sillegny and Pournoy.

While the 5th Division was fighting on the Seille, 90th, on the American left, had taken over the attack on the Amanvillers position, but it too failed. On 25 September Bradley, commander of 12th Army Group, ordered Patton to halt until enough fuel and ammunition arrived to permit resumption of the offensive. Patton was permitted to make minor adjustments by taking positions which would give him a springboard for his eventual attack and this was to lead to the third round of the battle for Metz – the attack on Fort Driant.

Patton reshuffled his divisions. He had been forced to

send 7th Armoured off to Belgium, and 5th Infantry could only concentrate on Driant by giving ground on the Seille east of Arnaville. On 27 September P-47s dropped napalm and 1,000-pound bombs on the fort, while artillery hammered pillboxes on the forward slopes of the hill. When two companies of 11th Infantry Regiment and a tank destroyer company advanced, they could not begin to dent the position and were speedily withdrawn. General Irwin had doubts about the feasibility of the task entrusted to his tired division, but both Patton and the corps commander, General Walker, decreed that the attack should go on.

On 3 October Irwin tried again, this time with a battalion of 11th Infantry reinforced with an extra rifle company, an engineer company and twelve tanks. The attackers got on to the fort, but suffered heavy casualties and were relieved on 5–6 October by another reinforced battalion, this time of 10th Infantry. The Americans held parts of the fort's superstructure, but although they began to fight their way underground with the aid of oxy-acetylene cutters they failed to make much impression on the garrison, who sallied forth after dark to inflict more casualties. The assistant divisional commander, who led the task force on the fort, argued that four battalions would be required to take it, and with this both 3rd Army and XX Corps agreed to a withdrawal, which took place on the night of 12–13 October.

Further north, 90th Division mounted a limited offensive against the industrial town of Mazières-les-Metz, on the railway north of Metz. It started well but then bogged down in house-to-house fighting in the town centre with the solidly built town hall – engaged from 135 metres by an American 155mm gun – forming the kernel of resistance. The 357th Infantry Regiment went about its task methodically, lavishly supported by artillery, taking a house or two at a time, and by the end of October the town was in American hands at relatively little cost. In late September XX Corps was able to extend further to the north when it received an extra infantry division, 83rd, and in October this cleared the bend of the Moselle north of Thionville, establishing the line the Americans were to hold for the winter.

While XX Corps used the October pause to improve its position, XII Corps, to the south, did much the same, chipping away at the ridge between the Moselle and the Nomeny–Nancy road (D913). It took the high ground around Mont St-Jean, pulverising 553rd *Volksgrenadier* Division in the process. On 21 October Bradley ordered his army commanders to resume their advance: the 1st and 9th were to begin on 5 November, and the 3rd five days later. Although Patton's men had profited from the lull by being re-equipped, reinforced and, in many cases, retrained, their task was made more difficult by the sort of weather that Bazaine's army had known all too well. It rained so heavily – often making effective air support all but impossible – that Eddy asked Patton to postpone the offensive. Patton characteristically told Eddy to get on with it or name his successor and when XII Corps advanced on the 10th it made slow progress through the wooded hills into which the French had attacked in August 1914. It took the whole of November for the corps to fight its way to the line of the Sarre, taking the Maginot Line forts at Welshoff and Le Haut Poirier in the process. With that XII Corps passes out of Lorraine and thus out of our story.

The painful advance of XII Corps was eclipsed by the bludgeon work endured by XX Corps. It was reinforced by two fresh divisions, 95th Infantry and 10th Armoured, before it began its advance on 9 November. General Walker planned to envelop Metz from both flanks. He already held a southern bridgehead at Arnaville, and in order to seize a northern crossing he ordered 95th Division to make a diverson at Uckange before 90th and 10th Armoured crossed at Thionville. A small bridgehead was established across the flooded Moselle south of Uckange on the night of 8–9 November and this helped the main crossing achieve surprise.

The 90th Division formed up in the Forest of Garche, west of Cattenom, and before daybreak on the 9th the attacking battalions of the 358th and 359th Infantry carried their assault boats the 600 metres to the two crossing sites at Cattenom and Malling. Once the first waves were over, the crossing sites were subjected to intense artillery fire. The 359th took Malling without resistance, and exploited to seize Kerling and the ridge line above it. To its south the 359th set about taking the Maginot Line fort at Koenigsmacker, but although combat engineer teams dealt with many of the observation turrets, the garrison still held the fort by nightfall. Nevertheless, 90th Division held a substantial bridgehead, although it had no armour across the river, and when the Germans counter-attacked on the 10th they made little headway.

Fort Koenigsmacker proved a thorn in the American flesh. There were three rifle companies on the fort on the 10th and they consumed explosives at an alarming rate: an American pilot was decorated for flying in only three metres off the ground to chart a safe route for supply drops. Aircraft – artillery spotters used in this unconventional role – could provide only a fraction of the division's needs, and engineers worked frenziedly under fire to bridge the torrential Moselle so as to enable tanks and trucks to cross. The bridge at Malling was completed on the night of 10–11 October, but was under constant fire and its approaches were one and a half metres deep in water. The river began to fall on the 11th and that day the garrison of Fort Koenigsmacker surrendered.

In the small hours of the 12th a battle group of 12th *Panzergrenadier* Division attacked the centre of the bridgehead, taking Kerling and making straight for Petit-Hettange down the D855, aiming for the bridging site. The 359th's G Company and its 2nd Battalion's heavy weapons company fired from woods south of the road. Sergeant Forrest E. Everhart, who had taken over the machine-gun platoon when its commander was killed, was awarded the Congressional Medal of Honor for charging the Germans alone and personally accounting for fifty of them. Lieutenant-Colonel Robert Booth, the 2nd Battalion's commanding officer, held the crossroads south-east of Petit-Hettange with a scratch force of headquarter personnel. Two tank destroyers, which had crossed the bridge before it was damaged by artillery fire, arrived in the nick of time and the thrust was stopped. A second attack, into the woods north of Hunting, fared no better. Over the next few days the bridgehead was expanded and linked up with the Uckange crossing, and on 15 November 10th Armoured crossed the river to begin the northern arm of the double envelopment of Metz. On the 17th General Balck, the new commander of Army Group G, decided to swing his northern flank back eastwards to prevent large sections of his 1st Army from being cut off, and on the 19th the leading elements of the 90th Division met tanks supporting 5th Infantry Division near Retonfey, east of Metz: the envelopment was complete.

The 5th Infantry, spearheading the southern prong of the operation, was fighting on familiar ground. On 9 November it attacked across the swollen Seille. The D5, which follows the west bank of the river, takes us along the line of the battle, from Cheminot, occupied without a fight, past the stout-walled farm at La Hautonnerie, which held out for some hours, up to Sillegny. The offensive developed well, with armour pushing up to secure crossings over the Nied, and infantry taking Forts Aisne and Yser on the southern edge of the Bois de l'Hôpital, near Orny. The wood itself was completely secured on 15 November and the Americans fought their way to the edge of Metz-Frascaty airfield.

Shortly before the pincers closed on Metz, Lieutenant-General Heinrich Kittel was appointed fortress commander. He had an excellent record as a divisional commander on the Eastern Front and was quick to realise the disadvantages of defending Metz against an enemy with absolute air superiority. The 462nd Division, recently upgraded to *Volksgrenadier* status, was the mainstay of his garrison, which also included an assortment of fortress infantry, artillery, engineer and machine-gun units. Kittel decided to retain the Moselle heights as long as he could, then to drop back on to Forts Jeanne d'Arc, Driant, Plappeville and St-Quentin. He was short of supplies, although one of the last trains into Metz station brought him extra rations and forty-eight light guns. Some of his troops were of the patchiest quality: unlucky *Volksturm* in civilian clothes and armed with old French rifles were marched up into the line between Forts St-Privat and Queuleu by police officials. One night in the open finished most of them.

On 14 November 95th Division advanced on the Moselle heights. It tried to infiltrate north of Fort Jeanne d'Arc and clear the 'Seven Dwarves' fortifications linking Jeanne d'Arc with Driant. The fighting that followed was predictably messy, with the Americans getting up amongst the forts, only to be accurately shelled from Driant and counter-attacked by German infantry who worked their way back into the Mance Ravine to cut off the companies fighting forward. It was not until 18 November that the 379th Infantry reached Moulins-les-Metz, across the river from Metz. It had accomplished this by bypassing Germans holding out in the forts, though they were well masked and unable to influence operations. The last bridge over the Moselle was blown with an American detachment actually on it, but on 19 November the leading elements of the division were in Metz.

Even before 95th Division entered Metz from the west, Task Force Bacon of 90th Division – from the northern prong of the pincer – had entered the city from the north. Bacon's men were checked by Fort St-Julien, whose garrison repulsed an infantry attack along the causeway to the main entrance. A tank destroyer failed to breach the gate, but a 155mm gun was run up and its heavy shells did the trick. To the south, 5th Division was stoutly opposed by fortress machine-gun troops and had to drop off a battalion to mask Fort Queuleu, but carved its way on into the city to meet patrols from the 90th Division near Vallières on 19 November. The defenders were now in a hopeless position and most had surrendered by the 22nd. Kittel, badly wounded and unconscious – he had gone forward to fight it out amongst his infantry – was found in an underground field hospital. The forts fell gradually, and Jeanne d'Arc, St-Quentin, Plappeville and Driant were still holding out by the end of the month. Driant surrendered to the 5th Division on 8 December, literally minutes before the division was relieved, and Jeanne d'Arc, last to fall, capitulated on 13 December.

With Metz at last in American hands, XX Corps drove on towards the Sarre, with XII Corps on its right. In early December it came up against the outposts of the *Westwall* and had wrested bridgeheads over the Saar when news of the Ardennes offensive compelled 3rd Army to redeploy for a counter-stroke into the German flank. Ironically, once the Ardennes offensive had been dealt with, 3rd Army based its operations on Luxembourg, not Lorraine: the hard-won ground between Moselle and Sarre was destined not to provide Patton with his expected springboard into Germany.

Patton's conduct of the Lorraine campaign has attracted much criticism. He and his senior commanders were initially over-optimistic and underestimated the effect of terrain and weather. American armour failed to achieve the results that might have been expected of it, in part because Patton and his corps commanders tended to parcel up their tanks amongst the infantry, and in part because the Sherman was outgunned by many of the German tanks it encountered.

It has sometimes been fashionable to criticise US combat performance in the Second World War and to suggest that the Americans only fought well when they had weight of metal on their side. The Lorraine campaign does much to rebut this and the Arnaville crossing site makes the point effectively. It is approached along the D952, which follows the west bank of the Moselle. A track leaves the road just south of Arnaville, between the village and a railway bridge, and takes the traveller across the canal – on the site of a bridge erected on 14 September 1944 and possibly even the same structure – down to the river bank, where there is safe parking by the village sports pitch. A path leads north-westwards to a stream – the Rupt de Mad – which rises in the Lac de Madine, just below the Butte de Montsec on the 1918 St-Mihiel battlefield. Across the stream is the Moselle Lateral Canal – which cannot be safely crossed on the dilapidated lock gates – then the railway and then the main road.

There were thus three separate obstacles – canal, Rupt de

Mad and Moselle – in the way of the American engineers. They first laid sections of treadway on the river bottom at a site well downstream, parallel with the southern end of the village of Novéant. This enabled infantry to ford, but was not suitable for vehicles. It was only after bridging canal and stream at the main crossing site, in the small hours of 11 September, that the engineers could approach the Moselle there, and bridging took place in filthy weather and under the fire of artillery and assault guns. On the night of 11–12 September a proper treadway bridge was erected some 200 metres downstream from the car park: a quarter of the two engineer companies involved became casualties that night. A heavy pontoon bridge was completed on 14 September. Voisage Farm, a prominent block of farm buildings, stands on the far bank, and the bridge lay just downstream of the track which still runs from the farm towards the river.

The vulnerability of the bridgehead can be appreciated from the high ground on the east bank. The road that leaves the N57 east of Arnaville and climbs up to Arry offers a ringside seat. Some German observers remained concealed up in the woods even after the Americans had taken the high ground and it was weeks before the site was free from observed artillery fire from front or flanks. The 5th Division, which crossed at Arnaville and fought at Fort Driant, lost 380 killed, 2,097 wounded and 569 missing in September alone. This does not include sick or psychiatric casualties and adding these, at the rate then prevailing, would bring the division's total casualties to about 6,000 – over one-third of its strength.

Lorraine, annexed by the Germans in 1940, became French once more in 1944. West of the Moselle its Frenchness is undisputed, but to the east there is a blend of cultures. Place names, with their -ing, -ach, -dorf and -strof endings, have a decidedly Germanic flavour. So too does the cooking: *choucroute* – *sauerkraut* – is a local favourite, and cabbage features in the rich salt pork and sausage casserole called *potée*. The 'high German' spoken in much of Alsace also appears in the triangle of French territory running from Sarrebourg, up the Sarre and across to the German border, and when inhabitants of this part of Lorraine speak French it is often with a thick accent, pronouncing their 'p's as 'b's.

Parts of Lorraine have little to recommend them. The Moselle valley between Metz and Thionville, and the iron country around Briey and Longwy, are best avoided. I find the uplands around Verdun so powerfully permeated by the events of 1916 as to be tolerable only in small doses. There is magnificent countryside in the Lorraine Regional Park north of Toul and around Château-Salins. Metz and Verdun both retain some of their original beauty. But neither can rival Nancy, with its fine eighteenth-century buildings around the wholly delightful Place Stanislas. The *Musée Historique Lorrain* has a good military collection, as well as almost the whole corpus of Jacques Callot's work. Callot, born in Nancy in about 1594, enjoyed the patronage of the Duke of Lorraine and Louis XIII. His engravings throw penetrating light on the manners of the seventeenth century, with his *Miseries of War* series emphasising the unhappy lot of soldier and civilian alike, and telling the hard old story of what happened in Lorraine for so much of history.

We should not be surprised that Patton, in common with several others who have tramped through Lorraine, failed to see its many good points. 'I hope that in the final settlement of the war', he wrote to the US War Department, 'you insist that the Germans retain Lorraine, because I can imagine no greater burden than to be the owner of this nasty country where it rains every day and the whole wealth of the people consists in assorted manure piles.' He does the country less than justice, but there *are* times, when the rain slices across the Moselle heights and black memories of 1870 and 1916 come howling in the wind, when it is not hard to agree with him.

VI

Sambre and Meuse

THE RIVERS SAMBRE and Meuse meet at the Belgian town of Namur. To the south, in the wedge between the rivers, is the long ridge of the 'entre Sambre-et-Meuse'. Northwards stretch the fertile uplands of Hesbaye and Hainault. The region is bounded by the flat lands of the Flanders plain and the Campine to the north-east and north, the hilly Ardennes to the west, and the wooded Argonne and rolling Champagne to the south.

Just as Lorraine offered a corridor to marching armies, so the land around Namur – now comprising the Belgian provinces of Hainaut, Brabant and Namur, and fringes of the French *département* of the Nord – had a military importance all of its own. The lines of Sambre and Meuse offered France the enticing prospect of an easily defended frontier, and did precisely the same for an independent Belgium: the area is as heavily splashed with fortresses as Lorraine. Moreover, the fact that the Ardennes form a natural obstacle to military movement forced armies to their flanks as the cutwater on a bridge's pier divides the torrent. It is only fifty kilometres across the open country of Hainaut from Charleroi to Brussels and one is scarcely out of gunshot of a battlefield all the way. A 1930-style monument on a crossroads in Fleurus, on the Gembloux road in Fleurus, commemorates the Duke of Luxembourg's victory in 1690, General Jourdan's in 1794 and Napoleon's in 1815. It fails to draw attention to another battle on the same ground: in 1622 a German army under Count Mansfeld and Christian of Brunswick was cut to ribbons by the Spaniards.

In the fourteenth century the provinces making up what

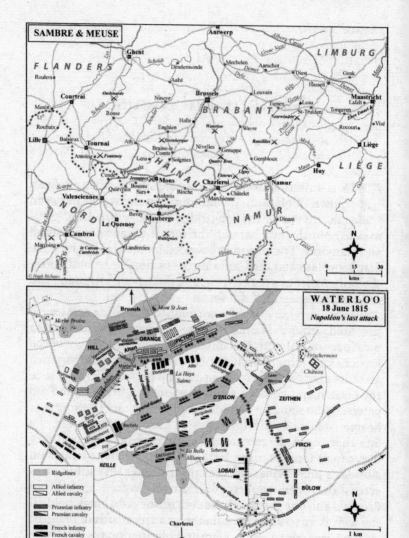

is now Belgium and Holland came under the dominion of
the Dukes of Burgundy, and in 1441 Philip the Good completed the unification of Burgundian possessions in the
north by his acquisition of Luxembourg. Liège, seat of a
powerful prince-bishop, revolted against the Burgundians
and in 1468 Charles the Bold destroyed it, leaving only the
churches standing. On his death the territories passed,
through his daughter Mary, to her husband the Emperor
Maximilian and on to Charles V and, in 1555, to Philip II of
Spain.

By the time the Low Countries became Spanish Brussels
had replaced Malines as centre of government and it was in
the palace of Coudenberg that the Emperor Charles V
abdicated in 1555. The city witnessed all the horrors of
religious conflict of the sixteenth century. In 1568 the
Counts of Egmont and Horn were beheaded in the Grande
Place. Egmont had served the Spaniards well, taking part in
their victories over the French at St-Quentin in 1557 and
Gravelines in 1558, but he became involved in resistance
against Spanish attempts to suppress the privileges enjoyed
by the provinces. The Treaty of Westphalia, which ended
the Thirty Years War in 1648, left the Spanish in possession
of Hainaut and Brabant – as well as Flanders and Artois to
the west – but confirmed the independence of the seven
United Provinces of the Dutch Republic.

The wars of Louis XIV saw the steady encroachment of
France on Spanish territory in the Low Countries. In 1675
the French took Dinant and Huy, securing the Meuse as far
as Maastricht – apart from the inconvenient Spanish fortress
of Namur. Although the peace of Nijmegen returned
Charleroi to the Spaniards, it left the French with a much
stronger frontier than they had previously enjoyed.

Campaigning on the Sambre and Meuse resumed with the
outbreak of the War of the League of Augsburg. On 1 July
1690 Luxembourg defeated Prince George of Waldeck at
Fleurus. The industrial sprawl around Charleroi has now all
but swallowed up Fleurus, making it as difficult to follow
Luxembourg's victory of 1690 as Jourdan's of 1794. This is a
pity, for he was a fine general whose achievements deserve

recognition. Although Luxembourg, with some 45,000 men, outnumbered his opponent, he was not content with an ordinary victory. By a classic combination of central infantry attack and cavalry envelopments on both flanks he inflicted 14,000 casualties on Waldeck's army and took forty-nine of his guns.

The following year saw the French out of their winter quarters before the Allies were ready for them. They sat down before Mons, capital of Hainaut, on 17 March 1691, and took it less than a month later. In May 1692 they descended on Namur and, after a siege sharpened by the fact that Vauban was taking on his rival, the Dutch engineer Menno van Coehoorn, the place surrendered in late June. This provoked William of Orange – now William III of England into the bargain – into attacking Luxembourg's field army. On 8 August 1692 he launched a dawn attack on the French camp at Steenkerque, between Soignies and Halle. He first made good progress, but Luxembourg rallied his army and William eventually withdrew with the loss of 7,000 men. The battle launched a new fashion, the wearing of the cravat 'à la Steenkerque', loosely knotted as if the wearer had thrown it on in haste.

In July 1693 the French took Huy, and Luxembourg then marched north to attack William between Landen and Neerwinden. William, with an Allied army of 30,000 men, had taken up a strong position on the Little Geet, but Luxembourg came close to repeating the double envelopment of Fleurus. His attack on the Allied centre forced William to weaken his wings to reinforce it and Luxembourg then jabbed in on both flanks. The Allies lost 18,000 men – many of them drowned in the marshy Geet – and over a hundred guns. Such was the caution instilled by Louis XIV into his commanders that Luxembourg did not pursue.

Acting on Vauban's advice, Louis then sent his great engineer off to besiege Charleroi. Vauban went about the operation with his customary precision, despite pressure from those who wished to hurry matters: 'Burn powder and spare blood,' he declared. The fall of Charleroi in October

1693 left Louis with what Vauban called 'the finest frontier which France has enjoyed for a thousand years'. He now held the line of the Sambre from Maubeuge to Namur, and the Meuse on up to Maastricht.

In 1694 the Allies began to chip away at this bulwark. Their first objective was Huy, whose citadel, stronghold of the prince-bishops of Liège, frowned down on the Meuse, covering the fine Gothic bridge and the little town where Peter the Hermit had preached the first crusade in 1095 and now lies buried. Its capture gave the Allies a handhold on the river. The present citadel was built in 1818–23 by the Dutch on the site of the prince-bishop's castle. During the Second World War it was used as a prison for Resistance workers and hostages, and contains a small military museum.

In 1695 the Allies went on to attack Namur. It had been extensively improved since its capture by the French in 1692 and the first attempt at formal siege had little effect. Coehoorn proposed a plan which combined the concentrated fire of guns and mortars with selective infantry assault, and the Elector of Bavaria, new governor of the Spanish Netherlands, gave him his head. Coehoorn's scheme led to Namur's unexpectedly rapid capitulation, and brought the redoubtable Dutchman the post of William's director-general of fortifications and the Spanish title of baron, an indication of the way in which the Spanish and the Dutch had been drawn together by the threat posed by France.

Some of the extant fortifications of Namur pre-date the siege. The citadel lies on the spur between the Sambre and Meuse. It is best approached along the Route Merveilleuse, which takes the visitor up from the car park where the Pont de France crosses the Sambre and past the Joyeuse tower, part of the Count of Namur's castle. The road winds upwards to pass between two more towers of the old keep, one of which houses an armoury, then crosses the ditch separating the keep from the Mediane Bastion, an early sixteenth-century work, strengthened by the Terra Nova bastion, built to its west by the Spaniards in 1640. Due west, beyond the Fabiola Park, is the detached Orange Fort, thrown up in 1691 to give more depth to the Terra Nova front of the fortress, and remodelled by the Dutch in 1816. Guided tours, which begin at the information centre at Terra Nova, are well worth taking, for they show this extensive citadel complex to good advantage.

In 1696 the Allies did not profit from the previous season's promise, for the French were in the field first. Vauban laid siege to Ath, at the confluence of the Dendre and Little Dendre. He had fortified the place himself, turning it into a neat little eight-bastioned fortress, but he took it in short order, using the new technique of ricochet fire – in which cannon were fired with a reduced charge so that their shot dropped over the parapet to bounce along the ramparts. Ath was too far north for the French to hope to retain it, but was a useful bargaining counter when the peacemakers met at Ryswick, and was handed back to Spain along with Courtrai, Charleroi and Luxembourg.

The strategic balance in the area altered dramatically with the outbreak of the War of Spanish Succession. In the 1690s Spain had made an important contribution to the struggle against France, but 1701 saw the Bourbonisation of the Low Countries as Louis occupied the Spanish Netherlands on behalf of his grandson Philip. Dutch troops left, garrisoning Spanish fortresses were swallowed up, and by the winter of 1701 the French held not only the whole of the Spanish Netherlands but also the bishoprics of Cologne and Liège. However, the new frontier was not easily defensible, for many of the newly acquired fortresses were in poor condition.

In order to strengthen the frontier the French threw up the Lines of Brabant. This continuous line of field fortifications – a curtain with bastions and redans, protected by a ditch – ran from Antwerp, through Lierre, to the River Demer in front of Aarschot. It swung south along the Geet from Diest and ran behind the Mehaigne to reach the Meuse west of Huy. There was little prospect of the Lines resisting deliberate attack, but they did protect the Spanish

Netherlands from raiding cavalry and made an invader's life generally more difficult.

In 1705 the Duke of Marlborough demonstrated that the Lines were not impenetrable when he responded to Marshal Villeroi's capture of Huy and attack on Liège by moving through the Lines south of Diest, reaching Tirlemont, then going on to face Villars near what was to become the battle-field of Waterloo. Difficulties with the Dutch prevented him from giving battle, but the Duke had at least been able to raze the Lines between Leau (now Zoutleeuw) and Merdorp, giving him an opening for future operations. In the following spring, anxious to bring Villeroi to battle before Marsin moved up from Metz to join him, Marlborough advanced from his camp at Maastricht through the demolished section of the Lines in an effort to draw Villeroi from the siege of Leau. The French took the bait and came down to Tirlemont to meet him.

The ground on which the armies met is typical of this part of Brabant, and though villages have grown larger and streams have shrunk in size, looks much as it did in 1706. Broad, flat-topped ridges, whose fields grow wheat or sugar beet, drop down into shallow valleys. Villages lie in the valleys and often show themselves only by church spires peeking above the crest line. Although the ground feels open, numerous gentle depressions offer hidden avenues of approach and there is no easily identified key terrain, which might invite a slogging match for its possession. In short, it places a premium on the commander's skill rather than the sheer weight of resources at his disposal. Its essential features are simply described. The River Mehaigne, marshy in 1706 but now more tightly confined, marks the southern edge of the battlefield. Two low Brabant ridges roll north-wards from it: to the east is the Plateau of Jandrenouille, divided from the Plateau of Mont St-André by the Geet and Little Geet.

Marlborough was about his business early on Whit Sunday, 23 May 1706. Its small hours found William Cadogan, his quartermaster-general, clattering westwards with a party of horse, looking for a suitable campsite. The rest of the army, in three great columns, was about two hours behind him. West of Merdorp, Cadogan brushed against a party of French hussars and saw strong bodies of cavalry further west. He sent word back to Marlborough, who joined him with the Allied staff, still uncertain whether he was facing the main body of Villeroi's army. All doubt was burnt off with the morning mist: the French army, its infantry resplendent in the beginning-of-campaign whiteness of its uniform, was drawn up in battle order to Marlborough's front.

Villeroi's right, under the command of the Elector of Bavaria, was anchored on the River Mehaigne, with five battalions in the villages of Taviers and Franquenée. The cavalry of the right wing, with brigades of infantry echeloned back in support, was draped across the shoulder of open ground to the north. The village of Ramillies, in the centre, was solidly held by infantry and guns. Villeroi took personal command of his left: more infantry stood on the west bank of the Little Geet, which ran through Ramillies and on through Offus to Autre Eglise, where the squadrons of the left wing took post. It was a well-sited position, with the bulk of Villeroi's army enjoying room for manoeuvre on the plateau of Mont St-André, the marshes of the Little Geet protecting its front and the villages acting as natural bastions.

We may doubt whether Villeroi's eye for the ground was as keen as Marlborough's. The plateau of Jandrenouille, where the Allies were to deploy, was creased by a shallow re-entrant running north-south opposite Ramillies. It was invisible to observers in Ramillies church and on the rising ground west of the Geet. Indeed, even today it is only by walking eastwards out of Ramillies, past the cemetery, that this fold of ground shows itself: it was to enable Marlborough to shift troops covertly from flank to centre. Both armies were more or less equal, at around 50,000 men apiece, though Marlborough enjoyed an advantage in artillery.

The battle began at about 1 p.m. when Marlborough's guns opened against the French centre. Shortly afterwards he struck at both flanks: in the south, the Dutch Guards took Franquenée and Taviers, while to the right Lord Orkney led twelve British battalions across the Geet marshes towards Offus. Both attacks contributed to what was to come. The Elector reacted to the fall of the villages by counter-attacking with fourteen squadrons of dragoons and two Swiss battalions. His attempt miscarried, for the dragoons were caught dismounting by the Dutch horse, and their collapse dragged away the entire extreme right. On the French left, Villeroi responded to Orkney's threat by moving troops from his centre. Thus, by about 3.30 p.m., the French centre had been weakened to support the flanks. But it still remained very strong and Schulenberg's infantry, with Overkirk's cavalry in close support, made little impression on it.

The remaining cavalry on the French right took up the challenge posed by Overkirk, and there was fierce fighting as attack and counter-attack swirled across the open ground north of Franquenée. This phase of the battle ended with Overkirk's cavalry somewhat pushed back and less able to support Schulenberg's infantry attack on the French centre. Marlborough was now sure that Orkney's attack – promising though it seemed to Orkney himself, with Autre Eglise all but in his grasp – could achieve no decisive result because the marshes would prevent exploitation. He could also see that his centre was not strong enough as things stood. Accordingly, he ordered Orkney to fall back and summoned the cavalry from the right to support Overkirk's horsemen. The re-entrant enabled the cavalry to move south unseen and their arrival tilted the battle on his left centre in the Duke's favour. After very heavy fighting, in which the arrival of the Danish horse proved decisive, the French were pushed back beyond a large tumulus called the Tombe d'Hottomont (east of the N91 and just south of Grand Rosière Hottomont), with their right wing arched back so that their position resembled a V with Ramillies at its point.

The attack on Ramillies continued, with Schulenberg being reinforced first by Scottish and English battalions, then by much of Orkney's force. Enough of Orkney's battalions remained in the north to persuade the French that his full force was still there – the process of deception was compounded by leaving regimental colours behind, just visible to the French over the crest line.

When the decisive attack began at 6 p.m. the French, with masses of unengaged troops facing the non-existent threat in the north, were badly off balance. Villeroi's centre was driven in and though he tried to form a new line to the rear, around Petit Rosier, the ground was cluttered with tents and baggage wagons, and his army was already beginning to dissolve. The Allied cavalry pressed the business and by dawn on the 24th perhaps half of Villeroi's force remained intact under arms. The Allies had suffered less than 4,000 casualties, small price for such a towering achievement.

Until recently one could cross the field of Ramillies and miss the fact that a battle was fought there at all. An obelisk outside Ramillies church commemorates a later conflict and within its railings stands a First World War German *Minenwerfer*, which has seen better days. One of the attractions of Ramillies is precisely its unexploited state and the recently erected orientation tables (there are two by the village cemetery, on the southern end of the re-entrant used by Orkney's men) do not spoil it. It is possible to drive along the French line, gaining an impression of the natural strength of Villeroi's position, although improved land drainage has greatly reduced the Little Geet and the marshes that once surrounded it. Ramillies is a pleasant, sleepy village. It is worth walking or driving up to the cemetery, on the slopes north-east of the church, to see the crucial re-entrant with the plateau of Jandrenouille stretching away behind it. On the other side of Ramillies the ground rises on to the plateau of Mont St-André, still remarkably open, with spires marking the locations of Offus and Autre Eglise to the north.

Ramillies had been fought early in the campaigning season, and Marlborough was able to exploit success by taking Louvain, Brussels and Antwerp, and going on to shepherd the French field army back to the Flanders frontier. Marshal Vendôme arrived in August to take command from

the discredited Villeroi, but could not prevent the fall of
Menin, Dendermonde and Ath. Marlborough's successes of
1706 ensured that for the next two years the fighting took
place in Flanders rather than Brabant, with the Duke's great
victory at Oudenarde in 1708 followed by his capture of
Lille.

By 1709 Marlborough was on the borders of Flanders and
Hainaut, dealing with fortresses outside the 'Lines of
Cambrin', which shielded the French field army, now under
Villars. In July he sat down before Tournai, stoutly held by
the Marquis of Surville-Hautflois. Tournai enjoyed a special
place in French affections. It was the birthplace of the
Merovingian King Clovis (465) and had been held in the
French cause during the Hundred Years War: loyalty to
France is commemorated by the fleur-de-lys in its coat of
arms. Henry VIII of England occupied Tournai in 1513–19
and the tower named after him – in the north, near the
station – is a remnant of the defences he built. Thereafter
the town passed to the Hapsburgs, and in 1581 it held out for
two months against the Duke of Parma. Tournai was in
French hands between 1667 and 1709, and its thirteenth-
century curtain wall – fragments of which remain at the
Pont des Trous in the north-west and in the parkland
running parallel with the Boulevard Walter de Marvis in the
south-east – was given bastions by Vauban. It became
Austrian after the Treaty of Utrecht, French after Fontenoy
in 1745, and Austrian until its capture by the French at the
century's close.

Marlborough conducted the siege while Eugène,
encamped on the Courtrai road, guarded against the arrival
of a relief force. The siege did not go well. Although the
Allies opened their trenches against three faces of the
fortress on the night of 7–8 July and established breaching
batteries opposite the Valenciennes Gate a week later, the
aggressive Surville mounted frequent sorties and on the 18th
his men exploded a large mine, which destroyed four of the
batteries. It was not until the 28th that practicable breaches
were battered into the curtain. Surville withdrew his
surviving garrison into the enormously strong citadel, a

five-bastioned work with masonry-revetted countermine galleries, under a temporary armistice. The garrison maintained the most obdurate defence, with mines and sorties, but at the end of August the Allies had fought their way to the lip of the ditch and Surville at last capitulated.

With Tournai secured, albeit at the cost of over 5,000 casualties, the Allies moved on to besiege Mons, which they invested on 6 September. Villars, under orders that allowed him to raise the siege even at the cost of a pitched battle, moved up to the south-west of Mons. The Sambre marked the southern limit of the area of operations. To its north the ground is gently undulating and in 1709 was covered by woods, large parts of which remain. The Forest of Mormal lay just north of the Sambre, with Tiry and Lanières Woods to its north-east. Across the Roman road from Charleroi to Le Cateau (the modern D932), which skirted the northern edge of the Forest of Mormal, lay another block of woodland, the Sars and Boussu Woods, extending almost as far as the River Haine, which joined the Scheldt at Condé and formed the northern border of the battle area. A number of streams – more significant in 1709 than they are today – rose in the woodland along the Roman road to flow northwards into the Haine.

There were three main routes to Mons from the southwest. One struck straight up from the Roman road near the village of Aulnois; the second cut through the gap in the woodland near Malplaquet and the third ran along the northern edge of Boussu Wood, from Quiévrain, through Boussu, to Mons itself. Villars decided to use this nexus of roads and woods to threaten the force covering the siege, and on 7 September he took position behind the Sars and Boussu Woods, with his cavalry well forward.

Marlborough reacted by sending Eugène off to watch the Boussu gap, while he covered the Aulnois gap. On the 9th Villars moved south-eastwards and set about preparing a wide defensive position centred on Malplaquet. In front of the village five redoubts were dug between the edge of Sars Wood and Blairon Farm. On the right, Marshal Boufflers dug a line of entrenchments, which folded right round into

Lanières Wood, with a twenty-gun battery carefully sited to command the Aulnois road: forty-six battalions under Generals d'Artagnan and de Guiche took post on Boufflers's flank. On the left, General Albergotti, with twenty-one battalions, threw up fieldworks running from the edge of Sars Wood due south to meet Villars's main line in a right angle north-west of Malplaquet. General Goesbriand, with seventeen battalions, dug further defences along a stream which followed the south-eastern fringe of Sars Wood. The cavalry were drawn up behind the foot and Luxembourg took a detachment to screen the right rear in case Marlborough attempted a turning movement. The position was well sited to deal with a frontal attack, for the ground between Blairon Farm and Sars Wood was a killing zone raked by cannon and musketry, with cavalry at hand to capitalise on the attacker's disorder. The flanks, which offered unappetising avenues of approach because of the woods, were also strongly held.

Marlborough enjoyed the advantage of numbers – with 110,000 men to Villars's 80,000 – and decided to repeat his familiar plan of attacking the enemy's flanks to induce him to reinforce them, then striking at the weakened centre. He originally intended to deploy a left wing, under the Prince of Orange, made up of thirty Dutch battalions, nineteen battalions under General Withers – marching quickly from Tournai to join the army – and twenty-eight squadrons of cavalry. The centre, under Lord Orkney, comprised fifteen battalions – eleven of them British – with 179 squadrons behind them and a twenty-eight-gun battery firing squarely into the redoubts. General Lottum, with twenty-two battalions, was to attack Albergotti's position, while, on the extreme right, Schulenberg led forty battalions and a detachment from Mons against Sars Wood. The late appearance of Withers's force, which arrived footsore on the 10th, persuaded Marlborough to deploy it on his extreme right – so that it would not have to make its way laboriously across the whole Allied army – thus extending his army well beyond the French flank.

Just after eight on the morning of 11 September the

British guns in the main battery opened fire into the mist that still hung over the field. Schulenberg and Lottum attacked the entrenchments in Sars Wood and were sternly received. With the armies locked in battle on the Allied right, the Prince of Orange led his Dutch battalions against the entrenchments on the left. The twenty-gun battery opened great lanes in the Dutch ranks and after repeated attacks the Dutch were forced to fall back, though they did so in good order. Hesse-Cassel's cavalry and two Hanoverian battalions, ordered up from the reserve, discouraged the French from pressing the retreating Dutch. The fighting on the Allied right, meanwhile, blazed on and on. Gradually numbers began to tell and the Allies gained a foothold in Sars Wood, inducing Villars to slide up battalions from the centre to prop up the threatened flank.

If Marlborough was making slow and costly progress on his right, his left was faring worse. Orange – because of disobedience, youthful enthusiasm or sheer misunderstanding – launched a second attack on Boufflers's guns. Marlborough reached him in time to prevent another attack and although Orange had actually pressed the matter more vigorously than had been intended, he did dissuade Boufflers from sending troops to help Villars on the left, where the fighting for Sars Wood was swinging against the French. Just before midday Villars had a line ready to deal with the Allies when they emerged from the wood, but he had been able to build it only at the cost of bleeding his centre white. Orkney's men were able to advance, almost unopposed, to take possession of the redoubts, and cavalry under Auvergne and Hesse-Cassel moved up to break the French centre.

The unfortunate Villars did not even reap the benefit of taking troops from his centre, for he was just about to launch them in a counter-attack when he was wounded. Albergotti, too, was hit and Puységur, who took command on the left, failed to organise an attack that might have rocked the already tired Allied infantry. Miklau's cavalry, which had accompanied Withers's wide outflanking movement, was vigorously attacked near La Folie and its discomfiture made the Allied right very cautious about exploiting beyond Sars Wood.

In the centre, meanwhile, the great mass of Allied cavalry began to move through the redoubts, now held by Orkney's infantry. Before they had completed their deployment on the flat ground beyond, facing Malplaquet, Boufflers – now in overall command – led his own horse forward and pushed his opponents back as far as the redoubts. A combination of musketry, the fire of ten cannon hauled forward from the battery in the Allied centre and spirited forays by Allied cavalry – in which the Scots Greys played a distinguished part – helped force Boufflers back on Malplaquet. Both sides maintained the contest with great determination and it was not until he heard that both his flanks had given way that Boufflers at last drew off.

On the French left Puységur, under renewed pressure from Eugène, ordered a retreat on Quiévrain, while d'Artagnan, on the right, had suffered severely from determined assaults by the Dutch Guards and retired on Bavay. Boufflers conformed to this movement in mid-afternoon by falling back on to the Hogneau, with Luxembourg and the cavalry reserve covering his retirement. The Allies were in no condition to pursue. They had lost the numbing total of 25,000 men; although they had captured some sixty-five cannon, killed or wounded perhaps 17,000 Frenchmen and taken a mere 500 unwounded prisoners, it was a shocking price to pay for so small a result and did Marlborough serious political damage. Villars, recovering from his wound in Le Quesnoy, was not overstating his case when he told Louis XIV that his men had 'done marvels' and Marlborough acknowledged that the French had fought better than ever before.

The field of Malplaquet straddles the Franco-Belgian border, with the Malplaquet–Mons road (the N543 in Belgium and the D932 in France) neatly bisecting it. Approaching from the south, the visitor passes through the tiny village of Malplaquet. The last building on the left, once the French customs house, was once a museum. It stands on the low ridge where the cavalry of Villars's centre formed

up. Further along the road an obelisk, erected by the French
to commemorate the two hundredth anniversary of the
battle, marks the area where the French and Allied cavalry
fought so hard in the early afternoon. The French redoubts
ran along the squat crest running across the road in the area
of the obelisk. The track to Blairon Farm, which leaves the
road just short of the border, more or less follows the line of
redoubts until it turns down to the farm. A monument,
erected by the *Navarre et Picardie* society – named after the
the oldest infantry regiments in the French Line – stands
alongside the track and bears a plaque in honour of the
British regiments that fought at Malplaquet. Blairon Farm is
a solid white-walled farmhouse, with an inscription
announcing that Marlborough spent the night of the battle
there. A more recently erected memorial on the edge of
Aulnois makes the tragic point that Swiss regiments fought
bravely on both sides and if the names of their colonels are
any guide, members of the same families found themselves
facing one another.

The woods have changed in size and shape since 1709, but
on both flanks they still give a good idea of the difficulties
faced by the Allies. Lanières Wood stretches out on both
sides of the Roman road and a line of trees west of Aulnois
marks the position of Thiry Wood. On the French left, Sars
Wood retains much of its original size and aspect, and the
road which crosses it from north to south is old *pavé*. The
streams are no longer significant obstacles, but they do help
establish the French position: the depression just south of
the modern wood edge marks the line of entrenchments
held by Goesbriand's men. Albergotti was deployed some
700 metres to the north-east, on the eastern edge of the
wood, in a line visible from the Malplaquet–Mons road
about 500 metres north of the frontier. The 11 September
1709 Museum, in the Rue des Juifs in Bavay, is run by the
helpful M. Arthur Barbera and is generally accessible in
the summer. He has written a booklet on the Swiss at
Malplaquet and his work on the 1709 topography of the
battlefield helps the visitor make sense of the place.

Having beaten the French army, though at such terrible

cost, Marlborough was free to resume the siege of Mons. The capital of Hainault, dominated, then as now, by its magnificent baroque clock tower, was held by the Marquis of Grimaldi and 4,000 men. Bad weather, raids by Villars's troops and sorties by the garrison slowed up the siege, but operations followed their usual course and on 20 October, with two of his gates knocked into rubble, Grimaldi asked for terms. He was allowed to march off to Maubeuge and Namur, and it was now so late in the campaigning season that the Allies were unable to go on to attack Maubeuge or Le Quesnoy.

Although there was to be no more significant fighting in the Hainault and Brabant for the remainder of the War of Spanish Succession, Malplaquet influenced events elsewhere because Allied politicians were now as reluctant as Louis XIV to authorise commanders to fight. The Treaty of Utrecht, which ended the war in 1713, resulted in the cession of the Spanish Netherlands to Austria, and the War of Austrian Succession again brought conflict to Brabant and Hainault. The Allied army, with its British, Austrian, Hanoverian and Hessian contingents, found itself opposed by French commanders of real ability. The sheer fighting quality of his troops enabled George II – commanding in person on the battlefield, the last time that a British monarch was to do so – to beat the French at Dettingen on the River Main in 1743. In 1744 the axis of the war shifted to the Netherlands, and Marshals Noailles and Saxe took a number of fortresses, including Menin, Ypres, Courtrai and Maubeuge. The Allies threatened Lille but lacked the resources to besiege it and then went into winter quarters around Ghent.

The French opened the campaigning season of 1745 by besieging Tournai. The Allied force of a little under 50,000 men, under George II's son, William Augustus, Duke of Cumberland, marched down by way of Halle and Soignies, and turned in towards Tournai on the Mons road in an effort to raise the siege. Saxe, with some 70,000 men at his disposal, left a strong detachment to continue the siege while he took the remainder, perhaps 56,000, forward to a strong natural position between the Escaut and Barry Wood, blocking the line of Cumberland's advance. The Duke reached the high ground between Maubray and Baugnies on the evening of 9 May, and camped there for thirty-six hours, in full view of the quadruple spires of Tournai cathedral and the lofty fourteenth-century clock tower. He initially considered sending his abundant cavalry to turn the French left, but a hussar reconnaissance discovered that Barry Wood was firmly held and this, as well as the broken ground north of the wood, persuaded Cumberland to concentrate on the French centre.

The pause gave Saxe ample time to strengthen his line with field defences. He had already concluded that French infantry was inferior to British in open field, and the redoubts thrown up on the ridge between Barry Wood and the village of Fontenoy propped up his centre. The wood was strengthened by an abatis of felled trees along its forward edge, and the villages of Fontenoy and Antoing – the latter marked by the slender tower of the Prince de Ligne's castle – were prepared for defence. The ground between the point of Barry Wood and Fontenoy was covered by two redoubts, and that nearest the southern point of the wood was known, after the *Régiment d'Eu*, which held it, as the Eu Redoubt.

On 10 May Cumberland cleared the French outposts from Vezon, in front of the main line: the Highland Regiment – the Black Watch – took part in this action, its baptism of fire. The army was then deployed to its positions for the attack – British and Hanoverians on the right, Dutch on the left – and its soldiers snatched a few hours' sleep before standing to their arms at two on the morning of the 11th. Cumberland was in the saddle at 4 a.m. and at once modified his initial plan by detailing a brigade under Brigadier James Ingoldsby to take the Eu Redoubt, securing the Allied right before the main attack began.

Ingoldsby found his task perplexing. He halted in a sunken lane on the edge of Vezon, well short of his objective, sent a messenger asking Cumberland for cannon, then asked his commanding officers how they thought the attack should be carried out. Sir Robert Monro of the Black

Watch suggested that his Highlanders could clear the wood while the rest of the brigade marched on the redoubt, but Ingoldsby took no action. Three cannon duly arrived and began to fire on the irregulars holding the wood, but there was still no move, despite a flurry of messages to and from Cumberland.

At about 6.30 a.m. Sir James Campbell led forward a strong detachment of British cavalry, ordered to screen the infantry as they formed up. He halted behind Ingoldsby, but could not get forward until Ingoldsby had dealt with the redoubt. Cumberland arrived shortly afterwards, expressed surprise that Ingoldsby had done nothing and, at about 7 a.m., ordered the general advance. At the signal of four cannon fired in quick succession, cavalry advanced to secure the ground so that the infantry could form up. Two columns of Dutch horse made for Antoing and Fontenoy, came under fierce fire and fell back, unmasking two batteries which engaged the French defences throughout the battle. Campbell led his cavalry forward but was engaged by cannon fire from the Eu Redoubt and Fontenoy. He was mortally wounded, and his men wheeled about and formed up behind the infantry.

The sharp rebuff to the cavalry did not unnerve Sir John Ligonier, the experienced officer of Huguenot ancestry who commanded Cumberland's foot. He ordered his infantry to advance, and successive battalions marched through Vezon and halted on the far side to cover the deployment of fresh units. The infantry waited, under constant artillery fire, for some two hours, until all was ready. Ingoldsby had still not budged, despite another exchange of messages, and told an aide-de-camp that he intended to advance when the first line of infantry on his left moved forward.

The Dutch infantry attacked first, advancing on Fontenoy and Antoing but making no progress in the face of murderous fire. It was now about 11 a.m. and Cumberland had been unable to drive the French from their flanking positions, leaving him with the alternatives of drawing off without further fighting or attacking the French centre under a fierce crossfire. He chose the latter option, sending

some of his British infantry – including the Black Watch –
to support the Dutch in their attack in Fontenoy, while
the remainder of his British and Hanoverian foot attacked
between Barry Wood and Fontenoy. The Dutch attack again
failed and the Black Watch lost heavily in the process.
Cumberland then put himself at the head of his infantry and
the whole body, some 16,000 strong, stepped out up the
slope towards the French, with twelve light field pieces
accompanying it. There was insufficient space between
wood and village for the Hanoverians to remain on the
British left and they were compressed into a third line
behind the British battalions. The advancing mass was
subjected to a fearful fire as it moved forward, but it reached
the ridge top to find itself confronting the main line of
French infantry.

There was a brief pause before the musketry began. Lord
Charles Hay of the 1st Guards stepped in front of his
company, doffed his hat and drank from his flask to his
adversaries of the *Gardes françaises*, shouting, 'We are the
English Guards, and we hope that you will stand till we
come up to you, and not swim the Scheldt as you did the
Main at Dettingen.' He called for three cheers, which were
lustily given; the French replied with cheers and fired their
first volley. Whatever the precise details of this incident –
Voltaire has the French asking '*Messieurs les anglais, tirez
d'abord*' – there is no doubt about its sequel. The sustained
volley firing of the British infantry caused great carnage in
the French ranks. Commandant Colin, the French military
historian, wrote that the *Gardes françaises* 'fled terrified'
and the *Régiment d'Aubeterre*, next in line, was half
destroyed. Despite fire from the Eu Redoubt and Fontenoy
the British pushed on over the crest line, but the combination of flanking fire and furious frontal attacks gave
even this redoubtable infantry pause for thought, and
Cumberland's men fell back to the ridge. There they were
reorganised into a huge hollow square and the advance
began again.

Saxe himself was suffering from dropsy and, too weak to
sit his horse, was taken forward to the Eu Redoubt in a litter.

However, once Cumberland had broken his centre, he mounted and rode out amongst the shaken infantry, trying to rally them, then galloped back to reassure the King and his suite, watching the battle from Notre-Dame-de-Justice – Gallows Hill – to the rear, between Antoing and Fontenoy. Saxe realised that until the Allied penetration had some solid fulcrum its gains were not irreversible, and both Fontenoy and the Eu Redoubt still held out. He animated the defence, sending fresh infantry and cavalry against the square.

Under Saxe's command was the Irish Brigade, composed of Jacobite exiles: its six regiments – Clare, Dillon, Bulkeley, Roth, Berwick and Lally – commanded by Charles O'Brien, Earl of Thomond and Viscount Clare, lieutenant-general in the French service. The Irish had their own reasons for hating the English and although their shout of 'Remember Limerick and Saxon perfidy' may owe a good deal to historical embroidery, there is no doubting the vigour of their contribution to the French counter-attack. It was certainly not cheap: the brigade lost 650 men that day, including Lieutenant-General Count Arthur Dillon.

There is a breezy fiction that the unsupported assault of the Irish proved decisive. In fact, the attack owed much to the fact that it made good use of combined arms. Numerous French cannon were hauled up, on the initiative of Captain Isnard of the *Régiment de Touraine*, to fire canister into the square from point-blank range, and several badly mauled regiments – notably *Royal-Vaisseaux* and *Normandie* – rallied to face the rolling musketry.

Cumberland's bolt was shot. He belatedly ordered up some of his cavalry, but their charges foundered in the face of fire from the Eu Redoubt and Fontenoy. His infantry at first held their ground, but the combination of artillery fire, musketry and attacks by the cavalry eventually persuaded him to order a general retreat. The retirement was well covered by the cavalry, and Cumberland's army formed up east of Vezon and retired on Ath. It had lost some 7,500 men, with the burden falling most heavily on the infantry of the right wing, and the twelve guns that had accompanied them were lost. Sergeant Wheelock of Bulkeley's Irish Regiment was commissioned in the field for the capture of 'an enemy flag' – allegedly a colour of the Coldstream Guards, although this remains disputed.

The battlefield of Fontenoy is a bizarre mix of ancient and modern. The A16 autoroute unhelpfully runs north-west to south-east right across it, the area around Ramecroix, in the French left rear, is disfigured by huge quarries, and Gallows Hill is now occupied by a vast sugar-beet plant and a waterworks. The N52 in turn crosses the battlefield from south-west to north-east, and the visitor can all too easily hurtle across it before he knows quite where he is. Come off the roundabout immediately south of the A16, then head east towards the sugar-beet factory. After a few hundred yards you will see the lane that leads down to Fontenoy and be able to halt just short of the cemetery. Suddenly the battlefield makes sense. The spires of Antoing scratch the skyline to the south-west, Fontenoy lies at the foot of the forward slope and Barry Wood sits on the northern horizon. Seen from this viewpoint, it becomes clear that the French position was like a reversed L, with its base running from Antoing to Fontenoy and its upright from Fontenoy to Barry Wood.

The cemetery wall bears two memorials, one to Colonel de Talleyrand and the officers and men of the *Régiment de Normandie* who fell before Tournai and on the day of Fontenoy, the other to 'The heroic Irish soldiers who changed defeat into victory' on 11 May 1745. Going on down into Fontenoy (a Celtic cross in memory of the Irish regiments stands in its square) and turning left, the traveller follows the foot of the slope as far as the autoroute. Ascending the little road that runs parallel with it takes one back up towards the N52. This was the centre of the axis of Cumberland's infantry attack and the French infantry line was just over the crest, where the ground falls slightly to the main road. The Maubrai–Baugnies ridge, where Cumberland camped until he moved up to the battlefield, shuts off the eastern horizon.

If, instead of climbing the slope parallel with the

autoroute, the traveller goes beneath it on the road which follows the foot of the slope, he enters Vezon. The village is much bigger than it was in 1745 and Barry Wood has shrunk somewhat. Nevertheless, by taking the lane that leads northwards out of Vezon and up to the wood, one can begin to grasp the problem facing Brigadier Ingoldsby: the wood is large, even today, and the ground screens it from direct observation from the east. The Eu Redoubt stood at the intersection of the old Mons road – which ran north of the line of the present autoroute – and the sunken road which followed the crest behind Fontenoy. Road-building has so altered the topography that it is impossible to be certain about the redoubt's site: my best estimate would place it some 300 metres due east of the bridge carrying the autoroute over the N52, just to the north of the former.

The garrison of Tournai, downcast by Cumberland's failure to raise the siege and by the well-directed fire that brought down the left face of the Orléans bastion, surrendered on terms on 20 June. Saxe's subordinate Lowendal took Ghent in mid-July, and Bruges and Oudenarde shortly afterwards. Saxe himself pushed Cumberland back towards Brussels and on 14 July the Allies fell back to Dieghem, then well outside the city but now between Brussels and its airport. There Cumberland court-martialled James Ingoldsby for failure to attack the Eu Redoubt. Ingoldsby claimed, with some truth, that he had been harassed by contradictory orders. The court found him guilty, but felt that his error stemmed from poor judgement and not from cowardice. He was sentenced to be suspended during the Duke's pleasure: Cumberland believed that he would 'never again be able to serve with honour and dignity' and suspended him for three months.

Over the weeks that followed Cumberland pushed reinforcements into Ostend, now besieged by the active Lowendal, and marched north to Vilvoorde to keep open his own communications with Antwerp. The French took Dendermonde in mid-August, threatening Cumberland's left flank, and on 24 August Ostend surrendered. Cumberland was under pressure to send troops to help deal with the rising in Scotland and was dismayed to discover that the Ostend garrison, granted liberty by the surrender terms, were marched to Mons for their release rather than being allowed to return straight to England. Cumberland had to mount an operation to shepherd them back from Mons and Sir John Cope's army had been defeated by the Highlanders at Prestonpans before the expected reinforcements arrived. By late October almost the entire British contingent with the Allied army had departed and it was with the powerful assistance of his Fontenoy regiments that Cumberland won the decisive battle of Culloden on 16 April 1746.

In Cumberland's absence the war went badly. Saxe took Antwerp in May 1746, Mons in July and Namur in September. In the autumn the Allied field army, now reinforced by a British, Hanoverian and Hessian force under Ligonier, concentrated around Liège, planning to winter there before advancing into Brabant in the spring of 1747. Saxe intended to clear them from the borders of the Netherlands and on 11 October he advanced on their strong position on the northern lip of the hills which ring Liège.

The Allied left rested on the village of Ans – now merely part of the ribbon-building on the N3 to St-Truiden – their centre held Voroux and Roucourt, and their right Liers. In mid-afternoon d'Estrées and Clermont led a strong force of infantry against Ans, which was duly carried. The attack on the centre, delivered by twelve brigades of French foot against twelve British, Hanoverian and Hessian battalions, was delayed by the inaction of its commander, Clermont-Gallerande, a capable officer who had a history of poor relations with Saxe, and the three assaulting columns were savaged by close-range musketry and canister shot. Saxe personally animated his men as the battle hung in the balance and the Allied centre was broken after a vicious fight. D'Estrées's force had now lapped around the Allied left rear as far as Vottem and a general retreat ensued. Only the fact that the battle had started late prevented the wholesale destruction of the Allied army, which made its way across three pontoon bridges over the Meuse at Visé. This little-known battle – called Rocoux in British accounts – cost the Allies at least 4,500 men and the French a thousand less. It was another remarkable achievement by Saxe, for the Allied position, still partly visible through the mix of surburban housing and industrial estate, was a well-chosen one: it is no accident that three of Brialmont's Liège forts – Loncin, Lantin and Liers – were sited to cover precisely that sector of high ground held by the Allied army.

In 1747 the Allied army took the field once more. Under the nominal command of the Duke of Cumberland, it had a theoretical total of 130,000 men, but only something over 100,000 concentrated at Breda, where the British contingent had spent the winter. There was some inconclusive jockeying before Antwerp and the Allies then fell back to cover Maastricht. Saxe, whose army had wintered comfortably in Hainaut, advanced on Maastricht via Brussels and on 2 July he attacked the Allies in front of Maastricht. The village of Laffeldt, held by British and German infantry, was the key to the Allied centre. It fell at midday and Cumberland failed in an attempt to recapture it. As the Allied army fell back on Maastricht, Ligonier charged the pursuing cavalry with the four regiments of British horse and bought enough time for the army to cross the Meuse, though he himself was unhorsed and captured. The bitter fighting in the village made Laffeldt a costly battle: the Allies lost 6,000 men in the fighting and another 2,000 prisoners, while French losses may have been as high as 8,000.

Although Saxe had bruised the Allies at Laffeldt, Ligonier's action had prevented the French from converting the retreat into a rout, and Saxe felt unable to risk besieging Maastricht with a largely intact Allied field army within striking distance. Accordingly, he left d'Estrées to mask the Allies at Maastricht while Lowendal marched westwards to besiege Bergen-op-Zoom. Cumberland, urged by the Dutch to relieve this important fortress, set off to bring Saxe to battle, but before he could arrive the French risked all on the desperate throw of a storm, and early on the morning of 4 September Lowendal assaulted the largely intact defences, and took the place in an operation marred by the wholesale slaughter and pillage that so often accompanied a storm.

The Allied army, based on Roermond, downstream from Maastricht, was preparing for the 1748 campaign when news arrived that preliminaries of peace had been signed at Aix-la-Chapelle. Cumberland at once concluded an armistice, surrendering Maastricht as its price. Although the war ended with the line of the Meuse in French hands, the settlement resulted in a return to the *status quo ante bellum*. The Austrian Netherlands were returned to Austria – much to the irritation of Maria Theresa, who would have preferred to regain Silesia and lose the Netherlands. Indeed, the fact that Silesia remained in Prussian hands made another war all but certain and for that reason the Peace of Aix-la-Chapelle, which ended so much bloodletting, turned out to be little more than a truce.

The Seven Years War (1756–63) saw no serious fighting in Hainault and Brabant. The Duke of Cumberland was badly beaten by his old enemy d'Estrées at Hastenbeck in 1757 and was forced to conclude the Convention of Klosterseven, agreeing to disband his army apart from its Hanoverian contingent. The Peace of Paris (1763) brought no changes in the Austrian Netherlands, and it was not until 1791 that armies again clashed along the Sambre and the Meuse. This time the adversaries were the threadbare warriors of revolutionary France and the pipeclayed forces of monarchical Europe.

Frequent changes of name – and even more frequent changes of commander – make it difficult to follow the activities of the armies of the First Republic. The task is a little easier as far as the northern frontier is concerned, for its defence was entrusted, on 14 December 1791, to the *Armée du Nord*, and this army wrote its colourful history across Brabant, Hainault and Flanders. Its offshoots, the *Armée de la Belgique*, the *Armée des Ardennes*, the *Armée de la Hollande* and the *Armée de la Sambre et Meuse* all flared into prominence in their day, but it is with the *Nord* that we shall have most of our dealings. Its first commander, the Count of Rochambeau, was a seasoned professional who had served in North America during the Revolutionary War, but his independent spirit crouched uncomfortably beneath the weight of political control which bore down on the armies of revolutionary France.

The *Nord* received its baptism of fire in inauspicious circumstances when Rochambeau, who had shrewdly advised that his troops were too shaky to do more than stand on the defensive, was ordered by Dumouriez, Minister of War, to push three columns into the Austrian Netherlands. One advanced from Dunkirk to Furnes, found nothing and retraced its steps without mishap. Another marched from Lille on Tournai, only to be checked midway at Baisieux and to dissolve in rout, murdering its commander, Théobald Dillon, whose ancestor had fallen at Fontenoy. The third made for Mons, fighting a successful battle at Quiévrain – just off the field of Malplaquet – only to fall victim to panic and bolt back to Valenciennes.

In mid-May the army commanders in the north and north-east met near Valenciennes to evolve a new strategy, which involved moving the *Nord* to Flanders in an effort to roll up the Austrian right flank, taking fortresses as it encountered them, while the *Armée du Centre* threatened the Austrian left. Perhaps Rochambeau only consented to this folly because he was ill and already intended to retire: in any event he left the army and was replaced by Dumouriez, who relinquished the post of War Minister in order to command in the field. No sooner had he arrived than there was another hare-brained scheme, the *chasse-croisée*, by which troops and commanders were redeployed in two main armies under Lafayette and Luckner between the northern and north-eastern frontiers, resulting in much predictable confusion. It left Dumouriez, for the moment, without an army to command, but Lafayette's flight to the Austrians put him back in command of the *Nord*. In September he swung down into the Argonne and decisively checked the Allied invasion at Valmy, then turned his attention to his old objective, the Austrian Netherlands.

The Austrians had been threatening Lille and Dumouriez moved against their main concentration, near Mons, to end the danger. He found the Archduke Albert established with about 14,000 men in a strong position on the high ground between Quesmes and Jemappes, west of Mons. The battlefield of Jemappes is now utterly unrecognisable, with high-density housing and an autoroute spur curling down from the E42 making it impossible to follow details of the battle on the ground. This is unfortunate, for Jemappes was a decisive action, which gave France control of the Austrian Netherlands and showed the Revolutionary army in a characteristic light. On 6 November Dumouriez intended to attack the Austrian centre with 40,000 men, while d'Harville, with 10,000, turned the enemy left. In the event the battle degenerated into a huge frontal assault. Dumouriez's regular regiments went forward with the war-cries that would have been heard at Fontenoy – '_En avant. Navarre sans peur_' and '_Toujours Auvergne sans tache_' – while the volunteers sang patriotic songs. The French artillery, less harmed by the emigration of officers than the infantry and cavalry – acquitted itself well and the unlucky Austrians were simply swamped. Mons surrendered the following day and the French had overrun the Netherlands within a month.

Early in 1793 the Allies profited from Dumouriez's preoccupation with pushing on to invade Holland by concentrating 40,000 men under the Prince of Coburg. Dumouriez was reluctant to abandon his goals in Holland in order to face Coburg, but he was ordered back by the government. He joined General Miranda, who had been commanding French forces around Maastricht, at Louvain and at once set off to meet Coburg. He took Tienen on 16 March and two days later attacked Coburg at Neerwinden.

Dumouriez's men went forward in eight columns. The battle was evenly balanced on the right and centre, but on the French left Miranda was repulsed and fell back, without orders, on Tienen. Dumouriez had been encouraged to seek a quick decision against Coburg by the fear that his own army would not stand a retreat or prolonged defensive operations and events proved him right. There was large-scale desertion from French ranks, and Dumouriez lost first Louvain and then Brussels. He had previously considered negotiating with the Allies in an effort to produce a new political settlement in France and after Neerwinden he entered into discussions with Mack, Coburg's Chief of Staff. He planned to march on Paris, overthrow the Jacobins and re-establish the monarchy, but hesitated too long and had to flee to the Austrians.

Dumouriez was succeeded in command of the northern armies by Dampierre, but the latter was mortally wounded in fighting between Quiévrain and Valenciennes and Lamarche took his place. The Allies intended to besiege Valenciennes, and in late May they captured the fortified Camp de Famars, just south of the town, and proceeded to take Valenciennes. Lamarche had already been replaced by Custine, who did his best to improve his army's efficiency. The strictness with which he applied himself to this task did not make him popular: he was recalled to Paris, arrested, tried and guillotined. Further commanders followed in quick succession. Kilmaine was suspended after the fall of Valenciennes and his successor, Houchard, tried hard to avoid being passed the poisoned chalice. He was quite right to do so. After a confused campaign in which he beat an Allied army at Hondschoote in Flanders but suffered a reverse near Cambrai, he too was arrested and guillotined.

With the departure of Houchard, the *Armée du Nord* fell into the firm hands of General Jourdan. He had served as a private soldier before the Revolution and become a linen draper on discharge. He rose rapidly in the army of the Revolution and if Colonel Ramsay Phipps is right to say that 'he cannot be placed higher than in, say, the third rank of commanders' his dogged obstinacy and tenacity of purpose was to stand him in good stead. He found Coburg besieging Maubeuge with 26,000 men, covered by a screening force of 37,000 deployed in a wide arc through the village of Wattignies, and expecting to be reinforced by the Duke of York, on his way from Menin with about 5,000 English and Hanoverians.

Jourdan concentrated on Guise, then set off for Maubeuge with some 45,000 men. His path was blocked by Clairfayt's covering force, spread across the close country of woods and hills on both sides of the Avesnes–Maubeuge road. Such was the spread of Clairfayt's deployment that he could bring only 20,000 men to bear on Jourdan's attack and the force besieging Maubeuge offered little assistance, although Coburg himself came forward to take command. As was so often the case with the armies of the Revolution, there was a deep gulf between planning and execution. The planning seems to have been the work of Lazare Carnot, member of the Committee of Public Safety and driving force behind the Republic's armies. There was to be an attack on both the Allied flanks, followed up with a decisive blow at the centre.

The first day of the battle, 15 October 1793, was every bit as confused as the partly trained state of Jourdan's army and the closed-in terrain of the battlefield suggested. Both flank attacks – by Fromentin on the left and Duquesnoy on the right – initially made good progress. Carnot then persuaded Jourdan to launch his centre division, under Ballard, at the village of Dourlers on the main Avesnes–Maubeuge road. This was an unwise move. The French artillery could not keep pace with the infantry attack and Ballard received a bloody repulse. At the same time Duquesnoy was checked in front of the hilltop village of Wattignies, while Fromentin unwisely debouched from the close country into the plain of the Sambre and was charged with grim effect by the Austrian cavalry.

Jourdan revised his plans that night and took 6,000 men from the centre to strengthen his right, where he intended to make the decisive effort. Fresh levies had come up from Paris, increasing Jourdan's overall superiority, but Coburg too had been reinforced and was confident of holding his very solid positions. Wattignies was key to the Allied left. Twice French columns came up the slope through the early-morning mist and twice the Austrian defenders threw them back. Jourdan himself rallied his men and led them on again, and the assaulting columns converged in Wattignies, pushing the Austrians over the crest and down the slopes behind.

French performance was decidedly uneven. On Duquesnoy's right, many of Beauregard's division ran away.

Well off the battlefield, beyond the French right, a column under General Elie, moving from Philippeville on Beaumont, bolted back to Bossu when it met the Austrians. So mixed were the results of the day that Jourdan was not certain if he had won, but on the 17th Coburg slipped away under the cover of fog. Jourdan marched into Maubeuge that afternoon and exchanged some harsh words with the garrison commander, whose 20,000 men had been singularly inactive while the battle was in progress. It was typical of the notion of justice then prevailing that the garrison's second-in-command was guillotined.

Maubeuge was no stranger to sieges. It had the unlikely distinction of being burnt by English troops of the Duke of Gloucester, husband of Jacqueline of Hainaut, in 1424, and changed hands several times before it became French following the Treaty of Nijmegen in 1678. Vauban fortified it in 1680–4 and it remained in French possession throughout the eighteenth century. It was successfully defended by Colonel Schuler in 1814 and the following year it surrendered to the Prussians at the end of the Hundred Days. Spared by the Franco-Prussian War, the town was given a ring of surrounding forts under the Seré de Rivière plan. These were similar to the forts we have already seen in Lorraine, although the unimportance of Maubeuge in French planning prior to the First World War meant that they were not modernised. The forts remain more or less intact, though most are difficult or dangerous of access: Fort Bourdiau is badly knocked about, with massive rents in its reinforced concrete, and almost completely overgrown, while Fort le Floricamp is Maubeuge Council's toxic waste dump.

However, Fort de Leveau, reached via the D105 running north-west from the town, can be visited on most weekday afternoons. There is some steel and concrete where its salient points out over the open country, but the remainder of the structure is unmodernised brick with earth covering. Large *traverses-abri* sit atop the barrack block with its arcaded brickwork. The fort was bombarded and taken by assault on 7 September 1914 and a memorial stands where many of the combatants lie buried beneath the ruins.

Maubeuge was declassified as a fortress in 1928. It was very heavily damaged by air attack in 1940 and after the war the fortifications suffered as the town grew: the *Porte de France* and the defences on the south bank of the Sambre were destroyed in 1958. The northern fortifications, based on five bastions, still remain, and the *Porte de Mons*, well preserved thanks to the efforts of the *Renaissance Vauban* association, is one of the most striking fortress gates in France, second only to that of the citadel of Lille. Approaching on foot from the north the visitor crosses a crisply maintained ditch (the covered way is especially clear) and goes over the drawbridge to the stately entrance. Immediately inside the gate is the drawbridge mechanism, and the arched entrance corridor leads to a courtyard with plaques in honour of General Schuler, defender of Maubeuge in 1814, and General Fournier, its governor a century later. Fournier deserves remembering, for his thirteen-day defence of Maubeuge in the face of impossible odds tied up German troops who might have added their weight to the finely balanced Battle of the Marne.

The courtyard opens on to the Place Vauban, rich in military memorials. The local regiments are commemorated by a plaque on the ramparts and a Renault light tank in the centre of the square. The square is dominated by the Wattignies memorial, its stone plinth topped by a jubilant bronze figure of a Revolutionary infantryman, the very essence of 1793, musket in one hand and bicorne hat waved by the other. There is a sadder footnote: at the rear of the plinth is another of the heroes of Wattignies, a drummer boy rapping out the *pas de charge* as death closes in on him. Wattignies itself, fifteen minutes away from Maubeuge by car, now rejoices in the name of Wattignies-la-Victoire and has an elegant obelisk celebrating Jourdan's triumph. The slopes around it are as open today as they were in 1793, and the narrow lanes and thick woods show just how difficult it was for the attacking French columns to remain in contact with one another.

The victor of Wattignies soon found himself in difficulties with his government. Ordered to advance down the Sambre, he protested that his army was exhausted and ill equipped, and was allowed to go into winter quarters. Summoned to Paris and accused of lack of zeal, he was lucky to escape with his head. Dismissed from his command, he went back to his drapery business in Limoges, but wisely kept his general's uniform hanging at the back of the shop. His successor, Pichegru, splashed the *Nord* across the northern frontier and in the early months of 1794 there were some Allied successes: a French division was routed at Troisvilles, near Cambrai, and Landrecies fell on 30 April. The tide began to turn in the French favour, and on 18 May Pichegru's men won a confused encounter battle at Tourcoing and went on to take Brussels.

The constant juggling of French armies threw the real burden of the 1784 campaign in the north on to the *Sambre-et-Meuse*, commanded by Jourdan, his draper's shop abandoned once more. He moved against Charleroi, left a force to besiege it and crossed the Sambre, only to lose a stiff battle against the Prince of Orange near the 1814 battlefield of Quatre Bras. Charleroi surrendered, rather prematurely, on 26 June and on the same day Jourdan fought Coburg, advancing to its relief, at Fleurus. Jourdan had some 75,000 men, dug in on the high ground north of the Sambre, and Coburg attacked with 52,000 in five columns.

It was a day of mixed fortunes. On the French left, around Courcelles, Kléber held his own and then counter-attacked, although his left-hand division gave way, allowing the Allies to reach the Sambre. There was a panic on the extreme right, in front of Lambusart, permitting the Allies to reach the Sambre to its south-east. Failure on the right brought the French centre back from Fleurus and St-Fiacre, and the battle focused on Lambusart, where Lefèbvre – a marshal under the Empire – inspired what Soult, another future marshal, called 'fifteen hours of the most desperate fighting that I have ever seen in my life'. Lefèbvre regained and held Lambusart, and the crest line to his left, above Heppignies and Wagnée, was retained by the French after another furious battle.

By the end of the day it was not clear who had won and

Coburg might have succeeded if he had resumed the attack the next day. However, he drew off to the north, leaving the exhausted French on the field. Fleurus is not an easy battle to follow on the ground. The villages, small and distinct in 1794, have expanded and, in some cases, merged, and the thick web of roads overlaying Charleroi does not help matters. There is an interesting aside to the battle. The French used a captive balloon – *L'Entreprenant* – at Fleurus. Manned by General Morlot and an engineer officer, the balloon hung over the battlefield, with the observers sliding messages (of remarkably little value) down the rope.

Coburg's decision not to press matters at Fleurus proved disastrous. A victory over Jourdan would have enabled him to turn on Pichegru and the *Nord*, on his right, and reshape the military situation in the Netherlands. As it was, Fleurus proved a more conclusive battle than its immediate results suggest. The *Sambre-et-Meuse* linked up with the *Nord* and cleared much of Belgium, although interference from Paris soon had the armies separated again. The discomfiture of Coburg's field army enabled the French to take the fortresses on their northern border at their leisure, then to push the Allies out of Brabant altogether, Pichegru striking up into Holland and Jourdan crossing the Rhine.

The French successes of 1794–5 took the war out of the fatal avenue: it was not to return in earnest until 1815. The Duke of Wellington spent a fortnight in Belgium on his way to take up his appointment as British Ambassador to France. He suggested that the damaged fortifications of Courtrai, Menin, Mons, Tournai and Ypres should be repaired, and looked hard at positions that might be of use to him in the future: these included the Mont St-Jean position, covering the southern approaches to Brussels. Wellington left Paris for the Congress of Vienna early in 1815 and had been in the Austrian capital for a month when he heard that Napoleon had escaped from the island of Elba. He set off to command the Anglo-Dutch army and arrived in Brussels on 5 April.

The Duke was not encouraged by what he found. Although Belgium and Holland were united in a single kingdom, Belgium had been in French hands since Fleurus and there was much local sympathy for Napoleon. Wellington's army was not the seasoned force he had commanded in Spain, but was a mixture of British and Dutch-Belgian units of uneven quality. It was the Allied intention to mount a combined invasion of France, with Wellington and Marshal Blücher's Prussians moving down from the north, while Barclay de Tolly's Russians and Schwarzenberg's Austrians marched up through Lorraine and Alsace, but the components of this master plan would not all be ready at the same time and Napoleon was raising troops fast in order to put 600,000 men into the field by early summer.

Wellington organised his army into three corps: the first under Lord Hill, the second under the young Prince of Orange and the third or Reserve under his own command. In each corps veteran units were mixed with the inexperienced so as to stiffen the whole and the Duke worked hard at improving his army's training. By the time operations began Wellington had 93,000 men, only about one-third of them British. He met Blücher at Tienen on 3 May, and the two commanders agreed to work in close co-operation, with the Roman road from Bavay to Maastricht – the Roman road that runs across the battlefield of Malplaquet – as the boundary between their forces.

Other details of the meeting are obscure, but both commanders probably agreed to lean in towards the centre to cover Brussels, and seem to have decided that in the event of a French advance through Charleroi or Mons the Anglo-Dutch would concentrate at Nivelles and the Prussians at Sombreffe. Liaison officers were exchanged, Baron Müffling joining Wellington and Colonel Hardinge, Blücher. The latter had some 117,000 men, in four corps, under his command. Ziethen's 1st Corps was around Charleroi, Pirch's 2nd at Namur, Thielmann's 3rd at Ciney and Bülow's 4th near Liège.

On the afternoon of 15 June Wellington heard that the French were on the move. They had attacked the Prussians south of the Sambre, around Binche, early that morning, but it had taken some time for word to reach the Duke and even then the report was incomplete. Wellington sent out preliminary concentration orders, but still did not know whether Napoleon would advance through Tournai, Charleroi or Mons.

The Emperor had stolen a march on his opponents and moved, with 122,000 men forming two wings and a reserve, straight on the Allied centre, hoping to deal with the enemy armies in detail. Early success veiled serious flaws within his army. Berthier, his trusted Chief of Staff, had died under suspicious circumstances and Marshal Soult, appointed to the post, was a man of action rather than a staff officer. Nor was the choice of commanders for the two wings much better. Marshal Ney, who led the left wing, was an officer of unquestionable bravery but flawed judgement and Marshal Grouchy, of the right wing, was an excellent cavalry commander but lacked experience of combined-arms command. Napoleon himself was past his best. Although aged only forty-six – the same age as Wellington – he was in poor physical condition and, as the events of the next days were to show, his mind had lost its sharpness.

The French had been on the move early on the 15th, infantry columns on the march behind a cavalry screen. Bad staff work led to tangles around Charleroi, where Ziethen's Prussians were pushed off the Sambre bridges. Blücher had received accurate information of Napoleon's movements from a French divisional commander of Royalist sympathies who chose this propitious moment to desert, but persisted in his plan for a forward concentration at Sombreffe and arrived there himself on the afternoon of the 15th. Wellington, as we have seen, was less sure of what was afoot and his initial orders – for a concentration to the west of Brussels – would have taken his army further away from the French and exposed the Prussians to even greater danger.

The early engagements went in Napoleon's favour. Late in the afternoon of the 15th Grouchy, spurred on by the Emperor himself, drove Ziethen out of the outskirts of Fleurus, while the leading element of Ney's wing took Gosselies. There was little sense of urgency on either wing and only Lefèbvre-Desnouëttes's cavalry, scouting ahead of

Ney, made real progress, moving up through Frasnes and bumping infantry and guns south of the road junction at Quatre Bras. These Allied forces were part of Perponcher's 2nd Dutch-Belgian division and held the important crossroads without difficulty against the cavalry probes. The fact that the junction was held at all owed much to the 'intelligent disobedience' of General de Constant Rebecque, the Prince of Orange's Chief of Staff, who had sent Perponcher there on his own initiative and ignored a direct order from Wellington to move the division to Nivelles, away to the west.

Wellington spent the night of the 15th at the Duchess of Richmond's ball in Brussels: failure to attend would have encouraged pro-French elements in the city and it was no disadvantage to have numerous senior officers readily available. He was up betimes on the 16th and rode forward to Quatre Bras. His reserve was also on the move early and marched down to the road junction at Mont St-Jean, where it could move to the south or swing west to the Halle–Tubize area as occasion demanded. By this time Wellington had a much better idea of what was afoot. While he was still at the ball news had arrived, from cavalry scouting towards Oudenarde, that there were no French on the western flank. He was therefore able to modify his orders and to call for a concentration on the inner flank, towards the Prussians. But it has to be said that without Constant Rebecque's foresight in holding Quatre Bras the Duke would have been in serious difficulties, for the central axis of the Allied army would have been lost.

Napoleon was also busy early on the 16th. At about eight that morning he sent dispatches to Ney and Grouchy. The former was told to halt at Quatre Bras, with one division to its north-west and another to the east to link up with Grouchy, and to prepare for an immediate advance on Brussels, which would commence once the reserve had come up. Grouchy was ordered to march on to Sombreffe and Gembloux, taking on the Prussians as he found them and being prepared to swing westwards to assist Ney in a battle against Wellington. The Emperor himself went forward to Fleurus, where it became evident that, far from falling back as had been expected, Blücher was concentrating in front of Sombreffe.

Three Prussian corps – 84,000 men and 224 guns – took up a position on the rough line of the Ligny brook. Wellington rode over to meet Blücher, at the Bussy windmill near the village of Brye. He was not impressed by the ground Blücher had selected: Prussian columns were drawn up on the slopes leading down to the marshy brook and Wellington would have preferred to edge back into the reverse slope so as to shelter them from cannon fire. At some ten kilometres wide, Blücher's position was rather broad for the strength of his army, although he was hoping that his IV Corps would come up in time to hold it and was also heartened by Wellington's assurance that he would come to his assistance if he himself was not attacked.

Napoleon believed that Blücher's stand offered him an excellent chance of smashing the Prussians. While cavalry masked the Prussian right wing, the corps of Vandamme and Gérard were to attack Blücher's right and centre. Ney would march in on the Prussian right rear, completing the rout and forcing Blücher to fall back on Liège with his surviving troops. At about 2 p.m. Ney was told to clear the enemy from his front and then to envelop Blücher, and at 2.30 p.m. the main attack on the Prussians began. Blücher's infantry was pounded by artillery, as Wellington had predicted, but when Vandamme's men assaulted Ligny and St-Amand they were stoutly resisted. The villages changed hands several times and Prussian reserves were drawn forward into the fighting exactly as Napoleon had hoped. This encouraged him to send a second message to Ney, at 3.15 p.m., ordering him to move on the Prussian flank. Shortly afterwards Napoleon heard that Ney was heavily committed at Quatre Bras, so modified his plan by telling Ney to clear the crossroads there and to send only d'Erlon's corps to attack the Prussians.

Napoleon was preparing to send in his reserve, the Imperial Guard, when Vandamme announced that a column of unidentified troops – presumably hostile – was approaching his left flank. This eventually turned out to be d'Erlon's corps, but no sooner had it approached the battlefield than it departed inexplicably. It took time to put fresh momentum into the attack and it was not until 7.30 p.m. that the Guard pushed through Ligny, breaking the Prussian line. Blücher, displaying energy that belied his seventy-two years, put himself at the head of thirty-two squadrons of cavalry and charged the Guard, but was beaten off. His horse was shot: French cavalry rode over him as he lay pinned beneath it, but a devoted aide rescued him and he was carried from the field.

French victory at Ligny was tantalisingly incomplete. Although the Prussians had lost 16,000 men to the French 12,000, they were to recover quickly and this resilience was to have dramatic effects. Napoleon undoubtedly erred when he decided not to pursue them that night, but he was concerned that the arrival of Bülow's fresh corps might obstruct the pursuit and disturbed not to have heard from Ney for several hours.

Ney had made a slow start on the 16th. His men were not on the move through Frasnes until just before midday and Wellington was given valuable hours to reinforce Perponcher's badly outnumbered division at Quatre Bras. Ney's leading corps, under General Reille, advanced to the attack at about 2 p.m. Reille himself had fought the British in Spain and he approached the Allied position, in the cornfields on the gentle slope leading up to Quatre Bras, with caution. Nevertheless, his leading division, Bachelu's, took Piraumont Farm, on the Allied left, at 3 p.m. and Foy's division captured Gemioncourt, in the centre, shortly after-wards. The larger of the two Pierrepont Farms, on Perponcher's left, proved a harder nut to crack and it was only when Foy's men were reinforced by part of Prince Jérôme Bonaparte's division that the place was at last taken. The French then began to work their way through Bossu Wood, on the Allied right front, in the face of stiff resistance from Prince Bernhard of Saxe-Weimar's brigade of Nassauers.

Perponcher's men were now bearing the weight of three

French divisions and were close to cracking. But Wellington himself was on hand, a Dutch-Belgian cavalry brigade arrived from Nivelles to butteress Perponcher and Sir Thomas Picton's division marched down the Brussels road. Wellington deployed two of Picton's brigades along the Quatre Bras-Namur road and placed the third in support. Thick French columns were coming on behind a skirmish line and Wellington sent the two brigades forward to face them rather than await their assault. The firepower of the British infantry drove the columns back, but as Picton's men reached the brook that crosses the battlefield through Gemioncourt they were exposed to the full weight of the French artillery and compelled to fall back.

At this juncture the Duke of Brunswick brought up further reinforcements and Wellington deployed them as they arrived, sending most of the infantry west along the Nivelles road to support Saxe-Weimar's men who had been all but forced out of Bossu Wood. Brunswick himself was ordered to take his cavalry forward on the axis of the Charleroi road and at once charged at the head of his hussar regiment. He made little impression on the French infantry and was rallying his men just south of the crossroads when he was mortally wounded. The Prince of Orange collected the remainder of Brunswick's cavalry and led them forward with most of Van Merlen's brigade, but there was French cavalry on hand at Gemioncourt and the Prince's men ebbed back up the slope. The French cavalry followed and Wellington found himself in danger. The 92nd Highlanders of Picton's division were at the crossroads, facing south and west, and the Duke jumped their ranks to safety. French horsemen swirled around the infantry line, forcing their way between regiments. Wellington managed to rally some of his own cavalry, and posted Van Merlen's men behind Picton's infantry and the Brunswickers just south of the Quatre Bras–Nivelles road.

At 4 p.m. Ney was jolted into activity by Napoleon's 2 p.m. order, at once summoned d'Erlon's uncommitted corps and urged Reille to make fresh efforts to capture the crossroads. We have seen the difficulties that Reille's assault caused and there can be little doubt that the arrival of d'Erlon's corps would have been decisive. However, the staff officer bringing Napoleon's summons to d'Erlon's corps met one of d'Erlon's divisions and sent it off towards Ligny, found d'Erlon himself – with some difficulty, for he was already viewing the Quatre Bras position prior to launching his attack – and finally told Ney what was afoot. The explosion of wrath that followed dissuaded another of Napoleon's staff officers, who had just arrived with the full text of the 3.15 p.m. order, from delivering his message. No sooner had Ney turned his attention to the battle than an Allied counter-attack pushed Reille's men back and Ney decided that he could not hold his ground without help: he sent a messenger ordering d'Erlon to return. D'Erlon was, as we have seen, within sight of the battlefield of Ligny when the order reached him: he left an infantry division and some cavalry to link Napoleon's left with Ney's wing, and returned to Quatre Bras.

While this marching and countermarching was in progress the fighting raged around the crossroads. As the French cavalry ebbed back after reaching the Allied line, fresh reinforcements – including Halkett's brigade – arrived, at last tilting the balance of forces in Wellington's favour. Ney ordered Kellermann – son of the hero of Valmy – to charge Wellington's centre. Only one of Kellermann's three brigades had so far arrived and the general asked Ney to clarify his order, but Ney peremptorily told him to charge at once. Kellermann took his *cuirassiers* straight forward on either side of the Charleroi road. Although they rode down some British and Brunswick skirmishers, they were easily dealt with by the volleys of the infantry at the junction and Kellermann had his horse shot beneath him.

In the lull that followed, French skirmishers worked their way up to engage Wellington's line and French guns, as far forward as Gemioncourt, were able to find targets more easily now that the corn had been trampled by successive waves of attack. Many of the defenders were running short of ammunition, and although Wellington and Picton shuffled battalions so that fresh units took some of the strain, the combination of artillery fire and sniping caused a steady drain of casualties.

In the late afternoon another cavalry attack, delivered by a mixture of hussars, lancers and *cuirassiers*, swirled around the squares running parallel with the Namur and Nivelles roads, and at the same time the 95th Rifles and Brunswickers who had held the hamlet of Thyle, on Wellington's extreme left, were forced back into Cherris Wood, north of the Namur road. The plight of the Allied infantry was serious enough and the Prince of Orange was to make it more so. Halkett's brigade was drawn up with its four battalions in square between the northern end of Bossu Wood and the Charleroi road, and Orange, believing that the cavalry threat had diminished, ordered the battalions into line despite Halkett's objections. French horsemen charged immediately, first breaking the 69th Regiment (just east of the Charleroi road), then driving Halkett's other battalions to the shelter of the Bossu Wood.

Wellington rode over from his left flank, helped restore order to Halkett's brigade, and sent two newly arrived Brunswick battalions to clear the ground between Bossu Wood and the Charleroi road. The Guards Division was also on hand, its leading battalions were sent piecemeal into Bossu Wood, driving out the French: the whole division was soon concentrated south of the wood. The arrival of more artillery and extra ammunition encouraged Wellington to launch a general advance, taking Gemioncourt in the centre, Pierrepont on the right and Piraumont on the left. By the time the fighting ended at about 9 p.m. the Allies had recovered all the ground lost and the French were back on the outskirts of Frasnes.

Casualties at Quatre Bras were more or less equal at just over 4,000. Neither Wellington nor Ney emerges without blame. The former had been slow to recognise the importance of Quatre Bras, though once he had done so he managed matters with his usual skill. Ney made a lamentably slow start, then never realised that the real business of the day lay at Ligny rather than Quatre Bras. Poor French staff work made its contribution to a disappointing day: it is hard to resist the conclusion that Berthier's presence at his master's side would have made a telling difference.

The field of Quatre Bras is as instructive as that of Ligny is disappointing. Ligny, like nearby Fleurus, has been so heavily built up that it is hard to follow the battle on the ground. There is a small Napoleonic museum there, in the *Centre Général Gérard*, as well as memorials to General le Capitaine, killed in 'Napoleon's last victory', and the Old Guard. Quatre Bras, in contrast, has suffered far less. Modern roads follow the line of the 1815 routes, although the Nivelles and Namur roads are now very much larger than they were, and the thunder of traffic down the main Brussels road impedes quiet contemplation. Gemioncourt Farm – so typical of this part of Belgium, with farmhouse and farm buildings forming an easily fortified enclosure – stands east of the Brussels road, with the Gemioncourt brook – now a decidedly sorry obstacle – just north of it. Looking towards the crossroads, some two kilometres away, the disadvantages of the Allied position become clear. There was no convenient reverse slope behind which Wellington could shelter his infantry, as was his wont, and once French guns had been moved forward to Gemioncourt early in the afternoon they were able to trundle roundshot through the Allied lines with ease. Just west of the crossroads in Qautre Bras, almost opposite each other, are a British memorial, erected in 2000, and a Dutch monument, some ten years older.

On the Allied left, down the Namur road, the hamlets of Piraumont and Thyle are comparatively unchanged. The Materne Lake, formed by damming the Gemioncourt brook, stood on the initial line of the Allied centre left and can still be seen, although Cherris Wood, which stood north of the road, has gone. On the right, Bossu Wood was cut down shortly after the battle, but the two Pierreponts – Large and Small – stand south of the old wood edge. A pillar topped by a lion marks the spot, just south of the crossroads, where the Duke of Brunswick fell, and east of the pillar, parallel with the Namur road, is the area where Picton's infantry stood its ground.

Wellington rode to Genappe to snatch some food and sleep, and was back at Quatre Bras early on the 17th. He knew that Blücher had been beaten at Ligny, and received further information from cavalry reconnaissance and a Prussian officer who came in with a message to Müffling, the Prussian liaison officer. Wellington told Müffling that he was prepared to offer battle in the Mont St-Jean position, between Quatre Bras and Brussels, if Blücher would march to his assistance with a single corps. He then set about withdrawing his men from Quatre Bras and getting them on to the Brussels road to march up to Mont St-Jean.

Gneisenau, Blücher's anglophobe Chief of Staff, trying to arrange matters in his commander's absence on the night of Ligny, had been in favour of falling back on Liège. However, the direct route to Liège was blocked and it seemed wiser to fall back to the north: he selected Wavre as his army's concentration point. When old Blücher arrived, battered by his experiences at Ligny, he dosed himself with a deadly mixture of gin and rhubarb, and took issue with Gneisenau's preference for independent withdrawal, arguing that soldierly honour and common sense demanded that the Allies should stick together.

The French made another slow start on the 17th. Napoleon was sure that the Prussians were making for Liège, but was surprised to find that Wellington was still holding Quatre Bras. It was not until 11 a.m. that he ordered Grouchy to move on Gembloux, keeping in contact with the Prussians and reporting their movements. Lobau's Corps and the Guard were to march on Wellington's left flank via Marbais, while Ney attacked Quatre Bras frontally. At 1 p.m. Napoleon arrived near Quatre Bras to find Ney's troops at the serious business of lunch, and it was not until 2 p.m. that d'Erlon spurred off in pursuit of the Allied rearguard. The pursuit was impeded by a torrential thunderstorm, and the Allies were able to draw back on to Mont St-Jean with small loss. Napoleon established his headquarters in the roadside farmhouse of Le Caillou, where he received reports indicating that Wellington intended to stand and fight.

The ground over which the battle of Waterloo was fought

is scarred by the twin diseases of road-building and tourism. An autoroute crosses the field, and a thriving collection of cafés and souvenir shops nestles under the shadow of the Lion Monument, a mound whose construction helped alter the topography of the centre of the battlefield. That said, the field of Waterloo has not been extensively built on and most of the buildings which formed key points in the Allied line survive.

Mont St-Jean was the best defensive position between Quatre Bras and Brussels. A low ridge, notched by re-entrants, ran at right angles to the Brussels road and the roads behind it were shielded from an attacker advancing from the south. Three groups of farm buildings formed bulwarks on the forward slopes. In the centre, beside the Brussels road, was La Haye Sainte. Hougoumont stood on the western edge of the battlefield, while on its eastern side were the farms of Papelotte and La Haye, and the nearby hamlet of Smohain. The boggy valleys of the Dyle and Lasne gave some protection to the left, but the right, beyond Hougoumont, was more vulnerable. Fears about his right induced Wellington to station an Anglo-Dutch force 15,500 strong at Halle and Tubize, while Chassé's Dutch-Belgian division was posted even closer, at Braine l'Alleud.

Hougoumont was strongly garrisoned, initially by the light companies of the Guards Division, a Nassau battalion and some Hanoverian riflemen. The remainder of the Guards were posted behind Hougoumont, south of the ridge so as to support the garrison but able to gain a little protection from the trees around the farm complex, far more extensive then than they are now. The Duke posted a brigade to the west of the Nivelle road, almost parallel with Hougoumont, with the Brunswick contingent and Clinton's division behind it in Merbe-braine. To the Guards' east, along the lateral road running just behind the ridge top, stood Alten's division with some Nassauers in support, their positions extending as far as the Brussels road. La Haye Sainte was held by the 2nd Light Battalion of the King's German Legion, from Ompteda's Brigade of Alten's Division.

East of the Brussels road, the lateral road ran on the forward edge of the ridge and Bylandt's brigade of Perponcher's division was deployed along it, with Picton's Division to the immediate rear, just behind the ridge. A battalion of Rifles was thrown forward into the sandpit beside the main road opposite La Haye Sainte. Another of Perponcher's brigades, less a battalion at Hougoumont, held the Papelotte complex.

Wellington placed two brigades of cavalry, under Vivian and Vandeleur, on Picton's left. Ponsonby's Union Brigade and Somerset's Household Brigade were astride the Brussels road, and three brigades of British and King's German Legion horsemen were behind the Guards Division. The Dutch-Belgian cavalry were behind Wellington's centre and centre-right. In all, the Duke had some 68,000 men and 156 guns.

Napoleon, with 72,000 men and 246 guns, had a slight numerical edge. He was also extraordinarily confident and was delighted to discover that Wellington intended to offer battle. On the morning of 18 June, after the soldiers of both armies had spent a wet night, Napoleon deployed on the slope facing Wellington's position. East of the Brussels road stood d'Erlon's four divisions, with the three divisions of Reille's corps on the other side of the road. Kellermann's Cavalry Corps and a division of Guard cavalry formed up behind Reille, while Milhaud's Cavalry Corps and the Guard light cavalry division stood behind d'Erlon. Lobau's corps, with two more cavalry divisions, took position to the rear and the Guard was further back, just north of Ronsomme Farm on the Brussels road.

Early that morning Napoleon met his senior officers at Le Caillou. He was as unimpressed by Soult's suggestion that Grouchy should be brought back as by Prince Jérôme's report that two British officers had been heard discussing the fact that Blücher was expected to join Wellington. He did, however, agree that the battle should not open until 1 p.m., to give the ground time to dry out so as to make it easier for the guns to move, and sent off an ambiguous order to Grouchy. Napoleon's plan was scarcely subtle: he intended to throw his full weight against Wellington's centre with only a scant diversionary attack on Hougoumont.

The delay worked in Wellington's favour. That morning Blücher confirmed his intention of marching to Wellington's assistance and set off from Wavre at the head of Bülow's corps at 11 a.m., leaving Gneisenau behind to watch Grouchy. The latter had moved slowly on the 17th, only reaching Gembloux at ten that night. His cavalry patrols reported that the Prussians were moving on Wavre and Grouchy deduced that at least a portion of Blücher's army was heading for Brussels. Accordingly, he ordered Vandamme's corps off towards Sart-à-Walhain early on the 18th, with Gérard's corps following shortly afterwards. Vandamme did not move fast, and although Grouchy reported to Napoleon that he intended to interpose his forces between the Prussians and Wellington, the opportunity for him to do so was passing quickly.

Hougoumont was a substantial piece of building even by the robust standards of local farms. It comprised a manor house and farm buildings within a walled enclosure, with woods and rough pasture to the south, and a formal garden and orchard to the north-east. A sunken road – the 'protected way' – ran along the back of the orchard up towards the crest line. The Guards light companies had thrown a French unit out of the woods on the evening of the 17th, and prepared the house and buildings for defence, loopholing walls and barricading gates. The bombardment began at about 11.50 a.m. and Prince Jérôme's men advanced into Hougoumont woods, suffering heavy casualties from the Hanoverian *Jägers* and Nassau infantry. The Allied infantry gave ground slowly until they reached the southern face of the farm complex, where the firepower of the Guards beat off an assault.

It was probably never Napoleon's intention that the attack on Hougoumont should be pressed home, but Jérôme had the bit between his teeth and redoubled his efforts, now supported by Foy's division. Wellington had, meanwhile, posted part of a Guards brigade well down the slope, close to the protected way, and when the new attack came four

Coldstream companies moved down to recapture the orchard and restore the situation east of Hougoumont. West of the enclosure the battle was more finely balanced. The French surged round to the back, reaching the main gate, which was barred rather than barricaded. The gigantic Second-Lieutenant Legros of the 1st Light Infantry weakened the bar with an axe, then led a charge that burst open the door. The men who entered the courtyard were killed or captured, and the gate was closed again by main force. It was once more broken open, but as the attackers rushed in they were cut off by another four companies of the Coldstream, sent by Wellington to strengthen the garrison.

This was certainly not the end of the fight for Hougoumont, for the French soon worked their way back into the orchard, only to be expelled by reinforcements from the 3rd Guards. Extra troops were added to the garrison as the battle went on and supports were moved down the slope to buttress Hougoumont's defenders. Although its buildings were set on fire, probably by French howitzer shells, Hougoumont remained in Allied hands throughout the day. French failure to take it was doubly significant. The attack enabled Wellington to tie down large numbers of good French infantry with a smaller number of his own, and by retaining Hougoumont he reduced French room for manoeuvre in the area between it and the Brussels road.

With the struggle for Hougoumont still in its early stages, Napoleon's main battery – eighty-four guns on a low ridge east of the Brussels road – opened fire to prepare the attack of d'Erlon's corps. The fire was of limited effect, for the ground was so wet that cannon balls did not produce much ricochet effect and most of Wellington's infantry was on the reverse slope of the ridge. Before his main infantry attack was launched, Napoleon received the unwelcome news that a column of troops sighted towards Chapelle-St-Lambert belonged not to Grouchy, as had been hoped, but to Blücher. The Emperor dashed off a note to Grouchy, ordering him to march to the battlefield without delay. He then sent off two cavalry divisions to check the Prussians, following them with Lobau's corps, which formed a defensive shoulder between Plancenoit and Paris Wood.

D'Erlon's attack began at about 1.30 p.m. Three of his divisions formed up in dense columns – each with a frontage of 200 men and a depth of about twenty-five. Only the right-hand division was able to use the handier battalion columns. Little attempt was made to use cavalry to force the defending infantry to form square, which would have reduced its fire. The French columns were especially vulnerable to Allied artillery and although they achieved some success, clearing the gravel pit opposite La Haye Sainte and breaking Bylandt's brigade, they were raked by fire as they did so. On the extreme right, the French drove the light Nassau garrisons from Frischermont and Papelotte. At this juncture Picton's division moved over the crest, and checked the attack by musketry and a bayonet charge: Sir Thomas, incongruously dressed in plain clothes and stovepipe hat, was killed as he led his men forward. Lord Uxbridge, Wellington's cavalry commander, then launched the cavalry brigades of Ponsonby and Somerset into the French infantry, most of whom were soon running down the slope in disorder.

Self-control had never the strong point of Wellington's cavalry and Ponsonby's men galloped on into the French gun line, cutting down gunners and drivers, and were already in bad order when Napoleon launched his lancers and cuirassiers against them. Ponsonby was killed and his brigade was cut to pieces, and Somerset's brigade suffered less severely; the light cavalry brigades of Vivian and Vandeleur helped their comrades to break clean from the mêlée.

During the ensuing lull Wellington took the opportunity to reinforce the garrison of La Haye Sainte, reoccupy the gravel pit and reorganise his left centre. The attack on La Haye Sainte was speedily resumed, but Major Baring's newly strengthened garrison repulsed it. A disturbance in the Allied centre – probably caused by ambulance wagons moving back – persuaded Ney that Wellington's army was on the verge of giving ground and he ordered up a brigade of

Milhaud's cuirassiers to accelerate the process. The great mass of the French cavalry was speedily sucked into the battle, and by 4 p.m. the whole of Milhaud's and Lefèbvre-Desnouëttes's corps were thundering up the slope.

The charge was badly co-ordinated, like the infantry attack that had preceded it. Little horse artillery accompanied the cavalry, so British battalions were able to form square without being vexed by close-range artillery fire and the attacking horsemen swirled between the squares, sustaining close-range musketry. British gunners scampered for the shelter of squares as cavalry overran their pieces and ran out again as the horsemen rode back. No attempt seems to have been made to spike captured guns – or even to break rammers and sponge-staves – so guns were not out of action for long. As usual, Wellington himself was on hand at the point of crisis, riding from square to square and often trusting to his horsemanship when the French cavalry came too close.

At the same time that the first of the charges swept up the slope, Napoleon became aware that dark masses of Bülow's Prussians were emerging from Paris Wood. Lobau was soon being forced back, Plancenoit was lost and Prussian cannon balls began to reach the Brussels road. The Emperor sent Duhesme's Young Guard Division to retake Plancenoit and swung d'Erlon's right-flank division, under Durutte, round to face eastwards and brush Lobau's flank. Plancenoit was duly retaken, but Pirch's corps had now come up, and forced its way into the village, only to be expelled by a sharp counter-attack by two battalions of the Old Guard.

While fighting see-sawed to and fro on the French right, the struggle reached its climax in the centre. Extra French cavalry were fed into the battle and though their repeated charges – Ney himself is said to have led twenty-three – made little impression on the squares, the day's fighting had taken its toll of the Allied infantry and one officer recalled that the interior of his square resembled a hospital. Shortly after 6 p.m. Ney tried again, this time with a combined-arms force of Donzelot's division with some cavalry and artillery. Although the defenders of La Haye Sainte had been reinforced, they were running short of ammunition when French infantry reached the enclosure and the French soon gained a foothold in the buildings despite the unflinching resistance of the Germans.

The Prince of Orange endeavoured to support the garrison by ordering Colonel Ompteda to counter-attack with the two remaining battalions of his KGL brigade. Ompteda, an experienced veteran, warned that there was French cavalry within striking distance, but the Prince repeated his order and Ompteda's men duly advanced. They may have reached the edge of the garden, but French cavalry swept up from a hollow south-west of the farm. Ompteda was killed and one of his battalions was all but annihilated: the other formed square and fell back, its retreat covered by a charge of the much depleted Household Brigade. La Haye Sainte was taken and the French went on to expel the British Rifles from the sandpit.

The capture of La Haye Sainte enabled the French to push swarms of *tirailleurs* forward to take on the main Allied line and to bring some light guns into action from close range. To the west, Wellington was able to deal with the *tirailleurs* by a combination of volley-firing and controlled charges, but to the east the situation was more serious, and Alten's Division suffered terribly from the combination of *tirailleur* and artillery fire. This was the crisis of the battle, for if Napoleon had taken heed of Ney's desperate demands for fresh troops by sending the guard forward, he might have won. But the opportunity passed. Wellington ordered up some Brunswick battalions and Vivian's cavalry brigade to help support his centre, and Müffling was leading Ziethen's advance guard on to the Allied left.

At 7 p.m. Napoleon played his last card and sent the uncommitted battalions of the Imperial Guard against Wellington's centre. In an effort to whip up support amongst the battered infantry of Reille's and d'Erlon's corps, he had it put about that the troops now visible to the right were French, but morale faltered once it became clear that they were Prussian. The Guard marched on undeterred. Two battalions were dropped off to face Hougoumont and the remaining seven advanced on Wellington's right centre, moving obliquely across from the Brussels road. As they advanced, the battalions slid out into two distinct columns, probably because of difficulties in maintaining pace and direction, with the grenadier battalions moving to the right, and slightly ahead, of the *chasseur* battalions. The grenadiers neared the point in Wellington's line held by the Guards, just north of the crest line, with the Duke himself on hand.

As the advancing Imperial Guard breasted the rise, Wellington shouted to General Maitland, commander of 1st Guards Brigade, 'Now, Maitland! Now is your time!' He then addressed Maitland's men: 'Up Guards! Make ready! Fire!' The British line overlapped the French column at both ends and adjacent units joined in. Brave though they were, the grenadiers could not stand this fire and fell back down the slope. The *chasseurs* met the Allied line further westwards, and were engaged by Adam's light brigade and by Colonel John Colborne's well-handled 52nd Regiment, which swung out to take the column in the flank: the *chasseurs* broke and recoiled in the face of such firepower.

A third column, composed of two Guards battalions, well supported by line regiments of d'Erlon's corps, struck the Allied line near the Brussels road a little earlier and made some progress, pushing back two British battalions and rocking the Brunswickers and Nassauers. Wellington rode over and helped restore the situation, and General Chassé, whose Dutch-Belgian division was heavily engaged, brought up a fresh battery to fire point-blank at the column. It too began to fall back and Wellingtion, sensing that the moment had come, doffed his hat and motioned forward with it: the Allied army was to advance. He rode over to where Colborne's battalion was facing some guardsmen who had rallied and ordered, 'Go on, Colborne! Go on! They won't stand. Don't give them a chance to rally.' The cavalry brigades of Vivian and Vandeleur charged between the Brussels road and Hougoumont, and the Prussians pressed in on the Allied left. Wellington met Blücher south of the inn of La Belle Alliance on the Brussels road and the two commanders agreed that the Prussian cavalry should take over the pursuit.

Napoleon had left the field in a square of the Guard and soon took to his coach, only to abandon it for a horse. It took him an hour to force his way through the panic-stricken fugitives blocking Genappe. He paused briefly at Philippeville, where he left Soult to rally the army, then departed for Paris. Grouchy, who had conducted a skilful withdrawal by way of Namur, reached Philippeville on the 19th. Napoleon did not regard the situation as impossible, but it was certainly bleak. He had lost some 43,000 men – killed, wounded, prisoners and deserters – on the 18th, and losses at Quatre Bras and Ligny brought his total casualties to 60,000. The Allies had suffered 25,000 casualties at Waterloo, and about as many at Quatre Bras and Ligny, so that their overall losses were, at around 55,000, only slightly less than those of the French, but with their larger forces this butcher's bill was not exorbitant. They were forced to detach troops to mask the fortresses of the north and when Blücher approached Paris on 30 June he was duly repulsed. However, Napoleon had been forced to abdicate on 22 June and there was no political will to continue the war: Waterloo had indeed proved its decisive battle.

Despite the souvenir shops at its foot, the Lion Mound, just west of the Brussels road at Mont St-Jean, is the best place to begin a visit to the field of Waterloo. The monument was built in 1823–6 on the spot where it was thought the Prince of Orange was wounded, on the area of the Allied line roughly between the points of impact of the two main French columns late in the battle. It stands 40.5 metres high and was built with 32,000 cubic metres of earth taken from the surrounding fields. The monument gives an unrivalled view of the field. Hougoumont, over the crest to the right as we look due south, is like a breakwater in front of the main line. Cooke's division stood to the west, to the right of the monument, and Alten's to the east, on its left: the road from the monument to the Brussels road marks the line of deployment. On the other side of the crossroads, the Ohain road runs towards a prominent wood, then jinks right towards the Papelotte–La Haye complex: Picton's division held this sector. The Brussels road runs, past the big farm complex of La Haye Sainte, down the slope in front of the position, disappearing over the far crest just beyond La Belle Alliance, with a gentle swell of ground to its left just short of La Belle Alliance marking the position of the French gun line.

When I was last on the Lion Monument it was high summer and tourists snapped one another with the empty stubble as a backdrop. These fields around the monument saw some of the heaviest fighting of the battle, with waves of French cavalry thundering up towards the squares and infantrymen plying musket and bayonet in the thick powder smoke. As we strive to grasp the tactical detail, it is easy to forget the human dimension. Ensign Gronow, in the square of the 1st Guards, very close to the spot where the monument stands, recalled the approach of the cavalry:

> You perceived at a distance what appeared to be an overwhelming, long moving line, which, ever advancing, glittered like a stormy wave of the sea when it catches the sunlight. On came the mounted host until they got near enough, whilst the very earth seemed to vibrate beneath their thundering tramp . . . When we received cavalry, the order was to fire low, so that on the first discharge of musketry the ground was strewed with the fallen horses and their riders, which impeded the advance of those behind them and broke the shock of the charge . . . There is nothing perhaps amongst the episodes of a great battle more striking than the debris of a cavalry charge, when men and horses are seen scattered and wounded on the ground in every painful attitude.

Over the past few years the battlefield of Waterloo has become increasingly seedy and, for a variety of reasons, long-awaited work on restoring the site – especially the area round the Lion Mound – has not yet taken place. In view of the proliferation of French memorials and the shortage of their British equivalents, an alien from the Planet Zorg might easily form the view that the French actually won and a critical commentator might attribute the recent destruction of a memorial to Captain Cavalié Mercer of the Royal Horse Artillery to anti-British animus. When all is said and done, this is no way to treat a site of such profound international significance.

The visitor who has time to tramp the battlefield on foot would be well advised to purchase the readily obtainable *Promenade* 1815 booklet, available in English. The car-borne traveller should get his bearings at the Lion Monument, then visit as many of the surrounding attractions as he can stomach – the Waterloo Panorama and the visitors centre are both worth the effort – before driving to Hougoumont, now known as Goumont. There are filled-in loopholes to be seen in the barn wall, old loopholes in the orchard wall (shockingly gapped at the time of writing) and several memorial plaques to the men who fought there. Captain J. L. Blackman of the Coldstream and Sergeant-Major Edward Cotton of the 7th Hussars lie buried in the garden. The former was killed during the battle, while the latter, author of *A Voice from Waterloo*, became a pro-fessional guide, living in some style at Mont St-Jean until his death in 1849. Although the Lion Mound is easily seen from Hougoumont, it is worth noting that La Haye Sainte is not: there is a longitudinal ridge running between the two farms towards the Lion Mound and it cannot have helped the co-ordination of Napoleon's attacks.

Rejoining the road and moving south and then south-west, we strike the main Brussels road. After turning right, in the direction of Quatre Bras, the visitor reaches the farmhouse of Le Caillou. Its little museum contains numerous pieces of Napoleonic memorabilia – including the Emperor's camp bed – and the skeleton of a French hussar, a sobering antidote to much of the drum-and-trumpet military history peddled thereabouts.

Retracing our steps down the Brussels road, with the Lion Monument to our left front, drawing our attention to the centre of the battle, we pass the 'Wounded Eagle', on the left.

This monument – a mortally wounded eagle gripping a French flag in its talons – is where the last square of the Imperial Guard made its stand on the evening of 18 June. La Belle Alliance, once a farmhouse but a tavern by the time of the battle, stands on the right of the road a little closer to Mont St-Jean. Although the popular legend that Wellington and Blücher met there after the battle is reinforced by a commemorative tablet inside the building, the meeting was probably closer to Le Caillou than to La Belle Alliance.

The minor road running westwards off the main Brussels highway at La Belle Alliance leads to Plancenoit. A spiky Gothic monument, in the area of Bülow's advance, honours the Prussian dead of the battle. Following the road on to Lasne, beyond Wellington's left flank, brings the visitor to the pillar erected in memory of Colonel Graf von Schwerin, killed when the Prussian advance guard met the French cavalry screen on the 18th.

Heading westwards, back towards the central crossroads near the Lion Monument, we pass the farms of Papelotte and La Haye, partly rebuilt after the battle. Further along, near the crossroads, is a cluster of monuments to Sir Thomas Picton, Colonel Sir Alexander Gordon, the officers of the King's German Legion, and the Belgians killed in action at Waterloo. More instructive from the tactical viewpoint is a small, recently erected stone, which pays tribute to the 27th Regiment (Royal Inniskilling Fusiliers) who held this part of the line. Here Wellington's infantry were not protected by the crest, which runs well to the north. When they formed square to repel cavalry they were horribly vulnerable to artillery fire and the 27th was described as lying literally dead in square.

The road from the crossroads to the Lion Monument follows the line of the old road, but work on the monument has significantly altered the ground level. In 1815 this was a sunken lane with high banks, offering good cover to infantry. The sandpit opposite La Haye Sainte has almost completely disappeared, but the farm, in contrast, has altered little. The Brussels road leads northwards into Waterloo itself, in fact some distance from the battlefield.

Wellington's quarters, where he wrote his dispatch and spent the night after the battle sleeping on a pallet on the floor while the dying Gordon lay in his bed, have been turned into a Wellington Museum, with a substantial emphasis on Freemasonry. Lord Uxbridge, commander of Wellington's cavalry, lost his leg in the battle. The limb is buried just north of the Wellington Museum and its wooden replacement is in the museum. Directly opposite is the Royal Chapel, consecrated in 1690, its foundation stone laid by the Marquis of Castagna, Governor-General of the Netherlands, in the vain hope that Charles II of Spain might produce an heir.

This relic of the Spanish occupation of the Netherlands is an appropriate reminder of the changes that followed the Napoleonic wars. The Kingdom of the Netherlands was established by the Treaty of Vienna in June 1815 and the second Treaty of Paris, signed in November that year, redrew France's northern frontier in favour of the Netherlands, with the 1792 border shrinking to that of 1790. The marriage of Holland and Belgium was not a happy one, and after a rebellion in Belgium a conference of great powers, held in London, recognised the dissolution of the Kingdom of the Netherlands and later accepted the election of Leopold of Saxe-Coburg as King of the Belgians. It took a war, with the French intervening to support the Belgians, to force the Dutch out of Belgium and in 1839 Holland at last consented to the twenty-four articles, established by the London Conference of 1831, by which Belgium became perpetually neutral, her status guaranteed by the great powers.

Declaring Belgian neutrality was one thing; maintaining it was quite another. As one Belgian historian observed, perpetual invasions had given the population such pronounced anti-military reflexes that it was difficult for the army to obtain a budget compatible with its task. In 1851 a Royal Commission studied the problem of national defence, and reported in favour of fortifying the lines of the Meuse and the Escaut, and setting up an entrenched camp in the centre of the country, where the government could shelter while the powers which had guaranteed Belgium's neutrality came to its aid. Antwerp was selected as the entrenched camp and a line of forts was built around it.

The experience of the Franco-Prussian War, when Belgium felt herself to be inches from invasion, persuaded its government that existing defences were inadequate. Only 50,000 men responded to the call-out, which should have given General Chazal's army of observation, watching MacMahon's Army of Châlons as it neared the frontier, a total strength of 80,000. On 14 June 1887, after impassioned parliamentary debate, a law was passed establishing two groups of forts, one around Liège and the other around Namur. These were intended to hold an invader – French or German – while the field army mobilised and the great powers intervened.

The new fortifications at Liège and Namur were to embody armoured turrets developed by Henri Alexis Brialmont for the defence of Antwerp. Brialmont, born at Venloo in 1826, came from a family of Walloon gentry, and his father had served under Napoleon before playing an active part in the 1830 revolution and rising to become Minister of War. The young Brialmont was commissioned into the engineers in 1841. After travelling extensively in France and Germany he returned to Belgium convinced of the merits of detached forts, and in 1863 he installed an English-made turret into Fort No. 3 at Antwerp, combining the detached fort and armoured turret in a way that was to be characteristic of his fortifications.

The fortified region of Namur was designed to protect Belgium against invasion from France across the plateau between the Sambre and the Meuse. The forts in the protective belt were within sight of their neighbours and the furthest were under 6,000 metres apart. Large forts – Suarlée, Cognelée, Andoy and St-Heribert – alternated with smaller works – Malonne, Emines, Marchovelette, Maizeret and Dave. Although local Tourist Information maintains that all the forts are on private land and not open to the public, Fort Dave appears to be still under military ownership and can be entered easily, though visitors should note my earlier caveats. The inscription on the gate dates the work at 1897 and the place bears significant damage from heavy shells, which seem to have arrived from its rear. It appears to have had three large turrets, one main observation cupola at its summit and another near its apex: all have now been removed.

Important though Namur was, Brialmont was far more concerned about Liège. He recognised that the attraction of bypassing French fortresses and attacking into northern France across the open country of Hesbaye and Brabant might induce the Germans to violate Belgian neutrality across that narrow tongue of land between the French border and the 'Maastricht appendix', the sliver of Dutch territory which reached down parallel with the Meuse. Unless he chose to take on Holland as well, an invader would find himself compressed between the Maastricht appendix and the Ardennes: Liège, in short, lay squarely in his path.

Brialmont had intended that the belt of forts around Liège would be supported by powerful outworks at Huy and opposite Visé, but budgetary constraint caused them to be removed from the programme. Liège lies in the valley of the Meuse, and Brialmont sited his forts so as to secure the high ground and prevent an attacker from forcing his way up the valley. Forts Pontisse and Barchon covered the river north of the city, while Flémalle and Boncelle secured it to the south. The smaller valleys of the Ourthe and the Vesdre were covered by the neighbouring forts of Embourg and Chaudefontaine. At Liège the pattern of large and small forts was less regular than at Namur, partly because the cancellation of the work opposite Visé persuaded Brialmont to strengthen Pontisse, when logic would have demanded that Pontisse should be a small fort and its westerly neighbour, Liers, should be large.

The dominant principles of Brialmont's fortification were simple: all fortress guns should be armoured and larger than contemporary siege guns, and garrisons should be sheltered by earth and cement. Brialmont arrived at the depth of earth and concrete required by calculation and experiment. No horse-drawn siege howitzer or gun could reasonably exceed

220mm, and Brialmont believed that his minimum of 2.5 metres of concrete and three metres of earth would resist such weapons: in trials it took nine rounds containing sixty kilos of dynamite to penetrate 2.5 metres of concrete.

Brialmont's forts were built to a standard pattern, with set ingredients added as required. They were all either triangular, pointing like arrowheads towards the enemy threat, or four-sided, with the forward face of the work slanting out to cover vital ground. Ditches and counterscarp galleries provided local defence, while the fort's offensive power lay in its turreted guns, most of them concentrated in the armoured redoubt in the centre of the fort. A large fort might mount a cupola with two 155mms, two cupolas each with two 120mms, and two cupolas each with a single 210mm howitzer. Four cupolas housing 57mm Nordenfeldt guns sat in the salient angles of the parapet, and further 57mms. were to be found in the counterscarp galleries and the casemate covering the entrance ramp.

Domestic arrangements within the forts were similar to those in Seré de Rivière's works. Generators provided electricity for armoured searchlights, but not, because of penny-pinching, for the rest of the fort's interior. Messages could be sent by using the searchlight as a signal lamp, which was as well because the telephone lines linking the forts were not buried deeply enough to resist shelling. Further economies meant that there was no covered route between the kitchens, latrines and medical aid post, all usually found within the counterscarp, and the central redoubt.

In August 1914 the Schlieffen Plan turned Brialmont's hypothesis of invasion between the Ardennes and the Maastricht appendix into reality. A detachment from the German 2nd Army, under General von Emmich, was given the task of dealing with Liège, but its attempt to take the forts failed miserably. On 6 August Major-General Erich Ludendorff, fortuitously on the scene, took command of a brigade whose commander had been killed, infiltrated between the forts and reached the citadel. This unmodernised work, on a bluff high above the Meuse, had little military significance, but even so it took real flair for Ludendorff to bluff it into surrender. General Leman, military governor of Liège, took refuge in Fort Loncin, where his ability to co-ordinate the defence was seriously reduced.

Capture of the citadel scarcely altered the task facing the Germans. The forts held firm, and the guns with the field army made little impression on them. But Brialmont's calculations were soon rendered invalid by the march of technology. Huge 420mm siege howitzers had been made in Krupp's works at Essen, and these were sent forward by rail and pulled into position by tractors. Assisted by Austrian-made Skoda 305mms, the 420mms cracked cupolas like eggs, and penetrated earth and concrete to explode deep within the forts. The first 420mm opened fire on 12 August and the last fort blew up on the 15th.

Most of the Liège forts, modified after the war, remain in good order. It is entirely appropriate that the last to fall – Loncin, General Leman's command post – has been restored to the state it was in when the Germans entered it, and is open to the public. It stands where the St-Trond road crosses the autoroute eight kilometres from the city centre. Loncin is one of Brialmont's large forts, with five heavy guns and thirteen 57mms, and was assigned a garrison of 350 gunners and 200 infantry, the latter responsible for local defence. At 5.20 on the afternoon of 15 August a 420mm scored its twenty-fifth hit: the shell fell just behind the armoured redoubt and burst in the main magazine, causing an explosion that wrecked the fort. Two hundred and fifty of the garrison still lie beneath the rubble: Leman himself was pulled unconscious from the debris. There is a fine memorial just to its south, built so that its grieving classical figures physically overlook the fort: 'ce fort en ruines est leur tombeau.' A Belgian memorial to the crew of a Second World War RAF bomber stands across the N3, immediately opposite the Loncin monument. Fort Flémalle, with its small museum, is open on the first Sunday of each month and is in a far better state than the battered Loncin.

Liège is a bustling town, the third largest in Belgium. It is the centre of the Belgian arms industry – indeed, *Fabrique*

Nationale, where the FN rifle was designed, is in its suburb of Herstal. Just off the Quai de Maestricht, in an eighteenth-century building which housed the prefecture of the *Département* of the Ourthe in 1800–14 when Liège was French, is the *Musée des Armes*, with its unrivalled collection of firearms. The American Ardennes Military Cemetery lies ten kilometres away, in the village of Neuville-en-Condroz: it contains the graves of 5,310 Americans killed during the Second World War, mainly in the 'Battle of the Bulge', the German Ardennes offensive of December 1944, and the murals in its chapel are especially noteworthy.

While the Liège forts were buying time, the Belgian army concentrated behind the Geet in the triangle Diest–Louvain–Gembloux. The German 1st Army tried to outflank it from the south and push it away from Antwerp, but although the Belgians fought hard on the Geet for a day they slipped away in time to reach Antwerp safely, and the Namur garrison extricated itself to reach the same destination. We have seen, in an earlier chapter, how many of them were able to escape from Antwerp to take up a position on the Allied left in the Dixmude sector, ensuring that a small piece of Belgium remained in Belgian hands.

Their wheeling march through Belgium brought the German 1st and 2nd Armies within striking distance of the Allied flank. Joffre's westernmost army, the 5th, had begun to advance towards the Sambre on 2 August, allowing the 4th, previously in reserve, to move up into line on its right. The commander of the 5th, General Charles Lanrezac, had profound misgivings not only about his task – for he suspected that there were more Germans to his front than GQG admitted – but also about his British allies, coming up into line on his left. Britain had entered the war as a direct response to violation of Belgian neutrality, and the British Expeditionary Force, two corps under Sir John French, concentrated in the area Maubeuge–Hirson–Le Cateau.

French and Lanrezac met at Rethel on 17 August and got on badly. Although Lanrezac's rudeness is hard to forgive, he had good reason to be preoccupied. As his army lined the salient of Sambre and Meuse, like a face-down L with its long leg along the Sambre and its angle at Namur, Moltke issued a directive for Bülow's 2nd Army to attack west of Namur while Hausen's 3rd advanced between Namur and Givet: Lanrezac faced frontal and flanking attack. The Germans missed the opportunity. Kluck of 1st Army, operating under Bülow's orders, was forbidden to hook wide to the south-west to find the Allied flank. Bülow launched his attack across the Sambre on 22 August, a day earlier than had been planned, driving Lanrezac back across Hausen's front: when 3rd Army attacked on the 23rd it was too late to cut off the French.

The BEF, advancing to conform with French movements, halted for the night of 22 August behind the Mons–Condé canal, with Haig's I Corps on its right, echeloned back to Grand Reng, and Smith-Dorrien's II Corps on its left, stretching almost as far as Condé. GHQ had just heard, from its liaison officer with 5th Army, that Lanrezac was in difficulties and when French met his corps commander at Sar-la-Bruyère early on the 23rd he announced that the BEF would retain its present position for the time being. Smith-Dorrien's men made some hasty defensive preparations in the close country of mining villages and slag heaps along the canal. Many of them reached Mons after dark, when there was too little time to prepare all the bridges over the canal for demolition and it was not clear that the BEF would not need them itself. The field of Malplaquet lay just behind II Corps's position, and 2nd Lincolns and 2nd Royal Scots Fusiliers had spent the night of 21 August overlooking the battlefield on which their ancestors had fought.

There had already been contact between the BEF's cavalry screen and Kluck's advance guard. Early on the morning of the 22nd Major Tom Bridges's C Squadron 4th Dragoon Guards, on outpost duty at Casteau, north of Mons, engaged a patrol of German lancers. Corporal Thomas fired the first British shot of the war and Captain Hornby was the first officer to kill a German with the new pattern cavalry sword. The action is commemorated by a memorial beside the main N6 in Casteau, just north-east of the massive NATO headquarters. By one of the war's many ironies another memorial, exactly opposite, remembers the last shots fired by Canadian troops on 11 November 1918.

On the morning of the 23rd the first German attacks struck the troops of 3rd Division holding the canal line north of Mons. Lieutenant Dease and Private Godley of 4th Royal Fusiliers were to win the Victoria Cross – the first of the war – for manning a machine-gun on the railway bridge over the canal at Nimy, while 4th Middlesex fought a difficult battle around Obourg station. The Nimy–Obourg salient proved increasingly dangerous as the day wore on, for the Germans had crossed the canal to the east and came close to cutting it off, and only the determined resistance of 2nd Royal Irish, 1st Gordon Highlanders and 2nd Royal Scots averted this.

The battle spread along the canal to the west, held by 5th Division and the independent 19th Infantry Brigade, with massed attacks withering before the searing fire of British infantry. At midday Smith-Dorrien recognised that the troops in the Nimy–Obourg salient were dangerously exposed and must be withdrawn, even though this would entail pulling back 5th Division, which was holding its own. The withdrawal to a line running through Frameries and Paturages was carried out with skill, and the day's British losses – 1,642 killed, wounded and missing – were surprisingly light: most were incurred by the Fusiliers, Middlesex and Royal Irish in the salient. There is no accurate record of German losses, but these were probably no lower than 6,000 and may have been as high as 10,000.

A plaque beneath the Nimy bridge remembers the bravery of Dease and Godley, and another, on a small brick wall in the middle of the soulless railway halt at Obourg (actually on the other side of the canal from the village) celebrates the courage of an unknown soldier of the Middlesex who climbed on to the roof and covered his comrades' retreat at the cost of his life. A memorial to the first and last battles of the war, and a Celtic cross commemorating the Royal Irish, stand at the junction of the N40 and the N90 just east of Mons. Looking north, chimneys mark the position of

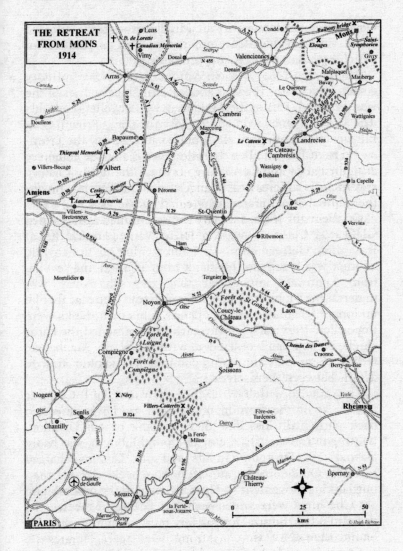

THE RETREAT
FROM MONS
1914

Obourg cement factory, in the Middlesex sector, while due south rises the wooded eminence of Bois la Haut, from which 49th Battery RFA supported the infantry.

Along the N90 lies St-Symphorien, whose military cemetery contains the first and last British casualties of the war. Private J. Parr of the Middlesex was a cyclist scout, his date of death amended to 21 August 1914, and Private G. E. Ellison of the 5th Lancers – buried just across the greensward from Parr – fell on 11 November 1918. The cemetery was opened by the Germans to contain the dead of Mons, and there is an assortment of German and British graves, including that of Lieutenant Maurice Dease VC.

On the night of the 23rd two important messages reached GHQ at Le Cateau. One, from Joffre, warned French that at least three German corps were threatening his front while another was feeling for his flank from Tournai; the second, from Lanrezac, announced that the 5th Army was undertaking a long withdrawal. French told Lanrezac that he no longer felt obliged to co-operate if his own flanks were imperilled. Very early on the 24th he summoned the corps chiefs of staff and ordered a withdrawal on Maubeuge, telling them that the corps commanders should arrange details between themselves.

The order to withdraw reached most units of II Corps at about 4.30 on the morning of the 24th. 3rd Division pulled back with little loss, but the left flank of 5th Division, fighting in the industrial clutter around Elouges, was badly clawed. 4th Dragoon Guards and 9th Lancers charged towards Quiévrain in an effort to check the German outflanking movement and lost 250 men in the process, and 1st Cheshires were wiped out after failing to receive the order to retire. Smith-Dorrien lost over 2,000 men that day – more than at Mons.

French considered falling back on Le Havre, where he could hold a bridgehead with the navy at his back, or on Maubeuge, where he might pause under its guns. Eventually he decided to withdraw on St-Quentin and Noyon, to refit behind the Oise. On the 24th he issued orders for retirement on Le Cateau. The Forest of Mormal, with no good

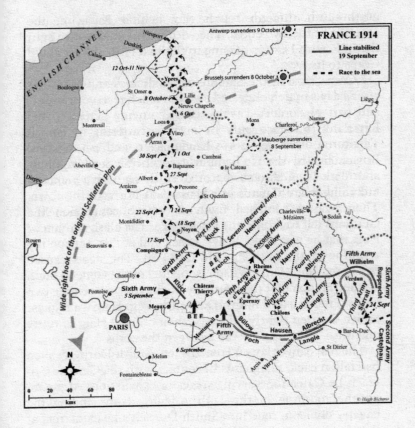

FRANCE 1914

Line stabilised
19 September

Race to the sea

ENGLISH CHANNEL

Nieuport
Antwerp surrenders 9 October
Dunkirk
Calais
12 Oct-11 Nov
Ypres
Boulogne
Brussels surrenders 8 October
St Omer
8 October
Lille
Liège
Neuve Chapelle
6 Oct
Montreuil
Loos
Mons
Charleroi
Namur
5 Oct
Vimy
30 Sept
Arras
1 Oct
Mauberge surrenders
8 September
Abeville
Bapaume
le Cateau
Cambrai
Dieppe
Albert
27 Sept
Amiens
Peronne
Charleville-Mézières
Sedan
St Quentin
22 Sept
24 Sept
Montdidier
18 Sept
Seventh (Reserve) Army
Heeringen
Rouen
17 Sept
Noyon
First Army
Kluck
Beauvais
Compiègne
Second Army
Bülow
Sixth Army
Manoury
B E F
French
Third Army
Hausen
Fourth Army
Albrecht
Fifth Army
Wilhelm
Chantilly
Kluck
Château
Thierry
Fifth Army
F. d'Espérey
Rheims
Sixth Army
Rupprecht
Verdun
Pontoise
Sixth Army
5 September
Ninth Army
Foch
Epernay
Third Army
Sarrail
24 Sept
Meaux
B E F
Châlons
Fourth Army
Langle
Second Army
Castelnau
PARIS
Montmirail
Fifth Army
Bülow
Hausen
Albrecht
Bar-le-Duc
Melun
6 September
Foch
Vitry-le-Français
Langle
St Dizier
Fontainebleau
Arcis

Wide right hook of the original Schlieffen plan

N

0 20 40 60 80
kms

© *Hugh Bicheno*

north–south through routes, lay behind Bavai, at the junction of the two corps. The withdrawal on the 25th split the BEF, with I Corps passing to the east of the forest and II Corps to its west.

I Corps crossed the Sambre at Landrecies and Maroilles after a day's march cluttered by two French reserve divisions and part of Sordet's cavalry corps, moving across to the left flank, screened only by a thin scattering of French Territorial divisions. Corps headquarters and 4th (Guards) Brigade settled down in Landrecies for the night and not long after dark the Germans appeared, colliding with a picket of 3rd Coldstream Guards at a crossroads north of the town. There was a confused fight, at whose conclusion the Germans fell back into the forest edge. The clash convinced Haig that he was under heavy attack and at 1.35 a.m. on the morning of the 26th he told GHQ so. The news caused consternation. Henry Wilson, sub-chief of the General Staff, warned French that the Germans might get between the two corps. French asked Smith-Dorrien to help Haig – a request Smith-Dorrien rightly refused – then ordered Haig to retire on Guise, widening the gap between the corps.

The real danger lay on French's left. Smith-Dorrien's men had fallen back, rearguards in contact with the Germans, to reach Le Cateau in varying stages of exhaustion. At 2 a.m. on the morning of the 26th Allenby, commanding the cavalry division, rode into Smith-Dorrien's headquarters at Bertry, south-west of Le Cateau, to say that his own brigades were widely dispersed and that he was pulling back off the ridge to the front of Smith-Dorrien's position. The latter asked Hubert Hamilton of 3rd Division if he could move at once, to be clear before the Germans moved on to the high ground. Hamilton, some of whose men were still coming in, replied that he could not start before 9 a.m., and Smith-Dorrien decided to stand and fight. Allenby agreed to act under his orders and Snow, whose newly arrived 4th Division was not part of II Corps, also put himself at Smith-Dorrien's disposal.

Smith-Dorrien intended to strike 'a stopping blow' and retire under its cover. He had no opportunity to prepare for a formal defensive battle and the best that most of his soldiers could do was to scratch a shell scrape in the hard ground. The northern edge of the battlefield was defined by the Roman road from Cambrai to Le Cateau, now the N43. The ground held by Smith-Dorrien's men slopes upwards, innocent of cover, south of the road. The Bavai road, the modern D932, crosses the right flank of the field. Between its junction with the Roman road and Le Cateau a shoulder of high ground, the onion dome of Le Cateau church visible from its crest, overlooks the valley of the Selle.

Fergusson's 5th Division held the right sector as far as Troisvilles, with 2nd Suffolks and several batteries of artillery up on the knuckle. The 3rd Division covered the centre and included the small town of Caudry within its lines. Thereafter 4th Division held a line edging south-eastwards in front of Esnes, the Warnelle ravine forming a natural barrier parallel with its front. 19th Infantry Brigade, at Reumont on the Bavay road, was Smith-Dorrien's only infantry reserve.

There was a thick mist on the morning of the 26th, and it was not until 6 a.m. that fighting began. German infantry, working its way into Le Cateau, pushed elements of two battalions out of the town and began to spill round the British right. Machine-guns came into action on the ridge line north of the Roman road, and a growing weight of artillery fire crashed down on the open ground south of it. Most of Smith-Dorrien's guns took post on the forward slopes, close to the infantry, and casualties from shell and machine-gun fire mounted. Fields of fire were far better than at Mons and the effect of British musketry was, if anything, more striking. The Germans made no progress on 5th Division's front, but percolated their way up the knuckle on its right. 3rd Division had little difficulty in holding its ground: Inchy was lost but quickly recaptured. 4th Division's battle began badly, when 1st King's Own was caught forming up south of Cattenières, but during the morning Snow's men held their ground well. At about midday he decided to pull back across the Warnelle ravine and took up the line Esnes–Harcourt–Ligny.

At about 1.30 p.m. Smith-Dorrien decided to break off the action and continue his retreat. It took time for his order to retire in succession from right to left to reach divisional and brigade headquarters and then to cross the bullet-swept ground to battalions. Withdrawing 5th Division's troops from the hillock proved taxing indeed. Teams were sent forward from the dead ground to bring out the guns, but ran into a storm of fire – German infantry was sometimes as little as a hundred metres away. Four of the six pieces of 37th Howitzer Battery, up with the Suffolks, were brought away safely and volunteers returned to save the other two. Both were hooked on to their limbers: one team was shot down but the other galloped clear in an exploit which earned the Victoria Cross for Captain Reynolds and Drivers Luke and Drain.

The order to withdraw never reached the KOYLI, just south of the Roman road at its junction with the Bavai road, and the battalion lost over half its strength. Up on the high ground to the right the Suffolks, latterly supported by 2nd Manchesters and 2nd Argylls, were lacerated as German infantry got up amongst them, though some survivors managed to fall back southwards. Covered by this self-sacrificing resistance, the remainder of 5th Division slid off through Reumont, well ahead of the Germans. 3rd Division broke clear with less difficulty – though 1st Gordons, with some of 2nd Royal Irish and 2nd Royal Scots, did not receive the withdrawal order and were engulfed the following day. 4th Division had a harder time of it, for the dismounted cavalry, who had put in the morning attacks on the left flank, had been replaced by infantry. Nevertheless, the division was away by nightfall, albeit at the price of leaving most of its wounded behind.

Le Cateau lacks the popular appeal of Mons, but was a more costly battle: II Corps lost 7,812 men and thirty-eight guns. German losses are again difficult to estimate with any certainty. Smith-Dorrien met his operational aim. Had he continued to retreat without fighting he would inevitably have been caught with his corps strung out on the line of march. He had hit the Germans hard enough to be able to continue his retreat virtually unmolested. Indeed, Kluck, released from Bülow's control on the 27th, drifted off the British line of retreat altogether and made for Amiens, taking the pressure off the BEF. Although French's official dispatch paid tribute to Smith-Dorrien's coolness and courage, the Commander-in-Chief was furious at Smith-Dorrien's decision to offer battle. When French wrote his own account of events in his idiosyncratic book 1914 he estimated the casualties of Le Cateau at at least 14,000 men and eighty guns. He privately regretted not court-martialling Smith-Dorrien for disobedience, and forgot that on that confused and confusing night of 25–26 August GHQ had, in an ambiguously worded message, given Smith-Dorrien permission to fight at Le Cateau. When French was woken by Henry Wilson to be told that Smith-Dorrien proposed to fight he agreed, but insisted that his exhausted CGS should not be disturbed. When the Field Marshal woke after more sleep he changed his mind, and told Wilson to speak to Smith-Dorrien on the telephone and order him to break off the action. By then, of course, it was too late.

The southern edge of Le Cateau Military Cemetery, alongside the Bavay road north-west of the town, gives a good view across the right front of Smith-Dorrien's position. It contains British, French, Germans and Russians – the latter died while working as prisoners of war. A British VC, Lance-Corporal John Sayer of The Queens, mortally wounded near St-Quentin during the German offensive of March 1918, died in a German hospital in the town and is buried there, as are two German nursing sisters and a female telephonist.

The Suffolk Memorial, a handsome white stone cenotaph, stands on the hillock and is best reached by a track leading up from the D21 as it leaves Le Cateau to the south: German infantry enveloping 5th Division's flank fought their way up re-entrants like that which the track now follows. From the Highland Cemetery, on the D12 Wassigny road, the entire British right flank can be seen in depth: looking due west, the water tower in Bertry – Smith-Dorrien's head-quarters – nudges the horizon, and Reumont – Fergusson's headquarters – lies just to its left. Amongst the dead in the cemetery are Lieutenant A. R. Skemp of the Gloucesters, professor of English at Bristol University, and Private Cleve Reuben Torgensen, who lived in the USA but had enlisted in a Canadian regiment, Lord Strathcona's Horse, long before America entered the war. But beyond doubt the most remarkable of the men lying there is James Blomfield Osborne (IX B 10). The son of an Australian colonel, he had served with the Australian Imperial Force from August 1914, was twice wounded in Gallipoli and was Mentioned in Dispatches. Invalided out as a captain, he voluntarily re-enlisted and was killed as a private in the Argyll and Sutherland Highlanders a month and a day before the war ended.

A track runs from Troisvilles, parallel with the Roman road, to join the Bavay road just south of its junction with the Roman road. A single prominent tree (*l'arbre rond* on maps) stands in the edge of the track – sunk beneath high banks at that point – a kilometre from Troisvilles. In 1914 it was the only tree in the area and stood just behind the line held by 1st Norfolks, providing German observers on the crest line opposite with a useful aiming mark. The Norfolks' commanding officer ordered his pioneers to cut it down, and they had almost done so when the local brigadier warned that he did not want it to block the track, which gave him useful lateral communication. The tree was to be guyed up with ropes and a team of horses from a nearby battery eventually tugged it down into the field behind. The present tree stands in exactly the same spot and a small concrete plinth, hidden in the grass at its base, once held a flagstaff.

The retreat took the BEF well away from the Sambre and the Meuse, down to the Marne, where it played its part in Joffre's swirling counter-attack in September. When the British returned to the north-east in 1918 it was on a scale that the men who fought at Mons and Le Cateau could scarcely have imagined. During October and November the British 1st, 3rd and 4th Armies advanced towards Mons and Maubeuge. Cambrai was liberated on 8 October, and on the

9th Clary and nearby Gattigny Wood – just behind the Le Cateau battlefield – were taken by the 66th Division and 3rd Cavalry Division after mounted charges by Lord Strathcona's Horse and the Fort Garry Horse. The 66th liberated Le Cateau the following day and its memorial drinking trough stands in the town centre. On 4 November the 3rd and 4th Armies fought the Battle of the Sambre, pushing the Germans back across the river, and at seven on the morning of 11 November the kilted pipe band of the Canadian 42nd Battalion marched into Mons. The armistice found the Canadians in action across the Condé Canal, clearing houses in Casteau, within easy range of the spot where Tom Bridges's squadron had fired its first shots a lifetime before.

The cemeteries of the area testify to the human cost of the last phase of the war: between 7 August and 11 November the British armies lost 13,603 officers and 284,522 men. The tiny Communal Cemetery at Ors, on the Sambre Canal north of Catillon, contains the graves of two holders of the VC – Second-Lieutenant James Kirk and Lieutenant-Colonel James Marshall – as well as that of the poet, Lieutenant Wilfred Owen. James Kirk, like Wilfred Owen, in the Manchester Regiment, had been commissioned from the ranks. He earned his VC on 4 November 1918, the same day as James Marshall, who also held an MC and bar. Marshall had served in the Belgian army early in the war, and although the cap badge on his tombstone shows that the Irish Guards were his parent regiment, he was killed commanding 16th Lancashire Fusiliers, striving to cross the Sambre Canal.

During the inter-war years the Belgians repaired eight of the Liège forts, putting modern guns into the old turrets, and refurbished all the others except Loncin for the storage of ammunition. They proposed to use the old forts as a second line of defence and built a new line, linked by concrete casemates, east of the city. Fort Eben Emael commanded the Albert Canal just inside the Belgian border. Roughly triangular in shape, it covered an area 900 by 700 metres, and mounted 120mm and 75mm guns. It was attacked on

10 May 1940 by glider-borne troops who landed on top of
the fort. Although the glider carrying their commander,
Lieutenant Witzig, was released early and had to be
relaunched, arriving three hours late, the remainder of the
force, under Sergeant Wenzel, used shaped charges to disable
the fort's guns. The attackers remained atop the fort for
twenty-four hours until they were relieved by ground forces;
the garrison surrendered early on the afternoon of the 11th.
The attack was part of a more complex plan, with other
detachments attempting to capture bridges immediately to
the north of Kanne, Vroenhoven and Veldwezelt. The
former was blown but the remainder were taken intact. The
fort is open for guided tours one weekend a month, but it is
possible to walk across the superstructure at any time. A
path ascending to the left of the entrance block gives access
to the exterior of the turrets and casemates, which bear the
marks of the shaped charges used by the attackers.

South of Eben Emael, near the village of Neufchâteau, lies
Fort Neufchâteau, which fell on 21 May 1940. It was later
used for target practice by the Germans and is badly
knocked about. Fort Battice, built to cover the main
Aachen–Liège road and railway, and within easy gunshot of
the modern A3, held out till 22 May. Fort Tancremont-
Pepinster, the most southerly of the new forts, covered the
Vesdre valley south-west of Verviers. Initially bypassed by
the Germans, it was the last of the Liège forts to fall.

Seven of the nine forts around Namur were also restored
and rearmed, and smaller works linked Namur to Liège.
Some work was carried out at Antwerp. An intermediate
line of pillboxes and anti-tank obstacles – the KW Line – ran
from Koningshooikt, near Lier, to Wavre, and defences
round Ghent were intended to help the city form part of a
last line of defence. All this was to do little to deflect the
German armoured onslaught in 1940.

On 10 May 1940 the Allied armies moved into Belgium,
with the main French and British forces taking up a position
on the Dyle. The BEF held the line from Louvain to Wavre.
There was heavy fighting along the Dyle on the 15th and on
the 16th Major-General Bernard Montgomery added lustre

to a growing reputation by his 3rd Division's defence of Louvain. German breakthrough to the south made it impossible for the Allies to stay on the Dyle, and as they fell back into Flanders and Artois, the panzer corridor opened out across northern France, with Rommel's 7th Panzer Division crashing through Avesnes, and on into Landrecies, Le Cateau and Cambrai.

In 1944 the Allied advance across the north was every bit as rapid: Paris was liberated on 25 August and Brussels on 3 September. The Ardennes offensive of December 1944 pushed a wedge into Allied lines south of Liège, but, although 2nd Panzer Division made good progress in its drive to reach the Meuse at Dinant, the attack was blunted well before it reached the river and, in view of Allied strength on the ground and in the air, it may be doubted whether the German objective of Antwerp was ever more than a hopeless dream.

VII

Champagne

CHÂLONS-EN-CHAMPAGNE, formerly Châlons-sur-Marne, lies in the centre of a dry chalky plateau, which curls like a crescent moon from Oise to Loire. To its east runs the narrower belt of the 'wet Champagne', where erosion has removed chalk to reveal clays and sand. South-east lies the Barrois and north-east the Argonne, and to the north the Ardennes rise across the Meuse. Westwards, towards Paris, the Brie and Tardenois plateaux are separated from dry Champagne by slopes planted with the vines whose grapes produce the inspiring wine that takes its name from the region. The rivers of Champagne flow westwards, the Yonne and Aube to join the Seine, and the Vesle and Oise to join the Aisne. The Marne, rising near St-Dizier in wet Champagne, flows through Châlons and Epernay to meet the Seine in Paris.

The counts of Champagne, who governed from the eleventh to the thirteenth centuries, recognised the importance of commerce, and six fairs – two each at Troyes and Provins, one at Bar-sur-Aube and another at Lagny – attracted merchants from far and wide. The fairs declined after 1284, when Jeanne, daughter of the last count, married Philip V of France. Nevertheless, industry continued to flourish, with Rheims, Châlons and Provins producing textiles, and iron being hammered out in wet Champagne and the Argonne, where rivers provided hydraulic power and forests fuelled the forges.

Prosperity ended in the Hundred Years War, when Champagne was pillaged by English and Burgundian war bands. In 1359–60 Edward III besieged Rheims, and in 1373

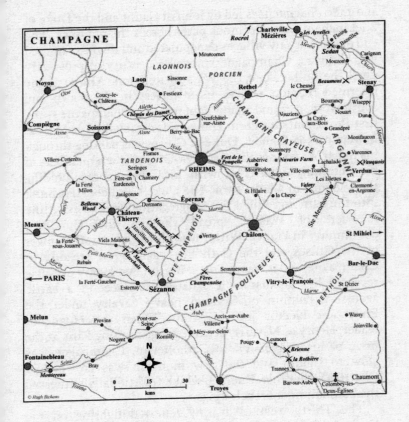

CHAMPAGNE

and 1380 *chevauchées* led by John of Gaunt and the Duke of Buckingham carved ruinous paths across the province. The end of the war saw villages depopulated and towns crammed with beggars. Champagne enjoyed a century of prosperity until the outbreak of the Wars of Religion. Although the reformed religion was at its strongest in the south and south-west, it found many adherents in Champagne – the first Protestant church in France was established at Meaux. In March 1562 Francis, Duke of Guise, was passing through his fief at Wassy, south-west of St-Dizier, when members of his entourage quarrelled with local Huguenots, holding their service in a large barn. The Duke's arquebusiers burst into the barn and killed the congregation, and news of 'the massacre of Wassy' spread through the Huguenot community in France. With the outbreak of war, the power of the Guises and the Catholic League grew steadily in Champagne, and it was at Port-à-Binson that Henry of Guise received his nickname 'le Balafré' (scarface) in a battle against a column of German heavy cavalry under the Protestant Elector John Casimir. When the Huguenot leader, Henry of Navarre, became king, entering Paris at the price of his conversion to Catholicism, Champagne was slow to submit and the new monarch was crowned at Chartres rather than Rheims, which traditionally witnessed the coronation of French kings.

The Thirty Years War saw renewed fighting on the borders of Champagne as the Spaniards sought to take advantage of the death of Richelieu and Louis XIII. Early in 1643 Don Francisco de Melo laid siege to the little fortress of Rocroi, north-west of Charleville, with an army 27,000 strong. The twenty-two-year-old Duke of Enghien marched from Péronne with 23,000 men to relieve the place. On his approach Melo took up a position on marshy ground two kilometres south-west of the town. Both armies formed up in the traditional way, with infantry in the centre and cavalry on the wings. Although Enghien charged and routed the horse on the Spanish left, on the other wing the Spanish cavalry beat back the French horse under L'Hôpital and La Ferté. Enghien swung left and cut his way through the enemy foot, separating the veteran Spanish infantry from their Italian and German auxiliaries.

Having dealt with the remaining Spanish cavalry, Enghien turned his attention to their foot. The Spaniards were drawn up in solid but inflexible *tercios*, whose pikemen and musketeers kept off the first French charges. When Enghien brought his artillery into action against the *tercios* the Spaniards asked for quarter, but mistakenly fired on Enghien when he rode forward to accept their surrender. The wedges of Spanish foot were overwhelmed by close-range fire and cavalry charges: 8,000 were killed and 7,000 captured. Melo himself was amongst the prisoners. The Spanish army was never the same after the destruction of its formidable infantry: Rocroi was an unmistakable sign of the decline of Spanish military power.

Rocroi was also a remarkable achievement, establishing the young Enghien – soon to succeed his father as Prince of Condé – as a general of uncommon ability. The scene of his victory is marked by a memorial *stele* on the little road from the centre of Savigny-la-Forêt to Rocroi, which runs parallel with the N51. A stream that flows southwards to join the Sormonne at Chilly marks the western edge of the battlefield and Melo's left flank probably extended as far as the modern main road. Rocroi was Vaubanised, like most other fortresses of the north, and its walls, their interpretation facililitated by some helpful orientation tables, are generally in good order. At Rocroi, like so many other fortresses, existing bastioned traces were modified, in the nineteenth century, by having big cavaliers thrown up atop the bastions, with massive traverses protecting bombproof defence works. A museum in Rocroi's *corps de garde* tells the story of the battle, and although the little town lies well off the beaten track it is well worth a visit.

Important though Rocroi was, it did not end the Spanish threat to Champagne at a stroke and the war with Spain went on even after the Treaty of Westphalia ended the Thirty Years War. In the instability accompanying the *Frondes* a Spanish army marched down from the Netherlands and threatened Rheims. The Treaty of the Pyrenees eventually brought peace and adjusted the northern border in France's favour. In 1642 France had already gained the principality of Sedan and this, together with the acquisitions of 1659, was to help keep war away from Champagne, although it was not until the Duchies of Lorraine and Bar at last became French in 1766 that Champagne was covered by the protective shield of Lorraine.

From the second half of the seventeenth century the local economy prospered, thanks largely to Jean-Baptiste Colbert, a native of Rheims, who replaced the disgraced Fouquet as Finance Minister in 1661. Colbert encouraged industry, helping revive textile production in Champagne and establishing a royal armaments factory at Charleville in 1688. This employed 500 workers in 1789, and the model 1777 Charleville flintlock musket was the standard infantry weapon of the armies of the Revolution and Empire, and remained in French service till the 1830s.

The military crisis of 1792 saw the works at Charleville producing weapons at top speed to equip the threadbare armies of the Revolution. When Dumouriez took command of the *Armée du Nord* he was attracted by the prospect of invading the Netherlands, but news that the Duke of Brunswick had taken Verdun drew him southwards. Dumouriez marched down from Sedan and disposed his army in the passages of the Argonne at Le Chesne, La-Croix-aux-Bois, Grandpré and La Chalade. Brunswick hooked north, setting off from Verdun towards Grandpré on 7 September, while an Austrian detachment under Clairfayt moved from Stenay on La-Croix-aux-Bois. A force of Austrians and Hessians, meanwhile, marched straight for the defiles at La Chalade and Les Islettes. Clairfayt forced La-Croix-aux-Bois on 12 September, and it says much for Dumouriez that he was able to pull his detachments back from the northern defiles and reassemble his army around St-Menehould, where he was joined by Kellermann, commander of the *Armée du Centre* previously based on Metz.

Brunswick had turned the French flank and the lightly held road to Paris stretched before him. He waited three days before resuming his march and when he set off at last he hoped that the threat of his columns, fanning south-west towards Châlons from the Argonne defiles, would force the French to retreat rather than have their communications cut. Emigré cavalry swung down from Le Chesne on the outer flank, Clairfayt and Kalkreuth marched from La-Croix-aux-Bois on Somme-Suippe, while Brunswick and Hohenlohe debouched through Grandpré and made for Somme-Tourbe and Somme-Bion respectively. Dillon still held the defiles at La Chalade and Les Islettes, facing the Allies advancing from Verdun, and Dumouriez took up a position on the high ground at Valmy, just west of St-Menehould.

Much was at stake on 20 September 1792. If the Allies overcame French resistance at Valmy and Les Islettes, they would be in a position to crush Dumouriez and Kellermann, and then march on Paris almost as they pleased. If, on the other hand, the French withstood the attack, Brunswick would be exposed to a counter-attack, which could cut his tenuous lines of communication. Sheer weight of numbers favoured the French, for Dumouriez and Kellermann between them had about 60,000 men at Valmy, while Brunswick had perhaps 30,000 under his hand, with another 16,000 Austrians and émigrés nearby. The weather was unseasonably wet and cold, and Brunswick's men were short of supplies.

The French held the Valmy position in two layers, with Kellermann's army at the front and Dumouriez's in reserve. Kellermann's troops stood on high ground, with Stengel on Mont-Yvron in the north, Muratel and Lynch across the ridge crowned by Valmy windmill, and Valence and Chazot south of the Châlons–Paris road, in the hamlet of Orbeval and the village of Gizaucourt. The Allies approached on the axis of the Châlons road, deploying on high ground between the road and the village of Somme-Bionne with the valley between them and the foot of Valmy Ridge. Most of the 200 guns with the armies of Dumouriez and Kellermann were at Valmy. Brunswick had a total of 230 in his columns and there was rough numerical parity when the artillery of the opposing armies opened fire on the foggy morning of 20 September.

After a long cannonade the Prussian infantry tramped down the slope in parade-ground order. Kellermann, aware that his men could not cope with elaborate manoeuvres, formed his centre into three huge columns with a frontage of one battalion and a depth of four, and prepared to charge with the bayonet once the Prussians were close enough. There was colossal enthusiasm in the French ranks, with shouts of *'Vive la nation'* and bursts of patriotic song. The explosion of three ammunition caissons caused a brief tremor, but order was soon re-established. It was painfully evident to Brunswick that the French would stand their ground. *'Hier schlagen wir nicht* [we will not fight here],' he concluded. Neither side had lost many men – there were some 184 Prussian and 300 French casualties – but the clash was as decisive as more costly battles. Brunswick's opportunity had passed and he had little alternative but to fall back. He abandoned Verdun and Longwy, and crossed the frontier on 23 October.

Thousands of motorists using the *Autoroute de l'Est* cross the battlefield of Valmy without knowing that they are hurtling across hallowed ground and many stop at the Valmy windmill service area with no idea of the mill's significance. The mill, reached on the D284 south-west of Valmy, was rebuilt in 1939 on precisely the same pattern as the mill that stood in the centre of Kellermann's position in 1792, and having blown down in 1999, this replica was itself rebuilt in 2005. Orientation tables describe the battle, and there is a superb view over Champagne and the Argonne. A suitably heroic statue nearby pays tribute to Kellermann, who became Marshal of France and Duke of Valmy under the Empire. We have already met his son, François-Etienne Kellermann, leading a desperate charge of cuirassiers at Quatre Bras. It is no coincidence that the *Voie de la Libération* passes through Valmy.

Champagne did not smell powder again till 1814. As the Allies approached the natural frontiers of France early that year, General Maison had 15,000 men in Belgium, Marshals

Marmont and Macdonald and General Morand commanded corps in Lorraine and northern Alsace, and Marshal Victor watched the Upper Rhine south of Strasbourg. Alsace and Lorraine fell quickly. Victor abandoned Strasbourg to Schwarzenberg's Austrians, and Marmont fell back through Metz with Blücher's Prussian and Russian army in hot pursuit. Blücher was across the Meuse by 22 January and his advance guard snatched a bridgehead over the Meuse at Joinville, between St-Dizier and Chaumont, on the 23rd. Napoleon found it hard to raise troops to meet the invader. Although he proclaimed a *levée en masse* along the eastern frontier, he was reluctant to take measures smacking of Jacobin extremism. The arrival of Cossacks in eastern France did not produce the hoped-for outburst of popular fury and the peasantry remained depressingly apathetic.

French fortunes soon revived. Schwarzenberg advanced gingerly, and although he reached the Langres plateau in mid-January, he was stoutly resisted by Marshal Mortier, sent there with the Guard at the beginning of the month. However, the danger of Schwarzenberg and Blücher joining forces impelled Napoleon to take the field himself and he arrived at Châlons on 26 January. The Emperor planned to make the best use of the area's good road system and numerous towns which could serve as depots to mount a high-tempo cut-and-thrust campaign against his more numerous opponents. Champagne is classic countryside for Napoleonic manoeuvre, mostly flat, with gentle contours, big woods and wide open spaces. It is a cavalry general's dream but an infantry colonel's nightmare, for things could go hard for unsupported infantry in this sort of landscape. Napoleon first stabbed at Blücher, who had pushed Victor back through St-Dizier. By the time Napoleon retook the place Blücher's main body had moved on, but he pushed forward swiftly and caught him at Brienne.

The little town of Brienne is dominated by its eighteenth-century château, now the *Centre Psychotherapeutique de Brienne*, sitting squarely on the hill that overlooks it. Napoleon knew Brienne well, for he had attended military school there between 1779 and 1794, and was to remember it with affection: he bequeathed the town the sum of 1,200,000 francs, part of which was used to build the town hall. The old military school, in the *rue de l'Ecole Militaire* south of the town, is now a Napoleonic Museum, and there is an attractive statue of the Emperor as a cadet outside the town hall.

Napoleon marched on Brienne from St-Dizier in three columns, sending a fourth under Marmont to Vassy to prevent Yorck's Prussian corps from supporting Blücher, and ordering Mortier to march on Arcis-sur-Aube. Unfortunately for Napoleon, his order to Mortier was intercepted by a Cossack patrol and Blücher, who had been marching on Arcis with Olsuviev's Russian corps in the hope of catching Mortier in the flank, was able to reverse his march and summon Sacken's Corps and Pahlen's 3,000 cavalry, part of Wittgenstein's corps.

Napoleon sent Grouchy's cavalry straight against Blücher, and supported them with the infantry of Ney and Victor as soon as they came up. Ney took two of his divisions against the town, while Victor sent a brigade to the château and swung round Blücher's left flank with the remainder of his corps in an effort to cut the Allied line of retreat. Though Napoleon's infantry, young conscripts for the most part, fought surprisingly well, the struggle was finely balanced: Victor was checked on the French left, but his detached brigade fought its way into the château. Blücher and Gneisenau left the courtyard through one gate as the French surged in through the other. Blücher tried a last counter-attack against the château but drew off when it failed, not long before midnight.

Brienne was an inconclusive battle. Although the French had inflicted more casualties than they had suffered, Blücher had not been destroyed. Napoleon followed the Allied retreat and halted at La Rothière, on the Bar-sur-Aube road (D396). He spent two days redeploying his forces in Champagne, moving Mortier back to Troyes, to cover his right, and ordering Macdonald to move most of his corps from St-Menehould to Châlons, and to reinforce the garrison of Vitry-le-François, covering his left.

The delay enabled the Allies to concentrate. Schwarzenberg joined Blücher at Trannes, south-west of La Rothière. It was snowing heavily and the weather helped blind Napoleon's reconnaissance, so that when the Allies advanced on 1 February they caught him at a disadvantage. Blücher led 53,000 men against his centre, while Wrede's Bavarians threatened his left. Mortier's report of enemy activity at Troyes suggested that the advance from Trannes might only be a feint and Napoleon ordered a general march on Troyes, Ney's corps leading. Marmont was told to move his corps from Vassy to Lesmont, on the main road north-west of Brienne, and Victor, destined to be the last to move, remained at La Rothière. At about midday Victor reported that enemy columns were approaching from Trannes and Napoleon hesitated, ordering him to stand his ground until the situation had become clearer. When Napoleon did at last react, ordering Ney to retrace his steps and bringing Marmont down from Lesmont, his plight was serious indeed. His own 40,000 men, deployed in an arc from Dienville, on the Aube on his right, through La Rothière in the centre to Chaumesnil on the left, were opposed by Blücher's 53,000 men, and Barclay de Tolly with the Russian reserve was on the move south of Trannes.

The Battle of La Rothière was fought in a howling blizzard. It opened with a see-saw cavalry fight around the village, with Nansouty's horsemen getting amongst the Russian gunners on the main road and the Russian cavalry lunging back to ride down twenty-four guns of the Guard horse artillery. When the infantry came up the French held their ground tenaciously, Gérard's men immovable in Dienville and Duhesme's division resolute in the northern half of La Rothière. In late afternoon, however, Wrede's Bavarians began to lever Marmont back down the Brienne road from Chaumesnil and even the reliable Duhesme wilted beneath the weight of Barclay's newly arrived troops. Napoleon averted disaster by putting one of Ney's divisions in to recapture La Rothière, while a handful of troops was sent off to support Marmont. This checked the advance and enabled the French to break contact in the snowy twilight.

Napoleon withdrew through Lesmont, hoping to join Mortier at Troyes. He had lost 6,000 men – as had the Allies – but fifty guns had also gone, and the battle and retreat damaged the mercurial morale of his conscripts. Another 4,000 men had disappeared by the time Napoleon reached Troyes, whose citizens were decidedly lukewarm in their welcome. The Emperor took steps to secure this attractive medieval town whose prosperity was founded on the manufacture of cotton nightcaps and stockings, and sent Marmont to cover the line of the Aube and secure Arcis.

Lack of unanimity amongst Allied commanders and the inevitable difficulties attendant on co-ordinating far-flung columns in a snow-covered countryside gave Napoleon a fresh opportunity. Schwarzenberg's natural caution was increased when the garrison of Sens rebuffed his attempt to take the town, a reconnaissance from Troyes nipped his right flank and reports arrived of fresh troops being raised at Lyons. Schwarzenberg responded by drawing Wittgenstein's corps to the south, weakening the link between himself and Blücher. The latter, in contrast, was supremely confident, and marched north-westwards in the hope of catching Macdonald and reaching Paris.

It took Napoleon some time to realise that Blücher was making for the capital, and the news forced him to abandon plans for dealing with Schwarzenberg and move northwards. This shift of balance showed Napoleon at his very best. Mortier covered the move by threatening Schwarzenberg's flank, inducing him to retreat on Bar-sur-Aube. Napoleon marched to Nogent, collecting a newly formed corps under Oudinot in the process, and stood poised, with 70,000 men, between the opposing armies. Bad news from Paris, where word of the Prussian advance had created a panic, from Belgium, almost entirely overrun by the Allies, and from a peace conference at Châtillon-sur-Seine, where the French had been offered peace on the basis of the frontiers of 1792, gave the Emperor pause for thought. However, he pushed reconnaissances to the north, trying to find out which of the three major routes west from Châlons was being used by Blücher. On 9 February Marmont reported from

Champaubert that Sacken was at Montmirail with at least 15,000 men.

This gave Napoleon his opening. Blücher was using the centre route, which runs from Châlons through Champaubert and Montmirail. Napoleon immediately sent his impedimenta to Provins, left Victor to cover Nogent and spread Oudinot's command along the Yonne to hold the bridges from Montereau to Auxerre, to give himself some breathing space if Schwarzenberg came back to life. Having poised the shield, he wielded the sword and marched for Sezanne to link up with Marmont before plunging into Blücher's flank. The Allies had some inkling of what was afoot. Blücher left Sacken to follow the retreating Macdonald and ordered three corps on to Sezanne. When he realised that Napoleon was ahead of him he changed his plan, sending Yorck to Montmirail and leading the other two corps towards Fère-Champenois in the hope of catching him in a pincer movement.

It was a dangerous game to play with Napoleon at the top of his form. Although he moved slowly over roads turned into quagmires by a thaw followed by heavy rain, he reached Champaubert on the morning of 10 February, smashing Olsuviev's corps and capturing its commander. He then turned west, leaving Marmont to watch Blücher, who had belatedly swung back towards Vertus. Sacken had been ordered to join Yorck at Montmirail. He declined Yorck's suggestion that he should move towards Château-Thierry so as to meet the Prussians before trying conclusions with the French, reached Viels Maisons, twelve kilometres short of Montmirail, and tried to hammer his way eastwards. He was within measurable distance of succeeding. Napoleon had only the Old Guard and Ricard's conscript division under his hand. Mortier was coming up fast along the main road, but Yorck, too, was approaching from Château-Thierry, his path watched by Michel's division and part of Nansouty's cavalry at Fontenelles. Sacken was at last making some progress and had driven Ricard out of Marchais, when Mortier arrived. Napoleon at once sent a strong force of the Guard against Sacken's left flank in front of Haute-Épine and pushed reinforcements under Lefèbvre – whom we last encountered at Jourdan's victory at Fleurus – into the battle for Marchais. Yorck had also arrived, but he committed only a small portion of his corps and recoiled as soon as he saw that the battle had turned sour. Sacken was bundled back on Viels Maisons, but Yorck's brief intervention prevented Napoleon from totally destroying Sacken's corps as he had Olsuviev's.

Napoleon hoped to reap further benefits by pursuing Sacken and Yorck, and catching them with the Marne at their backs. He depended on Macdonald having followed up the retreating Allies so as to secure the bridge at Château-Thierry. When Napoleon reached the river there on 12 February he found that the Allies had crossed and destroyed the bridge behind them: capturing the Prussian rearguard of 3,000 men was small compensation. By the time the Emperor's engineers had thrown a new bridge over the river the Allies were far away, and had crossed the Ourcq and blown its bridges to buy themselves more time.

Bad news from the south followed. Schwarzenberg recovered from his alarm, forced the line of the Yonne and drove Victor over the Seine. Napoleon would have to move south to deal with this developing threat, but before doing so he struck another deft blow at Blücher. Before setting off in his effort to catch Sacken and Yorck, Napoleon had left Marmont to watch his rear towards Vertus, and when Blücher began to move westwards once more on 11 February, Marmont fell back slowly before him. Napoleon left the bulk of his army concentrating near Montereau and sprinted off to meet Blücher with the Guard and Grouchy. On the morning of the 14th he found Marmont in action with Blücher west of Vauchamps. He immediately sent Grouchy round the Prussian right flank. Ziethen's division was cut to ribbons, but Blücher managed to pull his main body back to the east before the Guard could bring him to battle.

We last encountered Grouchy in the Waterloo campaign, when he was overmatched by his responsibilities. As a cavalry commander at Vauchamps, however, he was entirely in his element. He swung off the main road and found a route running parallel with it – the D11 to Janvillers and then a track which runs from Janvillers Cemetery back towards the main road – and put himself across Blücher's line of retreat west of Fromentières. However, the omnipresent mud prevented Grouchy from bringing up his guns and Blücher managed to tear his way past. Napoleon left Marmont to continue the pursuit, and the episode cost Blücher 7,000 men and much of his transport. Unfortunately for Napoleon, the Allies had the weight of numbers on their side. The Russian General Winzingerode, with his corps of 30,000 men, had captured Soissons on 14 February, but on hearing of Blücher's misfortunes he marched down to join him at Châlons, making good the losses suffered at Champaubert, Montmirail and Vauchamps.

Schwarzenberg's success in the Seine sector proved short-lived. Although he had seized crossings at Nogent, Bray and Pont-sur-Seine, Schwarzenberg was sensitive about threats to his communications and alarmed at the prospect of Napoleon descending upon him. After much hesitation he decided to concentrate on Bray while Barclay closed up on Nogent. Napoleon came thundering down on the 17th, his columns routing Pahlen at Mormant and Wrede at Valjouan. The operation developed on the 18th, with Macdonald chivvying Wrede towards Bray and Napoleon himself nearing Montereau with his main body. Victor had moved slowly that day, and by halting short of Montereau he had given the Prince of Württemberg ample opportunity to establish himself on a ridge north of the Seine. The exasperated Napoleon replaced him with Gérard, who attacked Württemberg the following morning.

Montereau, at the confluence of Seine and Yonne, had already been the scene of one dramatic episode. In 1419 John the Fearless, Duke of Burgundy, had been assassinated on the Yonne bridge, a tragedy that encouraged his son, Philip the Good, to conclude the Treaty of Troyes with Queen Isabel and Henry V of England. On 18 February 1814 Montereau again saw history in the making. Gérard's guns spent the morning pounding away at Württemberg's position on the ridge, and when the Guard artillery arrived he was able to dominate the enemy batteries and storm the ridge. Württemberg's withdrawal slipped out of control as Napoleon personally led his guns forward on to the ridge. General Pajol, already wounded, took his cavalry across the bridges before the Allies could blow them, and Lefèbvre exploited brilliantly by leading Napoleon's headquarters staff and his escort of guard cavalry in a charge along the Bray road. Württemberg limped off with the loss of 6,000 men and fifteen guns.

The Yonne bridge is especially rich in memorials. One tablet commemorates John the Fearless and another pays tribute to the 3rd Hussars, who charged there on 18 February. Another plaque commemorates 'Brave General Pajol, who, at the head of his cavalry, drove the Austrians and Württembergers into the Seine and the Yonne'. There is an equestrian statue of Napoleon a few yards from the bridge and its pedestal bears the words uttered by the Emperor when some of his escort warned that he was risking his life: 'The bullet which will kill me has not yet been cast.'

Schwarzenberg was alarmed by minor reverses and the shock of Montereau, a substantial defeat, had him in full retreat on Troyes, asking Blücher to join him at Méry-sur-Seine to offer a united front to the victorious French. Destruction of the Seine bridges at Bray and Nogent canalised Napoleon's movement through Montereau and enabled the Allied armies to link up. There followed a brisk debate, with Schwarzenberg arguing that the retreat must continue: he sent the exasperated Blücher back towards the Marne and resumed his own retirement on Bar-sur-Aube where, on 25 February, the Allied leaders met to consider strategy.

For all the unquestioned brilliance of Napoleon's conduct of the campaign so far, it was a war of missed opportunities. Montereau had jolted Allied self-confidence so severely that Napoleon might have gained generous peace terms and been permitted to remain as ruler of France, but by persisting in his demand for 'natural frontiers' he failed to draw political advantage from military success. And that success soon waned. The Bar-sur-Aube meeting formed the basis for the

Treaty of Chaumont, signed on 9 March, by which the Allies undertook to see the war through to a successful finish and to conclude only a single peace with France. The signatories offered Napoleon peace on the basis of the frontiers of 1791, an offer he scornfully rejected.

The campaign jolted on. Blücher turned towards Paris once more, inducing Napoleon to leave Macdonald at Troyes with 40,000 men to watch Schwarzenberg, while he tried to repeat earlier success by descending on Blücher's communications. He almost brought it off. The spirits of Blücher's men were depressed by their failure to defeat Marmont, supplies were short and Blücher managed to get across the Aisne only just before Napoleon caught up with him. Napoleon crossed at Berry-au-Bac on 6 March, after the bridge had been taken intact by his Polish lancers, and found an Allied force strongly posted on the Craonne plateau, covering the main route to Laon. Napoleon suspected that this was Blücher's rearguard and sent Ney off to deal with it, postponing his own advance and hoping that the delay would give Mortier and Marmont time to reach him.

The Battle of Craonne was mutually chaotic. Blücher had actually planned to hold the heights in strength with Vorontzov's and Sacken's corps, and to use his cavalry and Kleist's corps to envelop Napoleon's right once the infantry battle was joined, but the speed of Ney's advance threw his plan out of kilter. Ney was on the high ground before Sacken arrived and Napoleon then attempted an envelopment of his own. While Mortier threatened the Allied centre by attacking, with lavish artillery support, from the village of Hurtebise along the D18, Ney, supported by Victor and Grouchy, tried to turn the position from the north and Nansouty's cavalry swept up from the south.

The steep and wooded ground, which was to be the scene of heavy fighting in the First World War, made co-ordination of such an ambitious plan difficult. Ney, attacking between Vauclair Abbey and Ailles, went forward too soon, before Blücher's attention had been fixed by the bombardment of his centre. He was repulsed and Winzingerode, in overall command on Blücher's right, attempted to follow up, only to become bogged down in the valley of the Ailette. Blücher broke away and moved back on Laon, having inflicted slightly more casualties than the 5,000 or so that his army had suffered.

Napoleon remained convinced that he was only dealing with Blücher's rearguard while the remainder of his army was in full retreat and this encouraged him to pursue in the hope of crippling Blücher before marching southwards to restore the deteriorating situation on the Seine. It was a serious misconception, because Blücher intended to offer battle on ground of his own choosing south of Laon. His position, now crossed by the N44 as it bypasses Laon, was naturally strong and Blücher had posted his 85,000 men so as to take full advantage of it. Kleist and Yorck stood on his left, dominating the Berry-au-Bac road, Bülow held the centre and Winzingerode the right, out towards Clacy-et-Thierret.

Napoleon did not expect Laon to be seriously held, and sent Ney and Mortier straight down the Soissons road. When they reported that the Allies held the ridge line in strength, Napoleon supported their frontal attacks while he waited for news of Marmont, plodding northwards up the Berry-au-Bac road. For most of the day the fighting was inconclusive. Blücher, who thought that Napoleon was far stronger, suspected that the attacks up the Soissons road concealed an ugly surprise elsewhere, while Napoleon, for his part, lacked the resources to do more than simply keep Blücher in play. Winzingerode got above himself late in the afternoon and prodded Ney's flank, only to be sharply rapped for his pains.

Marmont arrived early in the afternoon, and although his men were tired they drove into Blücher's extreme left and captured the village of Athies. That, thought Marmont, was enough for one winter's day, and he sent a cavalry column under Colonel Fabvier to make contact with Napoleon and halted for the night. He had reckoned without Blücher's opportunism for, having deduced that Napoleon and Marmont were weaker than he had expected, the Prussian threw his two left flank corps, supported by Sacken,

Langeron and his cavalry, against Marmont. Marmont's men went reeling back down the Berry road. Total disaster was averted only by the bravery of Fabvier, who brought his detachment back to fall on the Allied flank and by the fortuitous appearance of a convoy with a tiny escort at the Festieux defile. The Allied cavalry had hoped to secure this natural choke point, effectively cutting off Marmont's corps, but the handful of guardsmen with the convoy beat them off and kept the road open. Even so, Marmont finished that terrible night just short of Berry with his corps in tatters.

On 10 March Blücher tried to complete the task he had begun so well on the 9th. Yorck and Kleist were ordered to finish Marmont while Langeron and Sacken swung westwards to Bruyères to cut the Soissons road. Had Blücher been at the helm the plan might have succeeded, but the old fellow was exhausted and command devolved upon his chief of staff, Gneisenau. Napoleon, who only heard of Marmont's calamity at dawn on the 10th, decided to remain facing Laon, and bluffed Gneisenau into calling off his victorious left. He withdrew after dark and when the Allies attacked Ney's rearguard at Clacy the following morning they received a bloody nose from that past master of the rearguard action.

Although Laon was not the cataclysmic defeat that had seemed so likely on the morning of the 10th, it was nevertheless a cruel blow to Napoleon. He had lost 6,000 men to his opponent's 4,000 and, with the balance of manpower firmly against him, could afford them even less. But the Emperor was dangerous even with the odds against him. While at Soissons he heard that General St-Priest, posted at St-Dizier to link the forces of Schwarzenberg and Blücher, had advanced to take Rheims. Napoleon descended on the city late on 13 March, brought up the Guard artillery to blow the Paris Gate to fragments and unleashed his cuirassiers into the streets. St-Priest lost 6,000 men, most of them prisoners, and the news brought Schwarzenberg and Blücher to a halt.

Victory at Rheims once again gave Napoleon the

initiative. Schwarzenberg represented the most serious threat to Paris and the Emperor sped off to deal with him. Marmont and Mortier were left to contain Blücher, and the rest of the army crossed the Marne at Epernay and Châlons. It was typical of Napoleon's gambling instinct that he chose the riskiest but potentially most profitable of the choices open to him. Instead of confronting Schwarzenberg frontally, he headed for Troyes to savage his communications. News that Napoleon was on the rampage produced something approaching a panic amongst the Allies, but Schwarzenberg drew back in time, marching hard for Troyes. When he heard this, Napoleon decided to shift his aim to St-Dizier and Joinville, where he could upset Blücher's communications as well as Schwarzenberg's, and perhaps relieve the garrisons of Verdun and Metz into the bargain.

Before marching on the Marne, Napoleon kicked southwards against the garrison of Arcis-sur-Aube with the aim of knocking Schwarzenberg further off balance. Schwarzenberg had, however, decided that Napoleon was making for Troyes, and concentrated between that town and Arcis in order to counter-attack. Instead of simply taking on Wrede's garrison, Napoleon was to find himself facing Schwarzenberg's entire army, outnumbering his by more than two to one.

Ney's corps and Sebastiani's cavalry, marching south of the Aube, entered Arcis on the morning of 20 March, repaired its damaged bridge and were preparing to join the remainder of the army north of the river when strong bodies of Allied cavalry appeared from the east. Neither Sebastiani nor Ney could hold back the torrent and only the arrival of Friant's Old Guard division propped up the French position. After heavy fighting around Torcy-le-Grand the Allied advance was checked, and the arrival of Lefèbvre-Desnouëttes with some fresh cavalry enabled Sebastiani to launch a charge through the dark, badly cutting up the Allied cavalry.

Napoleon presumed – as he had at Laon – that he was only dealing with an aggressive rearguard. It was a shock to discover, the following morning, that Schwarzenberg's entire army was drawn up in a half-circle to his front. Fortunately for Napoleon, Schwarzenberg hesitated to attack and the Emperor worked feverishly to get his army across the Aube, throwing a pontoon bridge over the river at Villette to supplement that at Arcis. Oudinot and Sebastiani covered the withdrawal, but when Schwarzenberg came forward in mid-afternoon, it was not easy to hold him off. Eventually the bridge was blown and Napoleon broke clear, making for St-Dizier. He paused on reaching it, then decided to head for St-Mihiel, collect the garrisons of Metz and Verdun, and ravage Allied communications.

His plight, had he but known it, was now hopeless. The Allied commanders met at Pougy to consider their position. The countryside was swarming with Cossacks and the Allies were able to feast on a rich diet of captured dispatches. In a letter to the Empress, Napoleon announced that he was moving on the Marne to threaten Allied communications. Blücher had beaten Marmont at Fismes, and Mortier had abandoned Rheims and marched westwards. Augereau, perpetually raising troops around Lyons, seemed to have fallen back. The Allied commanders decided to move Schwarzenberg north to join Blücher and to shift their lines of communication to the north. No sooner had this decision been taken than more captured dispatches told of the parlous state of Parisian morale and this provoked the Allies into staking everything on a united advance straight down the Marne to Paris. As a deception move, Winzingerode was sent off towards St-Dizier to deceive Napoleon into thinking that the Allies were indeed worried about their communications, luring him away from the capital.

Napoleon took the bait and beat Winzingerode on 26 March, but on the next day heard that the Allies had defeated Marmont and Mortier at La Fère Champenoise and must certainly reach Paris before he could intervene. Still undaunted, he collected all available men at Fontainebleau and it was there that Ney, acting as spokesman for the generals, told him that the army would not march on Paris. News that Marmont had gone over to the Allies was the last straw. On 4 April Napoleon abdicated in favour of his son, but the Allies insisted on unconditional abdication and on the 6th this was duly signed in what is now the *Salle de l'Abdication* in the Emperor's apartments in the Palace of Fontainebleau.

While the topography of the 1815 campaign is partly obscured by building and the field of Waterloo is heavily populated by tourists, Champagne is prettier, unspoilt by the growth of towns, and its battlefields are less oppressive. For the purpose of visiting they fall into two groups. Two comfortable days will enable the visitor to cover Laon–Craonne and Champaubert–Montmirail–Vauchamps. The town of Laon – 'the crowned mountain' – deserves to be better known. Its Gothic cathedral with statues of oxen at the angles of its towers, and lofty ramparts with stunning views over the countryside are worth the visit in their own right. Another two days will suffice for the battlefields of La Rothière, Brienne, Arcis-sur-Aube and Montereau. We have already seen that Brienne contains a small museum devoted to Napoleon. Arcis has fewer traces of his passing, although he spent the night of the battle at the château, now the town hall. The revolutionary politician Georges Danton had been born there in 1759. The site of his house is marked by a small memorial just north of the bridge and there is a bronze statue of him. He might have appreciated it, for he retired to his native Arcis at the end of 1793, trying to live in happiness with his second wife, but he was summoned back to Paris, tried and guillotined, facing death with exemplary courage. The bridge itself suffered the penalty of being on a military corridor: it was blown by the French in 1940 and by the Germans in 1944.

Fontainebleau is too firmly on the tourist route to be seen with comfort, but writes finis to the campaign with an elegant hand. The Emperor's apartments, which link two wings of the palace, were refurbished some time ago, with new silk brocade woven to match the original wall coverings. On 20 April 1814 Napoleon left Fontainebleau from the *Cour du Cheval Blanc* (named for the statue of a horse missing since about 1656), giving it the alternative title of *Cour des Adieux*. He descended the left-hand branch of the famous horseshoe staircase and paused to view his Guard drawn up in the courtyard before selecting the 600 men who were to accompany him to Elba. Then he bade the Guard farewell, embraced a regimental eagle and departed by carriage, leaving his veterans in tears behind him. Just north of the palace, in the Rue St-Honoré, is the *Musée Napoléonien d'Art et d'Histoire militaire*, which houses an assortment of uniforms from the First to the Second Empire, as well as a comprehensive collection of nineteenth-century regulation swords.

Champagne saw a military revival during the reign of the Emperor's nephew. In 1857 the Ministry of War purchased nearly 10,000 hectares of land about ten kilometres north of Châlons, between the villages of Mourmelon, St-Hilaire, Suippes and La Cheppe – further land, north of Suippes, was later added. In the summer of 1857 the Imperial Guard under Marshal Regnault de St-Jean d'Angély camped there under canvas. Huts were subsequently erected and a railway spur was built to link the camp with the main line in Châlons.

The camp enabled troops to carry out wide-ranging manoeuvres without the need to gain the permission of private landowners. Between 1857 and 1867 a full corps trained there each summer and from 1868 two corps visited the camp for two months each. Like so much of the training carried out by the army of the Second Empire, work at Châlons was overlaid by a thick veneer of formalism. Exercises followed a script and nothing was left to chance. Units marched from their campsite to the spot where the 'battle' was to begin and textbook tactical formations were adopted on the conveniently empty countryside with little regard for the effects of hostile fire. The battle over, everyone marched back to camp at top speed. Although Châlons did much for low-level training and *esprit de corps*, much more might have been achieved there. Not all its results were beneficial: regimental order books bear eloquent witness to the damage done to health and discipline by the enterprising ladies and roaring cafés of Mourmelon-le-Grand.

When France mobilised for the Franco-Prussian War, the divisions under training at Châlons became II Corps of the Army of the Rhine: we have seen how it bore the heat of the day at Spicheren and Rezonville before enduring the long agony of Metz. As an important military centre with good rail communications, Châlons was an obvious concentration point for France's second-line force, formed after the defeats in Lorraine, and it was there that Napoleon III arrived in a less than Imperial third-class railway carriage on 16 August. His army had always prided itself on its ability to triumph by *le système D: on se débrouillera toujours* – 'we'll muddle through somehow' – and the troops that streamed into Châlons bore witness to the regenerative powers displayed by the French army even in the downward spiral of a lost campaign.

Marshal MacMahon's Army of Châlons comprised the defeated right wing of the Army of the Rhine – I, V and VII Corps, which had come up by train from Chaumont and Belfort. These formations, filled out by new recruits of the 1869 class, were joined by the new XII Corps, under General Trochu. This contained one fresh division, another of good regular regiments from the Spanish frontier and a third of tough *Infanterie de Marine*. Overall, the army was of uneven quality: some of its regiments had not recovered from the effects of defeat at Froeschwiller in Alsace on 6 August and there were familiar shortages of artillery and equipment. Nevertheless, by 21 August it totalled 130,000 men and 423 guns, a respectable force by any standards.

Napoleon's principal difficulty was in ascertaining just what should be done with it. On 17 August he discussed future plans with senior officers and decided that Trochu should be sent off to Paris as governor while the rest of the army followed: it could then defend the capital with the powerful assistance of the Paris forts. No sooner taken, the decision was abandoned. News from Bazaine at Metz and advice from the Empress in Paris convinced Napoleon that he must march to the rescue of the Army of the Rhine which, according to sketchy information from Bazaine, would try to reach Châlons, probably by making a detour to the north. With its destination still uncertain, the army left for Rheims on 21 August, hustled on by the news that German cavalry was less than forty kilometres from Châlons. On 22 August plans for marching to meet Bazaine were cancelled, only to be revived again within hours on receipt of a message in which Bazaine declared that he would break out via St-Menehould or Sedan. On the 23rd the Army of Châlons lurched off to keep its appointment with destiny on the banks of the Meuse.

Following his untidy victory on 18 August Moltke had reorganised his armies, using the new Army of the Meuse – three corps under the Crown Prince of Saxony – as the northernmost prong of a two-tined fork, which stabbed hard into France on 23 August. The Meuse army made for St-Menehould and the 3rd for Vitry-le-François, but lacked accurate information on the whereabouts of MacMahon. Paris newspapers suggested that the Marshal might be making for Metz rather than Paris, and Moltke took the considerable risk of ordering both his marching armies to wheel to the right, plunging into the thick country of the Argonne.

MacMahon trundled on, drawn north by the need to pick up supplies from Rethel and able to swing eastwards once again only on the 26th. He ran straight into trouble, for that day the cavalry of Douay's VII Corps, on his right, met Saxon horsemen from the Meuse Army near Grandpré. Douay halted and deployed for battle, and MacMahon sent I Corps to Vouziers to support him. By the time MacMahon had realised this was only an affair of outposts valuable time had been lost and when the advance resumed V Corps met more Saxon cavalry at Buzancy. MacMahon may not have been a genius, but he was an experienced officer who recognised suicide when he saw it. The Meuse Army was between him and Bazaine, and 3rd Army had swallowed up Châlons and Rheims behind him. His only hope lay in escaping to the north and on the 27th he told the Minister of War that he proposed to withdraw on Mézières. The Minister replied with instructions as unequivocal as they were fatal. Paris would erupt if Bazaine was not relieved: the forces to

MacMahon's front were only a screen and he was to march
forthwith to the aid of Bazaine.

MacMahon took the order stoically and his army sloshed
on in the rain towards the Meuse south of Stenay, its steps
dogged by German cavalry. On the 29th, discovering that
the crossings at Stenay were firmly held, MacMahon
ordered his army to edge north towards Mouzon and
Remilly. Word never reached Failly's V Corps, moving along
the Buzancy–Stenay road, and on 29 August his advance
guard bumped the left flank of XII Saxon Corps, drawn up
ready for battle at Nouart. During the afternoon there was a
pointless firefight across the Wiseppe, and at nightfall Failly
withdrew to the north to Beaumont. Its route took V Corps
across the stunningly beautiful countryside of the Forest of
Belval, along roads and tracks not easily negotiated even
with daylight and good maps – and Failly's men had neither.
They reached Beaumont soaked and exhausted, and flopped
down in bivouacs south of the town, in the V between the
D19 and the D30.

As the Meuse Army fanned up through the Argonne on
the 30th, the vanguards of IV and I Bavarian Corps came in
sight of Failly's camp at almost the same time, and their
guns opened fire shortly afterwards. Despite chaos in
Beaumont, with guns and wagons jamming the streets, and
the faint-hearted joining the inhabitants in their flight
northwards, many of Failly's men took on the Germans
emerging from the wood line. But as more German troops
came up on either flank it became obvious that the position
was untenable and the French made a painful withdrawal,
aided by a handful of infantry and gunners who made good
use of the long fields of fire on the slopes to the north to
check the advance. By mid-afternoon Failly had his corps in
something approaching order in a position just short of
Mouzon, but this collapsed under the weight of German
attack.

MacMahon, aware that Failly was in trouble, recognised
that it would be unwise to bring other corps back across the
Meuse to help him. Lebrun, whose XII Corps had crossed at
Mouzon, tried to hold a bridgehead there through which V

Corps could retreat, but could not get men across the bridges, jammed as they were with fugitives. Failly sent his cavalry up the slopes to try to disengage the infantry, but the attempt met with predictable failure. The battle ended in a ghastly panic around the bridges, with men and horses drowned in the Meuse. The French had lost almost 7,500 men and the Germans rather less than half as many.

MacMahon had already decided to withdraw on Sedan. The principality of Sedan, once part of the prince-bishopric of Liège, had passed into the hands of the ferocious La Marcks, then into those of the La Tour d'Auvergne family. Viscount Turenne was born there in 1611 and gave early proof that he was destined for a military career by spending a snowy night on the ramparts of the *Château-Fort*, curled up on a gun carriage. Sedan became French in 1643, the year before Turenne attained the dignity of Marshal of France. The *Château-Fort*, begun by Everard de la Marck in 1424, was given bastions in the sixteenth century and formed part of the first line of Vauban's defences of the north-eastern frontier. Sedan nestles in the valley of the Meuse beneath surrounding hills and although the *Château-Fort* dominated the town it was itself commanded by the slopes around, and the place had been declassified as a fortress well before the Franco-Prussian war.

As MacMahon's army emerged from the tangle of the Argonne on 31 August he disposed it to make best use of the high ground around Sedan. Douay's VII Corps held the north-western sector, the long arm of ridge running from the Calvaire d'Illy, due north of the town, to the village of Floing, above the Meuse. I Corps, now under the capable Ducrot, took post on the north-eastern sector, covering the valley of the Givonne, a stream which runs through Givonne, Daigny and La Moncelle to enter the Meuse near Bazeilles. The lower Givonne and Bazeilles itself were in the steady hands of Lebrun's XII Corps. The debris of V Corps was in reserve in the town, apart from some units sent up to hold the Calvaire d'Illy, the hinge between VII and I Corps. Failly himself was no longer in command. Pressure from Paris had led to his replacement because of his earlier misfortunes on the frontier. His successor, General de Wimpffen, had arrived on 31 August, bringing with him – though he was the only man in the army who knew it – a dormant commission: if any misfortune befell MacMahon, he would take command of the army.

MacMahon did not intend to stay in Sedan for long, although he was not sure in which direction he might leave it. The newly formed XIII Corps was at Mézières, to the north-west, and MacMahon considered joining it by a new road north of the Meuse, which would not be on German maps. However, he had news of German cavalry threatening routes to the west and the Empress, from Paris, still urged him on to Metz. MacMahon intended to push out reconnaissances on 1 September to see where the Germans were weakest and act accordingly. He was already in deadly peril. Moltke, following his armies through the Argonne, recognised his opportunity and on 30 August he ordered the Meuse Army to advance on the right bank of the Meuse, brushing the Belgian frontier with its right wing, while the 3rd Army swept up on the left. On the 31st the Bavarians took the bridge at Bazeilles and the XI Corps that at Donchery. That evening, as German watch fires spread out across the hills, Ducrot summed up the army's plight in the apt remark: 'We are in a chamber pot, and tomorrow we shall be shat upon.'

The battle began early on 1 September. Before dawn I Bavarian Corps began to slip across the railway bridge and two pontoon bridges built by engineers on the 31st, and was soon locked in battle with Lebrun's *Infanterie de Marine*, fighting hard to hold the loopholed houses of Bazeilles. Some of the inhabitants turned out to support their countrymen and the infuriated Bavarians responded by shooting any civilians caught with arms in their hands. With Bazeilles in flames, the fighting spread northwards as the Meuse Army shook out to attack across the Givonne, the Saxon Corps coming into action on the right of the Bavarians. La Moncelle was taken quickly, but some of Ducrot's zouaves, posted in Daigny to cover its bridge, stabbed back hard when the Saxons appeared, and it was not until 10 a.m. that it was secured. German gunners, well established on the far side of the Givonne, caught the retiring zouaves as they struggled back up the slope.

Ducrot had other worries. MacMahon had ridden out to Bazeilles as soon as the fighting started, only to be hit in the leg by a shell splinter. He handed over to Ducrot, who immediately gave orders for a retreat to the west. He realised only too clearly that the Germans would soon envelop both flanks. Lebrun protested that his men were fighting well and Ducrot waited an hour before insisting that the withdrawal should start at once. Scarcely had he done so than Wimpffen appeared, announcing that he was in command and cancelling the retreat. It was now about 8.30 a.m. and Wimpffen knew – as Ducrot did not – that the Germans had crossed in strength at Donchery and cut the Mézières road: a withdrawal would have been more difficult than Ducrot imagined. Even so, a retirement might possibly have saved some fragments of the army; standing to fight would simply guarantee its destruction.

The Germans lapped around the western front of the French position as they had around the east. The advance guard of XI Corps reached Floing at about 9 a.m., while V Corps pushed up on towards Fleigneux. At about midday the leading elements of V Corps and the Guard met at Olly, on the upper Givonne, completing the encirclement. Some French cavalry, charging in an effort to delay the progress of V Corps, had forced their way out of the sack before it closed and a handful of soldiers found refuge in Belgium or Mézières. But most of Wimpffen's army was now bottled up, at the mercy of fire from batteries on the surrounding slopes.

Wimpffen spent much of the morning trying to collect troops for a breakout towards Carignan, and at 1 p.m. he asked Napoleon to come and place himself at the head of his troops. The Emperor declined. Not out of cowardice – indeed, he spent an agonising morning in the saddle seeking a death which would not touch him – but because he saw that it was hopeless. Few of the troops selected for the attack arrived at the start point and the operation produced only a brief flurry of activity at Bazeilles. Its dying embers were drenched by a flood of fugitives from the north-west: VII Corps, inert under shellfire for so long, had collapsed.

It was the constant hammering of German artillery that did the damage. Troops holding the apex of the position around the Calvaire d'Illy, where Douay's and Ducrot's men joined hands, were driven back into the doubtful safety of the Garenne Wood, between Illy and Sedan, by the fire. German infantry took the Calvaire at about 2 p.m., and it was all that Ducrot and Douay could do to hold the position together. Douay's left flank was in as much trouble as his right. His men were dug in in two lines of primitive trench on the edge of the plateau above Floing and German gunners soon battered his own guns into silence while their infantry seeped round the foot of the slope. Infantry counter-attacks bought some time and two squadrons of lancers a little more. Ducrot then ordered General Margueritte to use his cavalry division to smash a way through the Germans so as to permit a breakout to the west.

While his division prepared to charge, Margueritte rode over the edge of the plateau to reconnoitre. A rifle bullet smashed his jaw as he did so, but he had enough strength to ride back to his squadrons and point the way to the enemy before he fell from the saddle, mortally wounded. His troopers, good soldiers of the *Chasseurs d'Afrique*, charged with determination worthy of better fortune. The ground was steep and broken, and the German infantry before Floing stood their ground and slammed their volleys into a target they could scarcely miss. The *chasseurs* rallied to charge at least twice more: King William, watching from the heights of La Marfée across the Meuse, could not resist gasping 'Ah! The brave fellows!'

The collapse of the infantry followed. Douay's men went first, with Ducrot's behind them. As German infantry moved forward they encountered almost no resistance: Garenne Wood, raked by the Guard artillery, was packed with demoralised fugitives. Ducrot, Douay and Lebrun had all set off, independently, in search of the Emperor. By the time they reached him he had already hoisted a white flag in Sedan and it was agreed that Wimpffen should be found to sign a letter requesting an armistice. Lebrun tracked him down on the Balan road. He was still convinced that the game was not up and persuaded Lebrun to help him pull together an attack from the exhausted troops scattered about. They got it under way and actually retook Balan, before the troops folded with a finality which convinced even Wimpffen that the battle was lost. By this stage Moltke had been told of the white flag and had sent an emissary to the town: he returned with an officer of Napoleon's staff, bearing a letter in which the Emperor announced his surrender to King William.

The surrender negotiations, between the unwilling Wimpffen and General de Castelnau of the Emperor's suite on the one hand, and Moltke and Bismarck on the other, led to the signing of terms at the Château de Bellevue, on the river north of Frenois, on the morning of 2 September. The surrender brought the Germans 83,000 prisoners to add to the 21,000 they had already taken. Their own losses, at 9,000, were astonishingly light for a victory of such magnitude and demonstrated just how well the lesson of St-Privat had been assimilated: at Sedan German artillery conquered and German infantry occupied.

Sedan was badly damaged in the Second World War. Nevertheless, many parts of the battlefield remain little disturbed and easily accessible. On the high ground southwest of Floing, reached by a steep road with nerve-racking bends, is a row of 1870 French graves, then a 1940 military cemetery and finally a monument to the French cavalry, inscribed with King William's words: '*Ah! Les braves gens!*' There is a mass grave alongside it and all around are monuments to the regiments of *Chasseurs d'Afrique*, who played such a distinguished role in France's North African army. In Floing itself stands a statue of Margueritte, with an African neck cover on his képi. The Floing–Illy road follows the line of VII Corps' front and the Calvaire d'Illy, with Garenne Wood behind it, stands about 500 metres south-east of Illy on a tiny road. There are two crosses on a small knoll: one pre-dates the battle and the other commemorates it. The D129 runs along the Givonne valley, through Givonne,

Daigny – where the zouaves of Lartigue's division counter-attacked so briskly – to La Moncelle and Bazeilles.

On the western edge of Bazeilles – where a piece of spaghetti junction may confuse the traveller – is the *Maison de la dernière cartouche* – 'Last Cartridge House'. The artist Alphonse de Neuville who, with his contemporary Edouard Detaille, specialised in meticulously researched military scenes, painted *Les dernières cartouches* in 1873. It depicts the defenders of the house firing their last rounds from a window as the Bavarians close in. The heavy furniture is scarred by bullets, door torn from its hinges, mattress flung against the window. In the left foreground two figures search pouches for extra ammunition and a wounded infantry officer, face lit by a shaft of September sun, glares out towards the enemy. The first-floor room where Neuville set the painting remains as it was and the picture hangs on the ground floor. The house contains numerous relics of the battle, and in a nearby ossuary rest the remains of some 6,000 French and German soldiers. The exploits of the *Infanterie de Marine*, some of the best troops of the Army of Châlons, are remembered with pride. None of the four marine regiments there surrendered its eagle. Two were buried and the other two were carried by devoted NCOs across a countryside alive with Germans all the way to the corps' depot at Brest in Brittany. In Sedan itself the *Château-Fort*, recently refurbished in an effort to attract the tourists the town so desperately needs after the collapse of the local textile industry, invites a visit. Its ramparts offer wide views over the town and its museum has a section devoted to 1870, including a panorama which brings the battle to life.

There was little fighting in Champagne after the Battle of Sedan. Moltke gave his orders for the investment of Paris at Château-Thierry on 15 September and two days later his forces joined hands round the city. Champagne, on the line of communication between Germany and besieged Paris, endured the misery of occupation, sharpened almost daily as the activities of *francs-tireurs* brought an increasingly harsh response.

The peace terms, which amputated Alsace and dis-membered Lorraine, left Champagne intact. Some forts were built as part of the Seré de Rivière plan, three at La Fère, four at Laon and seven at Rheims. The latter were intended to block the advance of an enemy who had penetrated the Argonne. Fort de la Pompelle, lying alongside the N44 some five kilometres south-east of Rheims, was taken by the Germans in September 1914 but recaptured almost imme-diately. The heavily shelled exterior resembles that of Douaumont. Its interior can also be visited and houses a museum whose exhibits include an excellent collection of German military headdress. Amongst the other forts in this series at least one survives. Near the twin villages of Bruyères-et-Montherault, not far from Laon, is the well-signposted *Fort de Bruyères*. It is easy to find and to gain access, and is an excellent example of an umodernised (and unfought-over) Seré de Rivière work.

There were spiked helmets aplenty in Champagne in the summer of 1914. By 27 August the five marching armies on the German right were fanning out across northern France. That day Moltke ordered the 1st to march west of Paris, the 2nd to pass between La Fère and Laon, and head straight for the French capital, the 3rd to march on Château-Thierry, the 4th to march from Rheims to Epernay and the 5th to take the line Châlons–Vitry-le-François. This was very much in the spirit of the Schlieffen Plan, but on 29 August Lanrezac's 5th Army put in a sharp local counter-attack at Guise, north of Laon, which was to have important consequences. Bülow of 2nd Army called for help and his right-hand neighbour, Kluck, edged eastwards to assist. The crisis passed quickly as 5th Army resumed its retreat, but it left the German armies heading further east than had been intended. On 1 September Joffre ordered his armies to fall back to the line Verdun–Bar-le-Duc–Vitry-le-François–Arcis-sur-Aube–Nogent-sur-Seine, with the British at Melun and the newly formed 6th Army covering Paris. The government left Paris the following day and disaster loomed as the Germans threatened the railway lines linking the capital with the armies to the east.

Yet the Germans, too, were in difficulties. Not only had

the whole frame of their advance slipped eastwards, but the abrasive effects of nearly a month's marching and fighting reduced the effectiveness of commanders and troops alike. On 29 August Moltke's headquarters moved from Koblenz to Luxembourg, but Moltke himself was feeling the strain of his responsibilities, and his ability to give effective direction to his armies was drastically limited by the fact that telephones and couriers lay at the heart of his communications, and both were less than ideal in a mobile war on this enormous scale. On 2 September he continued the process of sidestepping to the left, directing his armies to drive the French south-eastwards from Paris, with 1st Army providing flank protection for the 2nd. Kluck should have marked time while 2nd Army wheeled past him, but his bellicose disposition impelled him to push on between the Petit Morin and the Marne, a move seen by reconnaissance aircraft reporting to General Galliéni, military governor of Paris – who, as a young man, had fought at Sedan. Bülow, meanwhile, crossed the Marne between Epernay and Condé-en-Brie, and Hausen moved up to the river west of Châlons. The 4th Army reached Valmy and the 5th was fighting hard in the Argonne on the line St-Menehould–Clermont. Despite the depth of his advance into France, Moltke was aware that he had won no decisive victory. In Lorraine the French were fighting furiously to hold Rupprecht and their new concentration around Paris threatened Moltke's right flank.

On the evening of 4 September Moltke issued a new directive. The 1st and 2nd Armies were to cover the German right against any threat from Paris, while the 4th and 5th Armies continued their drive south-east towards the rear of the forces engaged against Rupprecht's armies in Alsace-Lorraine. The 3rd Army was to march on Troyes, swinging east or west, as directed, to support either the Paris or the Lorraine operation. It took time for the directive to reach the field armies and when it was received by Kluck in his headquarters at La Ferté-Milon he was disinclined to obey. On the night of 5 September his men reached the line Montmirail–Rebais and a patrol advanced down the D403 towards Provins. No Germans were to get further during the entire war.

The Germans might yet have won had Moltke's failing resolve not been set alongside Joffre's phlegmatic determination. He did his best to stiffen his hard-pressed centre. Lanrezac of 5th Army was replaced by Franchet d'Esperey – soon christened 'Desperate Frankie' by the British. On the 5th Army's right the new 9th Army, commanded by Foch, who had made his reputation as a corps commander outside Nancy, battled on north of the Aube. But it was on his flanks that Joffre sensed the real opportunity. He conceived an ambitious plan for a concentric attack by his 6th and 5th Armies, together with the BEF, on his left, and the 3rd Army, on his right. On paper it offered the possibility of pocketing the right-flank German armies, but in practice the imponderables of war produced a battle that bore as little relation to Joffre's plans as it did to Moltke's.

The attitude of Sir John French, commander of the BEF, was an imponderable in itself. His confidence in the French high command had been shaken by the events of the past weeks and he had considered taking the BEF out of the line to refit. On 1 September Lord Kitchener, Secretary of State for War, travelled to France and met Sir John in the British Embassy in Paris. After an acrimonious interview French was told to conform with his allies' movements and he at once wrote to the French war minister expressing his willingness to participate in a properly planned battle on the Marne.

Joffre was dismayed to discover that this collaboration was not unconditional and when the *Instruction Générale No. 6*, which laid the foundations for the Battle of the Marne, reached GHQ at Melun French announced that he would study it before deciding whether or not he could participate. On 5 September Joffre drove over to see French in his quarters in the château at Vaux-le-Penil, overlooking the Seine outside Melun on the Chartrettes road. He went into the small room where French was waiting with Murray and Wilson, explained the situation and begged passionately for support. *'Monsieur le Maréchal,'* he urged, *'c'est la*

France qui vous supplie.' Tears coursing down his cheeks, French wrestled with his emotions and an unfamiliar language, and eventually declared, 'Damn it all, I can't explain. Tell him that all that men can do our fellows will do.' 'The Field-Marshal says "yes",' announced Wilson. Murray pointed out that the BEF could not start as soon as Joffre hoped, but the French Commander-in-Chief was well satisfied with his efforts. 'Let them start as soon as they can,' he said. 'I have the Marshal's word and that is enough for me.'

The Battle of the Marne began on 5 September when Manoury's 6th Army attacked, a day earlier than planned, into Kluck's flank north of Meaux. He was checked by IV Reserve Corps around St-Soupplets, and on 6 September Kluck shifted three of his corps from the Grand Morin to the Ourcq to join IV Reserve and eventually bring Manoury's attack to a halt. On the Grand Morin, meanwhile, 5th Army made little progress and the BEF was not yet ready to add its weight to the struggle. It spent the following day driving Marwitz's cavalry corps and a supporting infantry division across the valley of the Petit Morin around La-Ferté-sous-Jouarre. French's critics argue that this was the moment to strain every nerve, for Kluck's move to the Ourcq had left a weakly held gap between the bulk of 1st Army and Bülow's left flank north-east of Montmirail, but the BEF, frayed by its long retreat, did not move quickly enough to take the opportunity.

In the baking heat of early September the armies were locked in battle across the lush wooded country west and north-west of Paris. In truth the struggle still tilted in Germany's favour. On Joffre's left, Manoury's 6th Army had been halted by Kluck and was on the verge of falling back. The BEF had made slow progress and even d'Esperey seemed as concerned with holding ground as with taking it. But Joffre was undaunted. He had moved his headquarters back from Vitry-le-François to Bar-sur-Aube – where the Tsar and the Emperor of Austria had based themselves for part of the 1814 campaign – but was still well placed to put steel into his army commanders. A good lunch, eaten in single-minded silence, was the day's unmistakable landmark and during the key stages of the planning for the Marne battle Joffre sat quietly astride a chair under an apple tree in the school playground at Bar-sur-Aube. Moltke, in contrast, was at the end of his tether. A staff officer saw him weeping silently at his dinner table and letters to his wife betray awareness of his moral responsibility for the events of that terrible summer.

While Joffre was sped across Champagne by his racing chauffeur to deal personally with hesitant or incompetent commanders, Moltke was immured in the claustrophobic world of his headquarters. He had already sent Lieutenant-Colonel Hentsch, head of his foreign armies section, on a tour of the army headquarters and on 8 September Hentsch set off again. Although OHL was to deny that Hentsch had authority to order or approve a retreat, Hentsch himself certainly believed that the line St-Menehould–Rheims–Fismes–Soissons had been selected as the line to which the army should retreat if necessary. His journey began well enough. The 4th and 5th Armies were confident, and 3rd Army, its headquarters at Châlons, was making steady progress. But when he arrived at 2nd Army's headquarters in the neat red-brick Château de Montmort, overlooking the Surmelin between Montmirail and Epernay, all the talk was of retreat. The gloomy Bülow was concerned by the threat to his right flank and believed that only speedy withdrawal to the Marne could prevent his being outflanked. On 9 September Hentsch drove to Kluck's headquarters at Mareuil, on the Ourcq south of La Ferté-Milon. His journey had taken him across the back areas of armies engaged in mortal combat and his own confidence had been bruised in the process. Kluck's Chief of Staff admitted that it would be impossible for 1st Army to give immediate support to Bülow; this being the case, Hentsch ordered 1st Army to retreat 'because this was the only way one could bring it once more into co-operation with the 2nd Army'.

The retreat began at once. The Germans fell back on to the Aisne in weather that matched their mood as the blazing days of early September gave way to heavy rain. The Allied pursuit was not pressed vigorously, and the Germans took their stand in the excellent defensive country north of the Aisne and brought mobile war to an abrupt halt. We have already seen how the natural strength of the spurs rising above the Aisne had been well used by Blücher's men in 1814: the conditions of war a century later made them even more advantageous to the defender, machine-guns sweeping the slopes and howitzers dropping shells into the valley bottom. Although the Allies were able to cross the river, they could not break the German line on the far side, where the Chemin des Dames, running parallel with the river, provided a fine natural rampart.

Casualties were heavy. The British 2nd Infantry Brigade fought savagely for the sugar factory at the crossroads at Cerny-en-Layonnais, with one of its battalions – 1st Loyal North Lancashire – losing the Somme-like total of fourteen officers and 514 men. At Cerny an elegant pillar remembers the battalion. The British Military Cemetery at Vendresse contains many of the dead of this battle and more from the fighting of May 1918, including that rare bird, a Territorial brigadier-general.

It was soon obvious that the German hold above the Aisne would not be broken easily. French told the King that the Battle of the Aisne was typical of future operations: 'Siege operations will enter largely into the tactical problems – the spade will be as great a necessity as the rifle, and the heaviest calibres and types of artillery will be brought up in support on either side.' He pressed Joffre to agree to the BEF's move from the Aisne to Flanders and the move took place in early October, with GHQ leaving Fère-en-Tardenois for Abbeville on the 8th. The whole Champagne sector then became a French responsibility.

It remained important. By the close of 1914 the Western Front bulged out towards Compiègne in a salient which invited attack from Artois and Champagne simultaneously: we have seen how these two-pronged attacks affected the British in 1915 and 1917, producing the battles of Loos and Arras. The French Champagne battles were just as bloody and dismal. The Germans held the high ground and burrowed almost impenetrable dugouts in the chalk. The abortive Champagne offensive north-west of Rheims in March 1915 was the last of the old-style French attacks, with infantry going up to the assault in red trousers and blue greatcoats, to fall in huddles below Craonne. Thereafter horizon blue replaced the bright hues of yesteryear, although it took some time for offensive tactics to evolve to meet the changing demands of trench warfare.

German defensive tactics were already changing, partly because of the lessons of Neuve Chapelle and partly because of assimilation of a French document on the defensive battle which envisaged defence based on interconnected strongpoints, an idea that commended itself to several German officers. Not amongst that number was Colonel Fritz von Lossberg, posted to the Champagne front in September 1915 as Chief of Staff of 3rd Army.

Lossberg arrived at 3rd Army's headquarters in Vouziers on 27 September in the middle of a French attack. The offensive had been launched on 25 September with twenty divisions attacking between Auberive and Ville-sur-Tourbe. It made respectable progress, especially in the centre, north of Perthes-les-Hurlus and Les-Mesnils-les-Hurlus, owing to the effects of gas and the fact that shelling had cut most of the telephone cables linking German observers to their batteries. Complete breakthrough was averted only because the Reserve Line, on the reverse slope of the southern shoulder of the Py valley, had held. It was the sketchiest of positions, but the single shallow trench gave rapidly arriving reserve formations a grip on the ground.

Lossberg arrived after his predecessor had ordered a withdrawal. He cancelled this and departed for the battlefield, leaving his car at Sommepy and scrambling about in the shell holes of the Reserve Line at Navarin Farm. Lossberg decided that the Reserve Line should be held, but the position should be laid out in depth, with a rearward line on the northern bank of the Py brook and a second on the reverse slope of its valley's northern crest. This arrangement would give the Germans the benefit of holding reverse slopes, which would deny the French observed artillery fire, while at the same time permitting them to cover the valley with the fire of batteries behind the second rearward line. For an officer who had previously believed in the standard doctrine of holding one single solid line it was a volte-face indeed.

When the French renewed their attack the new system worked well and reserves echeloned back over ten kilometres behind the position enabled local counter-attacks to snuff out French gains. Lossberg spent much of his time visiting forward units and divided the front amongst staff officers at army headquarters, ordering all to visit their sector at least twice a week. This arrangement helped keep headquarters in touch with the realities of life at the front and 3rd Army's experiences of the battle established Lossberg as the leading authority on defensive doctrine.

Navarin Farm, on the D77 between Suippes and Sommepy, gives a good view over the ground taken by the French in September. Three figures top its monument, which commemorates General Gouraud, commander of the French 4th Army, and Lieutenant Quentin Roosevelt, son of President Theodore Roosevelt, killed in the Tardenois in 1918. Gouraud, a flamboyant officer of *chausseurs-à-pied* who had made his name in the French Sudan, lost an arm in Gallipoli in 1915 but went on to play a distinguished part in the Second Battle of the Marne and to serve as High Commissioner in Syria after the war. He died in 1946 and lies in the crypt beneath the memorial at Navarin Farm.

The farm stood on the 1915 Rearward Line – the original front line had been just north of Souain, in the area marked by military cemeteries on both sides of the main road. To the north, the road crosses the Py brook in Sommepy and quickly climbs the lip of the valley across the first rearward line, which clipped the northern edge of the village. The second rearward line was further north, in the area where the American Sommepy monument stands on the Blanc-Mont Ridge to the west of the road. The ridge was captured by the US 2nd Division in October 1918 and the ground beneath the pine trees around the monument bears evidence of the fighting.

There was little fighting in Champagne in 1916. At the beginning of 1917, as Nivelle, the new French Commander-in-Chief, surveyed his domain, Champagne began to feature more prominently on the maps on the walls of GQG in the Hôtel du Grand Condé in Chantilly. Nivelle soon moved GQG from this delightful horse-racing centre to the utilitarian *Institut Agronomique* in Beauvais. He was convinced that he could break through the German front 'on condition that we do not attack it at its strongest point, and that the operation is carried out by a sudden surprise attack, and is not extended beyond twenty-four or forty-eight hours'. Joffre had envisaged an offensive by the British at Arras and by the French in Champagne. Nivelle modified the plan so as to include the formidable German positions on the Craonne plateau. He proposed that the British should attack first, attracting German reserves to the north: his own armies would then assault. The attack was to be entrusted to the three armies of the *Groupe des Armées de Reserve* (GAR), delivering a violent blow, which would break the German defensive system and enable the attackers to push right on through it.

There were many who doubted if this would work. Nivelle lacked sufficient guns and mortars to achieve the concentrations deemed desirable, and it was hard to see how even the boldest attackers could unhinge a deep defence. While the Nivelle method had worked on a small scale at Douaumont, it had limitations when splashed across the vast canvas of Champagne. Nevertheless, Nivelle's political backers – British and French – desperately wanted to believe that the spell would work and in the first three months of 1917 the conjuring act went on. The British front was extended to Roye to free French troops for the offensive and at the Calais Conference in late February it was agreed that Haig should act under Nivelle's orders for the duration of the offensive.

Difficulties went beyond the essential impracticality of the Nivelle Plan. Those who knew the French army well detected an alarming brittleness. It had, wrote a British liaison officer, forgotten how to smile and had come to resemble 'a thoroughbred harnessed to a tumbril driven by a drunken carter who beat it till it bled'. German withdrawal to the Hindenburg Line in March wrong-footed Nivelle. He failed to attack while it was under way, but refused to acknowledge that it had altered the basis of his offensive. In early April GQG moved yet again, this time to Compiègne, as if to emphasise Nivelle's determination to press the attacks by being closer to the front.

French preparations came as no surprise to the Germans. On 4 April they staged a large-scale raid on Sapigneul, in the angle of the Aisne and the Aisne Canal north-west of Rheims, overran a section of the line and took numerous prisoners. One of them bore a copy of the VII Corps plan, a document that gave the Germans useful insight into the entire offensive. A meeting of senior military and political leaders at Compiègne on 6 April discussed the offensive: although some of the soldiers present spoke out against it – Pétain of the *Groupe des Armées du Centre* (GAC) was vehement in his opposition – Nivelle's plan was approved. There were still doubts within the army: the commander of I Corps was dismissed for saying that he was not confident of achieving a breakthrough. The Germans enjoyed air superiority over the Aisne battlefield and air observation enabled their gunners to disrupt Nivelle's batteries: despite this, between 5 and 16 April the guns of the French 5th and 6th Armies fired five and a quarter million shells.

General Micheler's GAR was to attack between Coucy le Château and Caurel, north-east of Rheims, with three armies: from the west, Mangin's 6th, Duchêne's 10th and Mazel's 5th. On the right, Anthoine's 4th Army from Pétain's GAC was to attack to the east, between Nauroy and Auberive, a day after the main battle had started. The Germans had turned the front into a veritable fortress. The chalk slopes had been honeycombed with caves and cellars – *creutes* – before the war and many of these had been turned into dugouts impervious to artillery. Only the German first line could be observed from the French trenches: two more systems, with a total depth of up to ten kilometres, lay out of sight. Counter-attack divisions were posted close to the front but concealed from French artillery. Along sections of the attack front the French were already across the Aisne and its parallel canal, but this had yet to be crossed on part of Mangin's front. Up from the river ran the Chemin des Dames – 'The Ladies' Road' – slashed with gullies and strewn with woods and undergrowth. It took its name from the road along its top – now the D18 – built so that the daughters of Louis XV could drive from Compiègne to the Château de la Bove, seat of the Duchess of Nemours. Behind it lay the marshy valleys of the Ailette, Bièvre and Miette, with yet more tangled ridges before the open country north of Laon and Sisonne was reached.

The weather piled the last straw on to the *poilu*'s overloaded back. It rained in torrents on the night of 15–16 April as the French moved up for the attack, their guns stepping up the bombardment as zero hour approached. At 6 a.m. the attack began, but it was soon obvious that the infantry could not keep up with the barrage as it roared and spat across the ridges. Some of Mangin's Senegalese of I Colonial Corps, attacking abreast of Vauxaillon on the extreme left, were so cold that they could not fix bayonets, but went over the top with their rifles under their arms, trying to keep their frozen fingers warm. XX Corps made some progress between Soupir and Paissy, and 5th Army gained ground on either side of Sapigneul, advancing up the valley of the Miette on its left and clearing a sliver of territory around Loivre and Courcy on its right. The weather made life difficult for the French tanks, floundering across no man's land: if they survived to breast the first crest they were taken on by German field guns behind it. Tank–infantry co-operation was sketchy: a group of tanks reached Amifontaine in the Miette valley, seven kilometres from their start line, but was destroyed there, well ahead of its infantry.

The infantry battle was hellish. In some sectors the French were machine-gunned as they left their front-line trenches; in others they got into the first position but had to fight like demons to stay there as Germans billowed up from deep dugouts. Command and control collapsed, and the worsening weather, as rain turned to sleet, added to the tribulations of the attackers. The battle staggered on despite growing losses and pitiful results. The French lost some 187,000 men. Though they took about 20,000 prisoners and 147 guns, first securing their hold on the Aisne valley, then compelling the Germans to withdraw from the Vailly salient in Mangin's centre and finally getting up on to the Chemin des Dames, the battle's gains did not approach the results touted by Nivelle. The lamentable performance of French medical services did not help and some of the wounded arriving at ill-equipped hospitals showed how deeply the defeat had entered the soul of the French army. 'It's all up,' they said. 'We can't do it: we shall never do it.'

In late April the army began to mutiny. The 2nd Battalion of the 18th Infantry Regiment was probably the first to go: its soldiers refused to return to the front and four were shot by firing squad. Severity failed to cure the fast-growing disease and by June mutiny was rife. In all, 113 infantry regiments, twenty-two *chasseur* battalions, a Senegalese battalion, two colonial infantry regiments, twelve artillery regiments and one dragoon regiment, from a total of sixteen corps, took part in the mutinies. Nivelle, already discredited, was replaced by Pétain on 15 May, and for the next few months the new Commander-in-Chief concentrated on restoring the army to health. He used both stick and carrot. In June, for example, a mutinous battalion seized Missy-aux-Bois, south-west of Soissons on the Villers-Cotterêts road. The area was cordoned off by a cavalry division and six of the leading mutineers were shot by firing squad. Yet if Pétain had little mercy on soldiers caught in the act of indiscipline, he devoted enormous energy to improving food and accommodation, stepping up the frequency of leave and distributing decorations more liberally. Above all, he put a stop to grandiose offensives and restored confidence by set-piece limited attacks. His political masters, meanwhile, clamped down on anti-military propaganda and parliamentary corruption, and in November Georges Clemenceau became Premier, declaring that he would 'conduct the war with redoubled energy'.

After the Aisne offensive of 1917 died away in mud and blood, the Chemin des Dames became a quiet sector, one of those backwaters where men slipped into the comfortable world of live and let live, developing more sympathy for their enemies a few metres to their front than the staff a dozen kilometres to the rear. Even after the German offensives of March and April 1918 had rocked the Allies it remained 'the sanatorium of the Western Front', where both sides could send battered divisions to recover. It was held by the French 6th Army, now under General Duchêne, and its natural strength encouraged complacency. Lieutenant-General Gordon's British IX Corps (8th, 21st, 25th, 50th and, later, 99th Divisions) had been sent to the sector after being knocked about in Flanders, and held a line from Loivre on the Vesle up on to the Californie plateau above Craonne. The British were disturbed to discover that their divisional sectors were wide but lacked depth. Duchêne argued that the character of the ridge prevented compliance with Pétain's 'Directive No. 4' of December 1917, which dictated that the front line would be thinly held and real resistance would begin in the battle zone to its rear.

The Chemin des Dames was the target of the third of Ludendorff's major offensives, *Blücher*, delivered by the six corps of 7th Army with assistance from two corps of the 1st: over 3,000 guns supported the attack. Although preparations were carefully concealed, on 26 May the Allies received hard intelligence that the attack would take place next day. Duchêne ordered his men into their battle positions, packed tight on to the ridge. At 2.15 a.m. on the morning of the 27th the German bombardment started and at 3.40 a.m., just before first light, the infantry loped forward into the murk. The bombardment had been so devastating that in some areas the Germans experienced their most serious opposition from the ripped-up slopes of the ridge. By 5.30 a.m. they had punched a gap through 6th Army's centre and by nightfall they were well across the Aisne. Fère-en-Tardenois fell on the third day of the offensive and on 1 June the salient included both Soissons and Château-Thierry.

The battle area is rich in relics and monuments. Behind it,

just north of Crépy on the Laon–La Fère road, is the site of a 210mm railway gun, which first fired on Paris in March 1918. The gun position is reached via the D26 and a track running from the main line into woodland marks the location of the siding from which the monster was fired. The western end of the Chemin des Dames was secured by Fort de la Malmaison, one of the group of Seré de Rivière forts around Laon. Its ruins lie beside the N2 Soissons–Laon road at its junction with the D18, just north of Aizy-Jouy. The fort had been used for experiments with the new high explosive in the 1880s. Captured by the Germans in 1914, it was retaken by the French 38th Division in April 1917. There is a large German Second World War cemetery within sight of it.

Suffering is etched deep into the Chemin des Dames. The French military in Cerny-en-Laonnois Cemetery contains 5,500 graves and the remains of 2,800 more soldiers lie in the ossuary. Further along the road to the east is the Caverne du Dragon, a huge *creute* expanded by the Germans. Capable of sheltering up to 6,000 men, the complex provided barracks, stores, armouries and command posts. It now houses a military museum, and near its entrance stand monuments to the 4th Zouaves, 41st *Chasseur* Battalion and 164th Infantry Division. The nearby Hurtebise Farm was in the centre of Blücher's position on 7 March 1814. It was contested again in September 1914 and April 1917. A bronze statue celebrates the valour of French soldiers a century apart and a plaque on the farm wall notes that another battle was fought there in 1940 by the French 4th Armoured Division.

An orientation table looking over the Californie plateau above the village of Craonne explains the 1917 battle and a nearby statue of Napoleon marks the position of the Emperor's command post in 1814. Craonne itself was devastated during the First World War, and an arboretum now stands on the site of the old village. The 'Chanson de Craonne', written in 1917 and sung to the music of Charles Sablon's 'Bonsoir m'amour' has been described as 'one of the world's great pacifist hymns.' It tells of the *poilu*'s hatred for

the *embusqués*, men with safe jobs, and the rich and powerful enjoying life in Paris, as he prepares to die at 'Craonne up
on the plateau'. Its bitter chorus begins: 'Goodbye to life,
goodbye to love, goodbye to all the women . . .'

By early June 1918 the German effort was visibly running
out of steam, and fresh Allied divisions arrived to contain
the break-in. Amongst them were the Americans. The US
3rd Division moved on to the line of the Marne on 3 June
and its machine-gun battalion shot up Germans attempting
to cross the river at Château-Thierry: the Quai Galbraith is
named for one of its officers. On the same day the German
bridgehead at Jaulgonne was snuffed out: the attack had
reached its zenith, and German troops were ordered to dig in
and hold their ground.

Allied attempts to regain lost territory began soon
enough. On the afternoon of 6 June 4th Marine Brigade
assaulted Belleau Wood, west-north-west of Château-
Thierry. The marines were inexperienced and began their
attack at a walking pace in straight lines. The village of
Bouresches, on the south-east edge of the wood, was taken
after hand-to-hand fighting, but it was not until 25 June that
Belleau Wood was secured. The Marine brigade had suffered
over 5,000 casualties, including half its officers. The
colonnaded American Château-Thierry Memorial stands on
Hill 204, overlooking the town from the west. Belleau Wood
is seven kilometres away. Within it is a memorial to 4th
Marine Brigade, with captured German guns in the clearing
around it. The American Aisne-Marne Cemetery is on the
northern edge of the wood, with a sombre German cemetery
nearby.

The last German effort, christened *Friedensturm* – Peace
Offensive – began on 15 July. The tactical mix was familiar
– lightning bombardment followed by storm troop assault.
The Germans plunged down on either side of Rheims. They
crossed the Marne east of Château-Thierry, pushing back
Berthelot's 5th Army on one flank. On the other they
attacked the robust Gouraud and his 4th Army, gaining a
little ground from the Vesle, through Auberive, to the high
ground above Tahure but at alarming cost. The attack did

little more than increase the size and vulnerability of the salient that bellied out to Château-Thierry, to which the Allies were about to devote their attention.

The *Friedensturm* came close to disorganising Allied plans for a counter-offensive. Pétain had planned an attack on the western face of the salient, but postponed it and ordered Mangin, whose 10th Army was chiefly concerned in the operation, to release reserves to help contain the Germans west of Rheims. Foch was already under pressure from Haig to strengthen the Lys front in Flanders, but had declined to do so, arguing that Mangin's attack across the base of the Marne salient would effectively undercut German offensive operations elsewhere. When he heard of Pétain's orders he decreed that Mangin's attack should take place as planned, and on 18 June it boiled out of the forests of Compiègne and Retz. It never came close to cutting off German forces in the salient, but reached the Soissons–Château-Thierry road, cost Ludendorff 25,000 men in prisoners alone, and led to a systematic withdrawal from the ground taken during *Blücher* and *Hagen*. In late July Ludendorff paused on the line south-east through Fère-en-Tardenois to Ronchères and was vigorously attacked as he did so. There was what one American account called 'hard fighting with little or no progress' on the 29th, with the villages of Sergy and Cierges taken at great cost. The capture of Seringes and Nesles, on the Allied left flank, led to the fall of Fère-en-Tardenois and on 2 August the Allies found that the Germans had slipped away under cover of darkness to take up a new line on the Vesle from Soissons through Fismes to Rheims. On the Vesle, Foch released American divisions from the Allied armies to form the US 1st Army, which we have already seen in action at St-Mihiel in Lorraine.

American participation in the elimination of the Marne salient is marked by the Oise-Aisne Cemetery in Seringes, with over 6,000 Americans buried beneath the tree-screened greensward. Just outside the nearby village of Chamery – down a track leading east off the D14 south of its junction with the D2 – two memorials mark the spot where Quentin

Roosevelt crashed after being shot down on 14 July 1918. A fountain to his memory has been erected in Chamery and bears the somewhat Spartan inscription: 'Only those are fit to live who are not afraid to die.'

Allied strategy in the late summer of 1918 gave the Americans pride of place in operations on the borders of Champagne. During August the concept of a convergent offensive evolved, with the British striking across from Cambrai and the Americans thrusting up through the Argonne towards Mezières. Foch outlined this plan to Pershing at his headquarters at Ligny-en-Barrois on 30 August. It meant that once the Americans had pinched out the St-Mihiel salient east of the Meuse their effort would shift to the Argonne west of the river. Pershing had been offered the choice of attacking between the Argonne and Rheims or taking the Argonne itself. He chose the latter, partly because it would be logistically easier for an army geared up for an advance on Metz and because he felt that no troops but his own would have the fighting spirit to undertake such a task.

As September went on the clarity of a two-pronged attack on the shoulders of the German position in France was replaced by plans for a more ambitious advance, with Foch's cry *'Tout le monde à la bataille'* as its watchword. The British armies would drive on through St-Quentin and Cambrai; the French centre would roll the Germans across the Aisne; and the Americans, with the French 4th Army on their left, would advance on Mezières. In the Argonne, the US 1st Army's 'jump-off line' ran from its junction with the French at La Harazée to Vauquois, then up to Malancourt and Bethincourt to the Meuse near Forges. Bullard's III Corps took the left-hand sector, Cameron's V the centre and Liggett's I the right. The nine divisions of these corps, most of them inexperienced, mustered a front-line rifle strength of a little over 100,000 men when they stood ready on the night of 25–26 September. Behind them were six more divisions, four of them with battle experience.

At eleven o'clock on the night of 25 September nearly 4,000 guns opened fire in unison, providing the high concentration of one gun for every eight metres of front. This weight of fire, together with the thin outer edge of the German defensive system, enabled the Americans and French to make good progress over an average of five kilometres on the first day. Over the next few days the advance was slower. It was difficult to get guns across the debris of the first day's battlefield, especially in the Argonne where smashed trees and resolute machine-gunners combined to impose a drag on the American advance. On 27 September the 79th Division of V Corps took Montfaucon Ridge, whose Hill 336 is now crowned with the Doric column of an imposing American monument. To the north, on the edge of Romagne-sous-Montfaucon, is the US Meuse-Argonne Cemetery and Memorial, with 14,000 graves on its gentle slope.

On 29 September American inexperience led to a serious check. The left flank was to hold fast between Binarville and Apremont, while the 35th Division pushed forward to secure Exermont. The Germans held the steep valley that crosses its line of advance and once the leading attackers had reached the edge of the village machine-guns firing down the length of the valley came back to life just as the garrison of Exermont counter-attacked. The division's forward elements came back apace: many officers were down and the retirement dragged the rest of the 35th back to a line just north of the Apremont–Eclisefontaine road. As a consequence of the day's reverse, the 35th, 37th and 79th Divisions were pulled out of the line, and replaced by the 1st, 3rd and 32nd Divisions, which moved up from the Verdun area that night.

Although Gouraud's army had made good progress across the open country on the American left, the setback in the Argonne was serious. The Americans were now up against the main *Kriemhilde* position and the need to make formal preparations to attack it imposed an operational pause. During this lull the 77th Division launched an abortive attack at Binaville. Most of the assaulting units were stopped by machine-guns in the dense woodland, but the 1st Battalion 308th Infantry under Major Charles S. Whittlesey managed to work its way up the valley east of the village to reach Charlevaux Mill on the D66. There the 'Lost Battalion' held out until its relief on 7 October.

A key role in the revived attack was entrusted to the regular 1st Division – which Pershing called 'the best damn division in any army'. It took the Bois de Montrebeau and Exermont, to its north, helping the 28th Division drive up the valley west of the Apremont–Flexeville road. Over the next few days there was a steady advance between this road and the Meuse, and by 10 October the Americans had reached the Grandpré gap, where the valley of the Aire divides the Argonne. That was the 1st Division's last day in action for the moment: during a yard-by-yard battle it had lost 9,387 men, the heaviest casualties suffered by any US division in the offensive.

There was another pause on 12–13 October while the Americans assimilated a major change in their order of battle. Pershing reorganised his command into an army group, with Hunter Liggett commanding the 1st Army in the Argonne and Robert L. Bullard the 2nd in Lorraine. On 14 October Liggett's men began the attack on the last *Kriemhilde* position. It was a disappointing day, with heavy losses and small gains, and this phase of the battle ended on 31 October with the Germans still secure on the line from Cléry on the Meuse through Aincreville and Champigneulle to the lower edge of the Croix-aux-Bois Forest.

When the attack resumed on 1 November the Americans profited from the policy of leaving a division's artillery in the line when its infantry were relieved, and the increased weight of fire assisted the infantry in breaking through the *Kriemhilde* position and getting beyond Bayonville and Remonville. On the 2nd the line crossed the Stenay–Vouziers road and on the 3rd another substantial advance saw the 2nd Division approach Beaumont, scene of Failly's defeat in 1870. The battle was now opening out, with succeeding days striking participants more as forced march than combat. 7 November saw the Americans on the Meuse as far as Wadelincourt and secure on the high ground above Sedan through Frénois, where the King of Prussia had

watched Margueritte's *Chasseurs d'Afrique* charge to glory
in 1870. They were there when the war ended four days
later.

It is another of the many bitter ironies bound up in the
history of the fatal avenue that the very slopes where the
Americans had halted their victorious advance in 1918 were
to witness the climactic events of 1940. The Maginot Line,
as we have seen, did not extend to cover the northern
borders of Champagne, but between 1936 and 1940 some
work was carried out on the line of the Meuse to shield
the Charleville–Sedan area against attack through the
Ardennes. The threat was not taken seriously, as the
Ardennes were believed to be unsuitable for large-scale
military movement, though the numerous casemates and
bunkers which survive on the left bank demolish the myth
that nothing was done to secure this sector.

In the spring of 1940 the Sedan sector was not allocated
great importance by the French high command. It was the
responsibility of Huntziger's 2nd Army, whose boundary
with Corap's 9th, on its left, ran just west of the Canal des
Ardennes, which joins the Meuse downstream of Donchery.
Both armies contained a high proportion of reservists,
allocated to what was expected to be a quiet part of the front.
Alas, the German operations plan developed as a result of
Manstein's dissatisfaction with *Case Yellow* aimed the
point of the German lance at precisely this sector.
Guderian's XIX Panzer Corps – 1st, 2nd and 10th Panzer
Divisions with the *Grossdeutschland* motorised infantry
regiment – were to cross at Sedan, with Wietersheim's XIV
Motorised Corps close behind. Downstream, the 6th and 8th
Divisions of Reinhardt's XLI Panzer Corps headed for
Monthermé, while the 5th and 7th Divisions of Hoth's XV
Panzer Corps were to cross at Dinant.

German spearheads moved into the Ardennes on 10 May,
incurring some delays as a result of Belgian demolitions and
sporadic resistance. Corap's 9th Army moved forward to
occupy the Belgian sector of the Meuse around Dinant and
sent patrols into the Ardennes. At Chabrehez the Belgian
3rd *Chasseurs Ardennais* fought a brisk action, forcing a

brief pause on Erwin Rommel's 7th Panzer Division. Huntziger's 2nd Army sent horsemen and light armour into the hilly woodland over the border, but its 2nd Light Cavalry Division was elbowed out of the Arlon gap. On the 11th the Germans continued their advance, pushing back the French light troops and that afternoon Guderian's men entered Sedan unopposed. Rommel did even better, getting a handful of men across a weir at Houx.

That night the rumble of tanks from across the river left the defenders of the Meuse in no doubt as to what was afoot. Grandsard's X Corps was responsible for the sector opposite Sedan. On its right, the 3rd North African Division held the line from Mouzon. In the centre, the 71st Division, a 'Series B' reserve formation, moved into position between Mouzon and Pont-Maugis on 12–13 May after much-needed training around Vouziers. The 55th Division, also composed of reservists, continued the line to the army boundary. The Germans burst across the Meuse on the 13th. Rommel secured a bridgehead around Dinant and Houx, and shrugged off a half-hearted counter-attack. Kempf's 6th Panzer Division found the going heavier at Monthermé, and by nightfall retained only a tiny bridgehead. The *Schwerpunkt* was at Sedan. At seven on the morning of the 13th Do.17 bombers began the process of softening-up the defences and at midday wave upon wave of *Stukas* joined in, screaming low to slam bombs on to French bunkers. Fighters swarmed above, keeping off occasional attempts of French aircraft to intervene. Great though the material effect of the bombing was, its moral consequences were even more serious. The reservists of the 55th and 71st Divisions were terribly shaken by the repeated attacks, and when Guderian's artillery joined in the deluge of fire was almost insupportable.

At 3 p.m. the barrage lifted to engage deeper targets and the infantry scrambled forward to the river. On Guderian's left, 10th Panzer crossed between Balan and Wadelincourt. Not all the defenders had been killed or demoralised by the bombardment and many of the rubber assault boats carried across the water-meadows were destroyed by machine-gun fire. Nevertheless, thanks to aggressive junior leadership the

Germans wrested a handhold on the south bank. In the centre, 1st Panzer crossed by the cloth factory at Gaulier, on the north-western edge of Sedan, with the *Grossdeutschland* leading the way and 1st Rifle Regiment following through to secure a large part of the heights of La Marfée. 2nd Panzer crossed at Donchery, about one kilometre downstream from the blown bridge. The first assault pioneers to make the attempt were hit in midstream, but eventually tank fire from the north bank helped silence the bunker opposite and by nightfall 2nd Panzer too had its bridgehead.

Many of the infantrymen holding the river line had recovered from the bombardment to fight well, but there was growing disorganisation behind the lines. Gunners abandoned their pieces and streamed off to the rear, and the confusion made it impossible for Grandsard to launch his armoured reserve in a counter-attack. This opportunity fast evaporated as German engineers threw bridges over the Meuse and the first tanks moved into the bridgehead. When the French did manage to counter-attack, on the 14th, they collided with elements of 1st Panzer on the D977 at Chéhéry. The French initially did well, but the Germans soon broke loose, chewing up the French armour and going on to ravage the unprotected infantry. By nightfall Grandsard's corps had ceased to exist as a fighting formation: the intact 3rd North African Division was transferred to another corps.

Help was on the way. General Jean Flavigny's newly formed XXI Corps, with two good divisions, 3rd Armoured and 3rd Motorised Infantry, was moving up to glimpse a golden opportunity. On the 14th Guderian decided, not without misgivings, to wheel right and break out across the Ardennes Canal, heading for Rethel. His left flank was vulnerable to a French thrust and he entrusted its defence to the *Grossdeutschland* Regiment. The Regiment headed for Stonne, on a long wooded east-west ridge with the steep-sided Butte de Stonne at its eastern end. If Flavigny was to have any chance of success it lay in acting quickly. In obedience to the cumbersome doctrine of the day, he decreed a methodical operation, which resulted in surprise and concentration being sacrificed. *Grossdeutschland* seized the initiative by attacking Stonne and although the French fought hard to break through, they could not do so, and late on the 15th the first of Wietersheim's motorised infantry arrived to support *Grossdeutschland.* The battle was won. The fighting at Stonne is remembered with increasing pride as, with the passage of time, the dark days of 1940 are no longer expunged from the national memory. When I first visited the village there was simply a cross up on the *butte.* Now there is a post-war French tank at its foot and memorials – as well as a *Char B* – in the village itself.

The other bridgeheads had also opened out, with armour boring through the defensive crust to plunge deep into Corap's vitals. On the 15th Gamelin replaced him with the more dynamic Giraud, but it was to little avail: the Panzer corridor opened up across northern France, building up a momentum that was to take it to the Channel coast on 20 May.

There were some delays. On the 17th Charles de Gaulle lunged out for the road junction at Montcornet, between Rheims and Vervins, with his embryo 4th Armoured Division. He advanced up the Laon–Montcornet road and succeeded in rocking 1st Panzer Division before being forced to withdraw. Elsewhere, however, it was the same depressing story of refugees and retreating troops jamming roads under the constant scourge of air attack.

The heartland of Champagne remained intact until the second phase of the battle of France. After the fall of Dunkirk the Germans redeployed for 'Case Red', the drive southwards. The left flank of this advance, from Laon to Montmédy, was the responsibility of Rundstedt's Army Group A, now including an armoured group of two panzer corps under Guderian's command. Bock's Army Group B, on the right, began its advance on 5 June and Rundstedt moved off on the 9th. There was heavy fighting as French troops defended the Aisne crossings against the infantry who were to secure bridgeheads for the armour and even when Guderian's tanks broke out on the 10th they did not have things their own way: the remnants of 3rd Armoured

Division, beaten at Stonne, went out with a blaze of glory at Juniville, on the little Retorne south of Rethel. Thereafter the advance gathered momentum across the excellent tank country of dry Champagne. Rheims fell on the 11th, Châlons on the 12th and only four days later the Germans were in Dijon, with Champagne inert behind them.

The last act of the drama was played out just outside the borders of Champagne, in a glade in the Forest of Compiègne near Rethondes. During the First World War railway spurs, built for the movement of heavy guns, met in the clearing. On 7 November 1918 Foch's train arrived there and on the following morning it was joined by the train carrying the German delegation tasked with negotiating an armistice. The agreement was signed just after 5 a.m. on 11 November and came into force at 11 a.m. the same morning. On 21 June 1940 Huntziger, head of the French armistice delegation, found himself taken to the same clearing by car after a tiring journey from the Loire. The *wagon-lit* in which the 1918 armistice had been signed had been taken out of its museum and moved to exactly the same spot.

Hitler was waiting with his service chiefs and, with a glare at the monument to the 1918 armistice with its inscription celebrating the end of 'the criminal pride of the German people', he strode into the carriage. After receiving the armistice terms, Huntziger returned to Paris where he relayed them over the telephone to Weygand at Bordeaux. Weygand had been Foch's aide at the 1918 armistice: the irony could scarcely have been more complete. The armistice was signed at Rethondes at 8.50 p.m. on the evening of 22 June, coming into force three days later. Hitler ordered that the monuments on the site should be destroyed – though he spared the statue of Foch. The *wagon-lit* was taken to Germany, where it was eventually destroyed in an RAF bombing raid. The armistice duly signed, Hitler set off on a tour of First World War battlefields.

The armistice site is reached by the D546 from the pleasant town of Compiègne, whose palace, largely built between 1780 and 1789, was a favourite residence of Napoleon III. The area has been restored and a slab marks the spot where Foch's carriage stood in 1918. Although Foch's *wagon-lit* was probably destroyed by a bombing raid in 1945, a similar coach, containing original fittings, stands in a nearby shed. This quiet glade in a beautiful forest is no bad place to end our journey.

VIII

Normandy

NORMANDY IS SLICED neatly in two. Upper Normandy, running from Le Mans in the south, across the Seine, through the city of Rouen to the borders of Picardy, lies on cretaceous limestone. A narrow band of jurassic limestone stretches down from Caen, through Falaise and on to Alençon, separating Upper Normandy from the granite and crystalline rock of Lower Normandy. Vegetation and agriculture march in step with geology. The traveller leaves the forests and rich fields of Upper Normandy for the naked ridges round Caen, before reaching the *bocage* of Lower Normandy with its orchards and cattle. Norman food is as lush as the countryside. Cream permeates the cooking and cheeses are unctuous: apples find their way into sauces and tarts, and provide raw material for cider and calvados (available in various grades from firewater to nectar) alike.

Normandy's military architecture reflects its history. There are abundant medieval castles, and a mass of Second World War steel and concrete. Its battlefield sites are part of a major 'Invasion coast' industry, with abundant signposts for the main attractions. Because it is so close to Paris by train or car, Normandy is also popular with French holidaymakers and can be a scrum in the summer months.

Normandy, like England, was subjected to a wave of Viking invasions, and in 911 the Viking leader Rollo secured a large tract of territory from Charles the Simple of France by agreeing to convert to Christianity and to become Charles's vassal. This agreement, the Treaty of Ste-Claire-sur-Epte, gave birth to the Duchy of Normandy and Rollo's

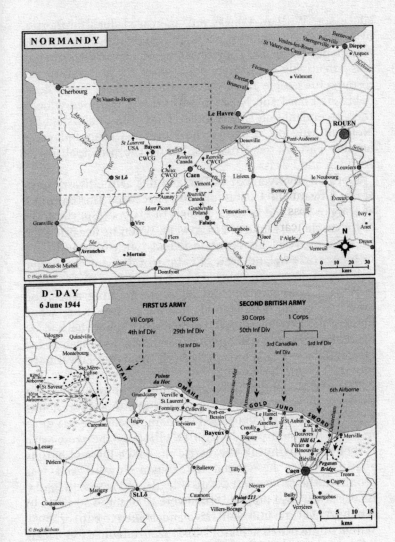

NORMANDY

Dieppe
Berneval
Pourville
Varengeville
Veules-les-Roses
St Valery-en-Caux
Anjecs
Fécamp
Etretat
Bruneval
Valmont
Cherbourg
St Vaast-la-Hogue
Le Havre
Seine Estuary
ROUEN
St Laurent USA
Bayeux
CWCG
Reviers Canada
Ranville CWCG
Colombelles
Deauville
Pont-Audemer
Louviers
Montdev
Douve
Taute
St Lô
Cheux CWCG
Seulles
Caen
Vire
Aure
Odon
Orne
Vimont
Breville Canada
Grainville Poland
Lisieux
Bernay
le Neubourg
Évreux
Ivry
Anet
Aunay
Mont Picon
Falaise
Vimoutiers
Chambois
Gacé
l'Aigle
Verneuil
Dreux
Granville
See
Avranches
Mortain
Vire
Flers
Sélune
Séc
N
Mont-St Michel
Domfront
Sées
0 10 20 30
kms
© Hugh Bicheno

D-DAY
6 June 1944

FIRST US ARMY   **SECOND BRITISH ARMY**

VII Corps   V Corps   30 Corps   1 Corps
4th Inf Div   29th Inf Div   50th Inf Div
                1st Inf Div   3rd Canadian 3rd Inf Div
                                  Inf Div

Valognes
Quinéville
Montebourg
82nd Airborne
Ste.Mère-Eglise
St Sauveur
101st Airborne
Pointe du Hoc
Grandcamp
Verville
St Laurent
UTAH
OMAHA
Arromanches
GOLD
JUNO
SWORD
6th Airborne
Formigny
Colleville
Port-en-Bessin
Le Hamel
Asnelles
St Aubin
Luc
Lion
Ouistreham
Merville
Tréviéres
Creully
Courseulles
Périer
Douvres
Hill 61
Isigny
Carentan
Bayeux
Esquay
Bénouville
Biéville
Pegasus Bridge
Lessay
Périers
Ballercy
Tilly
Noyers
Bully
Bourgebus
Vérrières
Troarn
Cagny
Caen
Mariguy
Caumont
Point 213
Villers-Bocage
St.Lô
Coutances
0 5 10 15
kms
© Hugh Bicheno

successor, William Longsword, increased ducal authority, bringing the Cotentin peninsula and the area around Avranches under his control. Eleventh-century Normandy was effectively an independent state, its prosperous and resourceful people ruled by ambitious dukes. The Viking habit died hard and Norman adventurers travelled widely: the descendants of Tancrede de Hauteville established the Norman kingdom of Sicily and Norman warriors played a notable part in the Crusades, ruling the principalities of Antioch and Edessa.

Norman involvement in England lies at the heart of our story. Edward the Confessor, who succeeded Harthacanute as King of England in 1042, had spent his youth at the court of his uncle the Duke of Normandy. When he became King he took many of his Norman friends with him and their influence in England grew, to the irritation of the native nobility. A power struggle led to the triumph of Earl Godwin of Wessex, who became chief adviser to the King, a position inherited by his son Harold. Harold became King when Edward died childless early in 1066. Although he had no royal blood, Harold was a natural candidate for kingship, and his candidacy was approved by both the dying Edward and the royal council, the witan.

There was another strong candidate for the throne of England. In 1027 a tanner's daughter in Falaise bore Duke Robert III a bastard son called William. Recognised as duke in 1035, William put down a rebellion in the west with French help, winning the battle of Val-ès-Dunes in June 1047. In 1051 he visited his cousin Edward and was promised the crown on Edward's death, and in 1064 Harold, in an oath whose precise terms were to become a matter of much conjecture, may well have undertaken to support him. When he heard that Harold had been crowned, violating Edward's pledge and arguably breaking the terms of the 1064 oath, William began preparations for the invasion of England, and on 14 October he defeated and killed Harold near Hastings, bringing Norman rule to England.

Of William's three sons, Robert Curthose ruled Normandy from 1087 to 1106, William Rufus ruled England from 1087 to 1100, and Henry Beauclerk, the youngest, was both King of England and Duke of Normandy. The sons of the Conqueror left no male heirs. Henry's daughter Matilda married Geoffrey Plantagenet, Count of Anjou, and their son Henry became the first Angevin King of England in 1154. Henry's marriage to Eleanor of Aquitaine brought him huge tracts of land in southern France, taking the Angevin Empire, at its height, from Cape Wrath to the Pyrenees. Henry, it was said, 'could rule any house but his own', and his sons Henry, Richard, Geoffrey and John had begun to divide the empire even before his death. Richard succeeded him as King of England, but spent only a few months in the country, devoting his later years to conducting an intermittent war against Philip Augustus of France. John, Richard's successor as King of England, was outmanoeuvred by the wily Philip, who took control of Normandy in 1204 – the duchy was 'reunited' to the French crown – and went on to make inroads into John's other possessions in France, leaving him holding Guienne.

Richard's castle of Château-Gaillard, blocking the route from Rouen down the Seine valley at Les Andelys, was built at tremendous speed in 1197–8. The castle became the king's favourite residence in the last years of his life and it is easy to see why. Richard jested that he could hold the castle even if its walls were made of butter and it made the best possible use of a bluff overlooking the Seine, its V-shaped outer bailey commanding the bluff's southern end, and a middle bailey separated by a moat from the inner bailey encircling a mighty keep.

At first Philip Augustus hoped to starve the castle into surrender and surrounded it with a double ditch in late 1203. In February 1204 he concluded that the process would simply take too long and decided to attack. He first dealt with the outer bailey, using a mine to bring down part of the tower at its salient, then turned his attention to the middle bailey. A handful of Frenchmen made their way in through the latrine chutes and managed to lower the drawbridge linking the middle and outer baileys, enabling the attackers to rush in. From the middle bailey they were able to bring

siege engines to bear on the inner bailey. Once the walls
were breached the garrison surrendered after a creditable
defence, opening the Seine approach to Normandy: Rouen
fell three months later.

Friction between the kings of England and France over
Guienne led to the outbreak of the Hundred Years War in
1337. On 12 July 1346 Edward III landed at St-Vaast-la-
Hougue with some 12,000 men. La Hougue, on the north-
eastern coast of the Cotentin, is a fine natural harbour, now
a fishing port with famous oyster beds. On 19–20 May 1692
an Anglo-Dutch fleet defeated the French under Tourville
off La Hougue, and the fortifications which now surround
the harbour – including the picturesque Fort de L'Ilet lying
offshore – were built by Vauban shortly afterwards. In 1346
Edward was able to land unopposed at La Hougue because a
local nobleman, Geoffrey de Harcourt, had thrown in his lot
with the English and in any case the French were preoccupied with campaigning in Guienne.

It took several days for Edward's force to complete
disembarking, and during this time detachments raided
Barfleur and Cherbourg. Several young men were knighted,
amongst them the Black Prince, and on 18 July the army set
off for Rouen with the intention of joining Edward's Flemish
allies in the north. Edward marched by way of Valognes,
Carentan and St-Lô. He reached Caen on 25 July. The capital
of Lower Normandy, Caen boasts a sturdy keep begun by
William the Conqueror in 1060 and improved by Henry I.
The Conqueror, who died at Rouen in 1087, was buried in
the Church of St-Etienne in the *Abbaye aux Hommes*. His
bones were scattered when the Huguenots sacked the
church in the sixteenth century, disturbed again by Second
World War bombing and all that remains of him is a femur,
which lies beneath a stone in front of the altar. Although
Edward's men respected the Conqueror's tomb they were
less kind to the rest of Caen. The town was methodically
pillaged and the plunder was shipped off down the Orne to
Edward's fleet.

King Philip had reacted to the English incursion with
considerable speed, recalling his son from Guienne and

concentrating at Rouen, capital of Upper Normandy. We have already seen, in the first chapter dealing with Flanders, Artois and Picardy, what happened next: after marching and countermarching on the line of the Somme, Edward crossed the river at Blanchetaque, won the battle of Crécy and went on to reach Calais.

The Duchy of Normandy loomed large in the political and military events of the 1350s. The Dauphin Charles had been created Duke of Normandy, but entered into treasonable discussions with his brother-in-law Charles of Navarre. Charles, a descendant of the Capetians, resented the Valois succession to the throne and intrigued with England in an effort to improve his position, gaining fiefs in Normandy by playing off King John off against the English. In April 1356 John surprised his son and son-in-law at Rouen, apparently in the act of plotting, and Charles of Navarre was imprisoned.

John's bold stroke aroused a storm of anti-Valois feeling in Normandy. Charles's brother Philip asked for English assistance, and the Duke of Lancaster landed at La Hougue and marched as far as Verneuil, only to retreat when John's army approached. The French King spent some time dealing with dissident nobles in Upper Normandy before marching off to Guienne to face the Black Prince. The campaign ended in disaster, for when the armies met at Poitiers on 20 September the French were utterly defeated and John himself was captured. While an English prisoner, he concluded the Treaty of London with Edward, making sweeping territorial concessions, which the Dauphin felt unable to honour. In the autumn of 1359 the English mounted a destructive raid across Artois and Champagne, inducing Charles, who had succeeded as Charles V on his father's death, to come to terms at Bretigny in May 1360.

During the nine years that elapsed between Bretigny and the resumption of hostilities Charles prepared for a renewal of the conflict and when war broke out again the new French policy of caution, exemplified by the tactics of Bertrand Duguesclin, paid dividends, paving the way for the Treaty of Paris, signed in 1396. When war was resumed in 1415 the English invasion, led this time by the young Henry V, followed the familiar pattern of a landing in Normandy. It came as no surprise to the French and the Constable, Charles d'Albret, concentrated at Rouen, while Marshal Boucicaut covered the coast around Honfleur. On 14 August Henry's fleet appeared off the north shore of the Somme estuary and the King immediately set about besieging the port of Harfleur.

The town of Le Havre, now the major port on the Somme estuary, was not founded till 1517, on the orders of Francis I. Until then Harfleur enjoyed pride of place. It is now a suburb of Le Havre, where the N15 enters the town, and is cut off from the sea by the reclamation work, which provided dry land for the huge industrial and petrochemical complex south of the Tancarville Canal. Some traces of the old town remain, including St Martin's church with its lofty clock tower. In 1415 Harfleur was strongly fortified, its wall buttressed by numerous towers and protected by a moat. Its energetic governor, the Sieur d'Estouteville, had dammed the Lezarde to flood the surrounding countryside and reinforced the gates with wooden barbicans.

Henry's force comprised some 2,000 men-at-arms, 8,000 archers and sixty-five gunners. The latter manned up to twelve cannon, the three heaviest called 'London', 'Messenger' and 'The King's Daughter'. The town held out despite heavy bombardment, and the besiegers suffered badly from dysentery in their cramped and unhealthy siege lines. On 17 September they took a barbican covering the south-west gate and a general assault was imminent when Harfleur surrendered on the 22nd. Henry had initially intended to march on Paris, but the siege had consumed time and resources. His council of war recommended garrisoning Harfleur and returning to England with the main body of the army, but Henry decided on a raid across Normandy and up to Calais. The venture took him across the Somme on 19 October and on the 25th he beat the French at Agincourt.

In 1417 Henry led another expedition to Normandy, taking Caen and Falaise. Over the next two years the English tightened their hold on the dukedom, with the Duke of

Gloucester subduing the Cotentin, the Earl of Huntingdon taking Coutances and Avranches, and the Duke of Clarence clearing the Seine as far as Rouen. The siege of Rouen was long and painful, and the town was eventually starved into submission on 20 July 1418. Alain Blanchard, who had inspired its defence, was hanged, having declined to furnish a ransom, saying: 'I have no goods, but even if I did I would not use them to stop an Englishman from dishonouring himself.' By the end of February 1419 most of Normandy was in English hands and of the surviving castles all save Mont St-Michel were eventually captured.

The granite outcrop of Mont St-Michel, defended by Louis d'Estouteville, lies just off the coast in the Couesnon estuary. Shifting sandbanks made it difficult for ships to attack and tides rendered conventional siege works impossible. A church had been built atop the rock in 1017–1144 and in the early thirteenth century the Gothic Merveille buildings were added. Subsequent work has produced the masterpiece that now crowns the mount, with its stylish spire, finished in 1897, pointing skywards. The medieval defences are intact. The Outer Gate, at the top of the ramp and steps leading up from the car park, opens into a fortified courtyard. A sixteenth-century guardroom contains a group of English bombards captured during the Hundred Years War. A walk round the ramparts takes the visitor through a series of defences, including drum towers and the fine arrowhead Boucle Tower Bastion. The mount's history is admirably covered in the Maritime Museum and the multimedia Archeoscope.

Normandy remained in English hands until the close of the Hundred Years War. Joan of Arc, whose example had helped inspire her countrymen, was captured by the Burgundians and sold to the English. In early 1431 she was tried for heresy and sorcery by an ecclesiastical court presided over by Pierre Cauchon, Bishop of Beauvais. Condemned to death, she made a wild recantation when taken out to be burnt, but subsequently abjured and was burnt at the stake on 30 May in the Place du Vieux Marché in Rouen. The site of her martyrdom is marked by the Cross of Rehabilitation and St Joan of Arc's church stands across the square. Another link between Rouen and the Hundred Years War is the Gothic Cathedral of Notre-Dame, where John, Duke of Bedford, third son of Henry IV and brother of Henry V, was buried in 1435. Bedford was a seasoned commander whose most noteworthy accomplishment was the defeat of a vastly superior French and Scottish force at Verneuil on the borders of Normandy on 17 August 1424. Verneuil was a repetition of Agincourt on a smaller scale. Bedford's 9,000 men took up a position blocking the French advance from Orléans and the 10,000 or so French under the Earl of Buchan, supported by 5,000 Scots under his father-in-law the Earl of Douglas, duly launched a frontal attack. The Duke of Alençon was more subtle and assailed Bedford's flank, but the attack failed and he was captured. Buchan and Douglas were both killed, and total French casualties may have been as high as 7,000.

The Duke died in time to be spared hard news. The Burgundians came to terms with the French in 1435 and Charles entered Paris the following year. In 1337–41 Lord Talbot fought a brilliant series of campaigns, which averted the sudden collapse of the English position in Normandy, but England's hold over the duchy was unravelled in 1449–50. There was probably less local enthusiasm for the French than some sources suggest, for the duchy had been in English hands for a generation. However, this was small reason for last-ditch defence, and the powerful French siege train under the competent Jean Bureau made up many a castellan's mind for him. Rouen fell in October 1449, Harfleur the next month and Honfleur in January 1540. Soon the Duke of Somerset, English commander in France, found himself threatened in Caen, and Sir Thomas Kyriel and Sir Matthew Gough marched to his relief with an army made up of detachments from garrisons across the whole of Normandy, reinforced by some 2,500 men sent across from England. Kyriel and Gough took Valognes and crossed the Douve and Vire. They were marching along the Carentan–Bayeux road when they encountered the Count of Clermont with perhaps 3,000 men and two small guns near the village of Formigny. The English force, some 4,500 strong, consisted of archers and billmen with a few hundred men-at-arms, and Kyriel and Gough decided to fall back on the brook still crossed by the road just west of Formigny, where their archers would give a good account of themselves. They posted the archers in two bodies forward of the brook, with a force of billmen drawn back to the stream on either flank and the few horsemen on the wings.

Clermont approached with great caution, avoiding the once-obvious temptation of frontal assault and instead feeling for the flanks. Inconclusive skirmishing had gone on for about three hours when the French brought up their two guns to engage the English line. The archers surged forward and took the guns, but while they were hauling them off they were attacked and driven back by dismounted French men-at-arms. Unable to break contact and open a clear field of fire, the archers were rolled back towards the brook and soon both armies were locked in battle along it. At this moment the Counts of Richemont and Laval, coming up from St-Lô with 1,200 men, appeared in the English rear. There was no reserve available to deal with them, and the English force was soon split into fragments and over-whelmed. Gough got away with the cavalry and Kyriel was captured, but most of the archers and billmen perished at the hands of Frenchmen who had little inclination to offer quarter. There is a florid 1903 memorial to the battle at the crossroads by the Mairie and a much more dignified column, erected in 1834, about two kilometres further west down the old main road, closer to the site of the battle. Formigny settled the fate of Normandy: Caen fell in June and on 11 August Cherbourg, bravely held by Thomas Gower, at last opened its gates.

The castles of the Hundred Years War still guard Normandy. Gisors is the capital of the Norman Vexin, that wedge of ground nestling between Seine, Andelle and Epte south-east of Rouen. Its castle was built in 1097 by the Conqueror's son William Rufus and improved by Henry II of England. Philip Augustus took the place in 1183 and it changed hands several times during the Hundred Years War, before becoming definitively French in 1449. The castle walls lour down on the town and are themselves over-shadowed by a keep, which sits atop an artificial mound twenty metres high. Robert the Devil's castle stands above the great loop of the Seine that swoops down south-west of Rouen. The original castle of the Dukes of Normandy was destroyed by Richard I's brother John, rebuilt by Philip Augustus and slighted by the Rouennais to prevent its capture by the English. It now houses a Viking museum with a reconstructed longship as its centrepiece.

Domfront holds the sandstone ridge above the River Varenne on the frontier between Normandy and Mayenne where the roads from Alençon, Laval and Avranches meet. The town coalesced around the castle built by William Count of Alençon, which passed to Henry Beauclerk in 1092. Henry II of England and Eleanor of Aquitaine stayed there frequently, and in 1170 the castle was the scene of an attempt to reconcile Henry with Thomas à Becket, Archbishop of Canterbury. Taken by the French, Domfront was recaptured by the English in 1356, relinquished ten years later and taken again, finally passing into French hands in 1450.

Domfront was soon under siege again. In 1559 Gabriel de Montgomery, captain in the King of France's Scottish Guard, mortally wounded Henry II in a joust in Paris when his lance pierced the King's visor. In 1574 Montgomery held Domfront for the Huguenots against a Catholic army under the Count of Matignon. Montgomery surrendered on terms, but Catherine de' Medici, Henry II's widow, ordered his execution. The castle was demolished in 1592 but sub-stantial traces of the keep remain in an attractive public garden giving a spectacular view over the valley. A large part of the old town walls survive, surrounding a number of attractive timber-framed houses.

When the French took Domfront in 1450 the Hundred Years War was all but over: its last battle, at Castillon in the Bordelais, took place only four years later. In 1469 Charles, the last Duke of Normandy, was dispossessed and Normandy became a French province, with the Exchequer at Rouen, housed in the fine Renaissance Palais de Justice, becoming its *Parlement* in 1514. As friction between Catholics and Huguenots deepened in the middle of the sixteenth century, Normandy grew increasingly disturbed, containing as it did some of the major Huguenot centres outside the strongly Protestant south-west.

In the First War of Religion, Anthony of Navarre, brother of the Huguenot leader Condé, was mortally wounded when the Catholics took Rouen in October 1562. Two months later, on 19 December, there was a pitched battle outside Dreux, on the fringes of Normandy between Chartres and Evreux. It was typical of the scrambling cavalry actions of its day. The Constable Montmorency charged the Huguenots, only to be beaten and captured. Condé pursued with his customary panache, but failed to keep his men in hand, was deftly counter-attacked by the Duke of Guise and was himself captured.

The next major battle in Normandy was not till 1589, when Henry of Navarre, who had recently succeeded to the throne as Henry IV, was fiercely opposed by the Catholic League. The new King, with only 7,000 men, was outnumbered by more than 25,000 commanded by the Duke of Mayenne. However, Henry threw up field fortifications on the boggy valley of the River Béthune, and when the Leaguers attacked on 21 September 1589 the ground prevented them from bringing their full weight to bear and Mayenne drew off discomfited. The battle took place in what is now Arques Forest, but its site, marked by a monument erected by the restored monarchy, can be reached on foot after a short drive out from Arques along the D56, turning left on to a small *route forestière*, which climbs up into the forest. The site gives a good view across the valley of the Béthune to Arques – its prominent castle was held by Henry's men in 1589 and cannon in it opened great lanes through Mayenne's ranks as he attempted to deal with Henry's army in the valley. The castle itself is an interesting mix of military architecture, with a medieval keep taken by the Conqueror in 1053 and rebuilt by his son Henry Beauclerk in 1123. It was strengthened by the addition of new towers in the fourteenth century and these were modified to take artillery in the sixteenth. Although the castle itself is currently closed, its site can be visited.

The adversaries met again on 14 March 1590 at Ivry, on the Eure near Dreux. Again the odds – 16,000 against 11,000 – favoured Mayenne, but again Henry's tactics were superb. He led the decisive cavalry charge that broke Mayenne's centre, but also deserves credit for massing his arquebusiers so as to produce concentrated infantry fire and in posting his cannon with great care. The League lost some 4,000 men to Henry's 500 in this, the decisive battle of the Wars of Religion. An obelisk erected on Napoleon's orders in 1804 stands on the battlefield north-west of Ivry, reached by the D833 and the smaller D163. It is hidden at the end of an avenue of mature trees and is easy enough to miss. Ivry itself – now, like Arques, bearing the title 'la Bataille' after its name, retains several attractive half-timbered buildings: Henry stayed in one at No. 4 rue de Garennes. There is another royal connection in St Martin's church: it was founded by Diane de Poitiers, the influential mistress of Henry II of France. Diane's château is at Anet, a few kilometres south-east of Ivry. Only the entrance front, part of the *cour d'honneur* and the chapel now remain, but they testify to Diane's artistic taste.

With the end of the Wars of Religion the fighting that had marked so much of Norman history was followed by a long period of peace. In the 1790s a combination of Catholic sentiment and resistance to the conscription law of 1793 produced a counter-revolutionary outburst by the royalist irregulars known as *chouans*, but the insurrection was neither as strong nor as prolonged as the similar disturbance in the Vendée. There was no fighting in the area during the Napoleonic wars, although the sight of British blockading warships – and occasional landing parties – reminded Norman seafarers that their country was at war.

There was war of another sort in 1864. The Confederate commerce raider CSS *Alabama*, which had enjoyed a long run of success under its captain, Raphael Semmes – it had sunk sixty-eight merchant vessels and the gunboat USS

Hatteras – put in at Cherbourg after a long cruise. The USS *Kearsage* waited just outside the harbour, and Semmes came out to fight on 19 June. *Alabama* went down after a hard-fought ship-to-ship action, but Semmes and many of his crew were rescued by French vessels. The episode is the subject of a sea shanty and often, when I near Cherbourg harbour in the ferry, a snatch of the song runs through my mind:

> Off the three-mile limit as the sun went down
> Roll, Alabama, roll . . .

Although later stages of the Franco-Prussian War saw Chanzy's threadbare army at Le Mans on the borders of Normandy, the war ended before Normandy shared the miseries of occupation with the north-east. Normandy was again spared during the First World War, though her harbours were busy, especially Le Havre, which served as the main British port of entry. Most British soldiers who served on the Western Front landed there and went thence by train to Rouen, then on to the great base area around Etaples ('Eat Apples') on the River Cance just inland from Le Touquet. The picture darkened dramatically in the Second World War. We have already seen how the German advance over the Somme in June 1940 brought fighting to the province, with considerable damage inflicted on many of its towns. When the armistice was signed the whole of Normandy came within the Occupied Zone of France.

Many of the defences built by the Germans to protect the Normandy coast against assault from the sea still remain. Most were constructed after 1942, and with good reason, for it was not until after that date that the possibility of invasion was taken very seriously. The raid on Dieppe, launched in August 1942, was of great importance, not only in encouraging the Germans to strengthen their coast defences, but also in showing the Allies what not to do when the time came to invade Europe on a large scale.

Dieppe has long enjoyed a close, if not entirely friendly, relationship with England. The Arques estuary makes a fine natural port. It is sheltered by cliffs topped with the church of Notre-Dame de Bon-Secours to the east and a long shingly promontory, now bearing the parallel Boulevards du Maréchal Foch and de Verdun and most of the town's hotels, to the west. Corsairs and privateers operated from the harbour against British and Portuguese shipping, and the elegant manor house built in nearby Varengeville by Jean Ango, who owned a fleet that ravaged Portuguese commerce in the early sixteenth century, can still be seen. The Florentine Verazzano set off from Dieppe in 1524 and discovered the site of New York, naming it 'Terre d'Angoulême'. He is remembered in New York's Verazzano Narrows Bridge.

Dieppe was besieged by Lord Talbot during the Hundred Years War and bombarded by a British fleet in 1694. After the defeat of Napoleon regular ferry services to Newhaven were established and an English quarter developed. Some of its inhabitants had left England because of debt or disrepute: Oscar Wilde lived there after his release from prison. Dieppe was the European centre of ivory-carving and the town's museum, housed in the fifteenth-century castle at the end of the Boulevard de Verdun, contains a large selection of carved ivories.

Winston Churchill knew Dieppe well, for his mother-in-law, Lady Blanche Hozier, had lived there. However, this appears to have had no influence on the fact that Dieppe was selected by Combined Operations, headed by the energetic Lord Louis Mountbatten, as the objective of a large-scale cross-Channel raid in 1942. The task was entrusted to Major-General J. H. Roberts's 2nd Canadian Division, whose lively behaviour had made it an object of some concern to the inhabitants of the Sussex towns and villages near which it was camped. The plan for what was initially code-named Operation *Rutter* called for infantry landings on the flanks of Dieppe, followed by the landing of an armoured regiment and more infantry in the town. The navy was unwilling to risk anything heavier than destroyers with 4-inch guns in the Channel, and it was eventually decided not to bomb Dieppe as the rubble would make it difficult for tanks to move through the streets.

Most of the troops earmarked for the raid – the Canadian 4th and 6th Brigades – did their work-up training on the Isle of Wight and rehearsed by landing on beaches near Bridport in Dorset. However, with the force embarked ready for departure, a combination of bad weather and a German air attack on converted ferries in Yarmouth roads led to the cancellation of *Rutter*. It was shortly resuscitated, as Operation *Jubilee*, partly because Churchill was under pressure from Stalin to open a Second Front.

The new plan entrusted the flank attacks to seaborne Commando units. To the east, 3 Commando was to land on beaches Yellow 1 and Yellow 2 at Berneval and Belleville to destroy the battery north-west of Berneval-le-Grand. To the west, 4 Commando was to land on Orange 1 at Vasterival and Orange 2 at St-Marguerite, destroying the battery on the main road just west of Varengeville. The Royal Regiment of Canada was to land east of Dieppe at Puys (Blue Beach), take a small battery on the clifftop and secure the eastern headland overlooking the harbour. The long stony beach in Dieppe itself was divided into Red and White Beaches, to be secured by The Royal Hamilton Light Infantry and The Essex Scottish before the 14th Canadian Army Tank Regiment (The Calgary Tanks) came ashore. The French-Canadian *Fusiliers Mont-Royal* formed a floating reserve and Royal Marine Commandos were on hand to seize German invasion craft, which were to be towed back to England. Finally, the South Saskatchewan Regiment, followed by the Cameron Highlanders of Canada, would land at Pourville (Green Beach), whence they would advance on the airfield on the D15 south of Dieppe and what was believed to be a divisional headquarters at Arques-la-Bataille. The attacks on Yellow, Blue, Green and Orange Beaches were to go in just before dawn, and the remaining assaults would follow half an hour later, when there was enough light for warships to deliver supporting fire.

The defenders of Dieppe came from the 302nd Infantry Division, thinly spread from Veules-les-Roses to the Somme estuary. A network of pillboxes covered the barbed wire protecting the seafront in Dieppe, and the caves on the eastern and western headlands housed artillery and machine-guns. The batteries at Berneval and Varengeville were the responsibility of the navy, and there had been repeated suggestions that the Varengeville battery was badly sited.

Canadian bad luck began early on the night of 18–19 August, for some of the defenders carried out their weekly defence alarm drill that night and were ready for the attack when it came. Worse still, German E-boats fortuitously bumped 3 Commando's convoy just offshore, causing casualties and compromising surprise. Nevertheless, Major Peter Young led a small party ashore on Yellow 2 and another detachment landed on Yellow 1. The Berneval battery had by now opened fire, but Young's men took it on and briefly silenced it. Assailed by German reinforcements they made their way back to the beach, where they were re-embarked and drew off under a hail of fire. The Commandos who had landed on Yellow I were less lucky and all were killed or captured apart from one who managed to swim out to a passing ship.

On the western flank, Lord Lovat's 4 Commando landed successfully, Major Derek Mills-Roberts's detachment at a little inlet west of Varengeville and Lovat himself near the mouth of the little Saane. A mortar bomb from Mills-Roberts's party put the battery out of action and Lovat's men overran it in a sharp hand-to-hand battle in which Captain Pat Porteous won a Victoria Cross. The guns were destroyed and the Commando's survivors were taken off by sea.

The Royal Regiment of Canada strove to land at Puys, between 3 Commando's attack and Dieppe itself. The tiny beach is dominated by cliffs and houses, and in 1942 its exits were blocked by wire and covered by pillboxes. Although some of the attackers managed to get ashore, they could make little progress in the face of murderous fire and those left alive surrendered at about 8.30 a.m.; the battalion had effectively ceased to exist, losing twenty-four officers and 465 men. A monument overlooking the beach, its foundation stone laid by the prime minister of Canada in 1946, pays eloquent tribute to its sacrifice.

The assault on Pourville's Green Beach by the South Saskatchewans was altogether more successful. The battalion poured ashore from its landing craft exactly on time and although some of the attackers arrived on the wrong side of the River Scie, surprise was total. Pourville and the high ground to its west were soon secured, but it was more difficult to exploit to the west because the bridge over the Scie was swept by fire. Lieutenant-Colonel Cecil Merritt, the South Saskatchewans' commanding officer, won the Victoria Cross for inspirational leadership in getting his men across the bridge and the battalion had secured a useful bridgehead by the time the Cameron Highlanders came ashore. The Camerons lost their commanding officer as he landed and they were soon fighting hard to evict determined Germans from the eastern edge of Pourville.

The South Saskatchewans were accompanied by Flight-Sergeant Jack Nissenthal, an RAF radar expert, and the special platoon escorting him tried to fight its way into the radar station east of Pourville, but failed to penetrate its defences. Some of the Camerons, under their second-in-command, pushed on up the Scie valley and crossed the main road, only to be halted by fierce fire from Quatre Vents farm. The survivors of both battalions pulled back on to the beach, expecting evacuation at 10 a.m., but owing to a misunderstanding the landing craft did not arrive till 11 a.m., by which time the Canadians had given up the high ground around the beach. The approaching landing craft were met by vicious machine-gun and mortar fire, and the evacuation was achieved under the most harrowing circumstances, and well under half those landed that morning survived to return to England.

The assault on Dieppe itself was spearheaded by the Essex Scottish, landing on Red Beach, near the harbour entrance, and the Royal Hamilton Light Infantry, attacking White Beach, opposite the casino. Hurricanes strafed the defences as the landing craft neared the beach, but little damage was done and although the two battalions reached the beach with few casualties, the Essex Scottish were stopped by intense fire on the sea wall and the Hamiltons were checked by the casino. The Calgary Tanks began to come ashore just after 5.35 a.m., the craft carrying them running the gauntlet of heavy and accurate fire. Although fifteen managed to get off the beach, they could get no further inland because engineers had not been able to deal with anti-tank obstacles blocking the exits from the Boulevard de Verdun. Another twelve were stopped on the beach, most because their tracks were jammed by the shingle. The casino was eventually taken and some of the Hamiltons struck out into the town, but the Essex Scottish remained thwarted by the sea wall. However, Major-General Roberts, commanding from his headquarters ship *Calpe*, believed that these battalions had made better progress and he committed his floating reserve, the *Fusiliers Mont-Royal*, to exploit it. The French Canadians shared the fate of their predecessors and were soon reduced to scattered groups. The commanding officer of 40 (Royal Marine) Commando, ordered to reinforce White Beach, waved his men back when he saw the task was hopeless, but was killed soon afterwards and some of his men, who had not seen his signal, landed anyway.

Withdrawal from Red and White Beaches began at 11 a.m., screened by smoke laid by Boston bombers. As the smoke cleared, the landing craft were riddled by fire from the cliffs. John Foote, the Hamiltons' chaplain, won the Victoria Cross collecting wounded on White Beach and declined evacuation to stay on with those who would inevitably become prisoners. The destroyer HMS *Berkeley* was sunk by bombing during the evacuation. Of the 5,000 Canadians who had gone into battle 3,367 became casualties; the Royal Navy lost 550 men and the Commandos 270. German losses came to a little over 600.

The Dieppe raid remains a contentious issue, but there is no doubt that valuable lessons were learnt. Defences would have to be reduced by prompt and accurate firepower, proper combined-arms training was essential and tight security was vital. The evidence of Dieppe suggested that it was unlikely that the Allies would be able to capture a harbour early on in their invasion, and this helped inspire the *Mulberry* mobile harbour. Hitler was encouraged to believe that an invasion could be smashed on the beaches and this contributed to his assertion that every inch of ground must be held – an order which was to cost German commanders dear in 1944.

Wartime improvement of coast defences and post-war demolitions make it difficult to relate the pillboxes and bunkers at Dieppe to the events of August 1942. However, the visitor who makes for the château will get a good view across the town's beaches, and there are a number of monuments to Canadians and Canadian units in the park just below the château. Embrasures staring out from the cliffs above the harbour mark the position of guns, which caused carnage on Red and White Beaches, and there are armoured observation cupolas above them. Near the golf course above Pourville was the bunker complex that housed the *Freya* radar, which was Flight-Sergeant Nissenthal's objective. All the assault beaches illustrate, all too graphically, the problem of landing on them in the face of a determined enemy: standing by the Royal Regiment's memorial at Puys, in the shadow of a pillbox, one is struck not by the fact that most of the attackers perished on the beach, but that any fought their way even a short distance inland. In Dieppe the casino, demolished after the raid, has been rebuilt on the same site, and the long shingly expanse that was once Red and White Beaches feels exposed even on the most balmy day. Most of the Canadians killed at Dieppe are buried in the Dieppe Canadian War Cemetery, near the N27's junction with the D915. Some died at sea and lie in the Brookwood Military Cemetery, north of Guildford in Surrey, and those who have no known graves are commemorated on the Brookwood Memorial.

A memorial to an earlier and more successful raid is to be found further down the coast, at Bruneval, near the Cap d'Antifer, south of Etretat. On the night of 28 February 1942 Major John Frost (later to win fresh distinction at Arnhem bridge in 1944) led a force that parachuted in to capture a *Würzburg* radar site. The set was dismantled and taken back to England when Frost's men were evacuated by sea.

Planning for the invasion of Europe was in hand even before the Canadians landed at Dieppe. The first project, code-named *Roundup*, was complete by the end of 1941: it bore little relation to the eventual plan, but was an important recognition that operations in Europe would eventually be essential. A subsequent plan, *Sledgehammer*, considered an invasion in the Pas de Calais – closest to airfields in England, but where the coast defences were strongest. General George C. Marshall, US Army Chief of Staff, produced a memorandum of his own, envisaging an assault astride the Somme with some forty-eight divisions, and the build-up of US troops in England, code-named *Bolero*, was begun. The decision to press ahead with the *Torch* landings 'swept the larder' of US troops concentrating in England, but at the Casablanca conference in January 1943 it was decided to set up a planning organisation to prepare small-scale raids, secure a bridgehead on the Continent in 1943 and mount an invasion in force in 1944.

In the spring of 1943 detailed planning was entrusted to Lieutenant-General Sir Frederick Morgan, COSSAC – Chief of Staff to the Supreme Allied Commander (designate) – and his staff. The COSSAC staff were able to profit from work done on *Roundup* and, as the US *Official History* declares, 'the most important practical experience came from the Dieppe raid in August 1942.' Other planners, whose task was now transferred to the COSSAC team, had felt that Normandy offered better prospects than the Pas de Calais, and their *Skyscraper* plan provided for landings on beaches north of Caen and on the eastern Cotentin. This, too, was to influence the COSSAC plan for Operation *Overlord*. In June 1943 Morgan's staff recommended landings in Normandy, going on to decide that the vicinity of Caen offered better prospects than Le Havre or the Cotentin. Even if Cherbourg could be captured early it would not be able to cope with the supplies required by the twenty-nine divisions to be pushed into the beachhead, and the planners therefore demanded the construction of *Mulberry* harbours, while recognising that some supplies would have to come ashore over open beaches.

The COSSAC outline plan was accepted at the Allied Quebec conference in August 1943, although there was subsequent debate over the relative importance accorded to France and Italy. In December President Roosevelt nominated General Dwight D. Eisenhower as Supreme Allied Commander for *Overlord*. Although Eisenhower had no experience as a battlefield commander, he was an accomplished staff officer and had served as Commander-in-Chief, Allied Expeditionary Force, in the landings in North Africa and was to prove an outstanding alliance manager. Air Chief Marshal Sir Arthur Tedder, who had worked closely with him in the Mediterranean, was appointed Deputy Supreme Commander. Admiral Sir Bertram Ramsay was naval Commander-in-Chief, and Air Chief Marshal Sir Trafford Leigh-Mallory commanded the Allied air forces. For the first phase of the campaign ground forces were to come under the command of General Sir Bernard Montgomery's 21st Army Group, but it was envisaged that Lieutenant-General Omar Bradley of the US 1st Army would take charge of an all-American Army Group in due course.

The defenders of Normandy were commanded by Field Marshal Gerd von Rundstedt, *Oberbefehlshaber West*, from his headquarters in Paris. They formed two army groups, Blaskowitz's G, south of the Loire, and Rommel's B to its north. Dollman's 7th Army held Brittany and Lower Normandy, while Salmuth's 15th Army watched the Channel coast from Upper Normandy to the Low Countries. Even in June 1944 the Russian front was the major German military concern, and 165 divisions were stationed in the east as opposed to fifty-nine in France and the Low Countries; armoured divisions were more evenly balanced, with eighteen in the east and fifteen elsewhere. Many of the divisions in the west were of poor quality, with obsolescent equipment and elderly soldiers. German weakness in the air was every bit as crucial and poor German air reconnaissance was to contribute to the success of Operation *Fortitude*, the Allied deception plan. This helped confuse the Germans as to the time, place and weight of attack. There was also a difference of opinion between Rommel and Rundstedt over the best tactics for meeting the invasion when it came. Rommel, with his recent experience of Allied air power, recognised that unless the Allies were stopped on the beaches they would not be stopped at all: other officers, used to fighting under the *Luftwaffe*'s umbrella, favoured the classic tactic of identifying the landings and moving armour to snuff them out. Hitler himself kept strings on Panzer Group West, the armoured reserve in the western theatre. Only a single armoured division – 21st Panzer – stood within striking distance of the beaches when the Allies landed.

For all its weakness, the German position was not impossible. A huge amount of work had been done on coastal defences since the Dieppe raid, and anti-invasion obstacles speckled beaches and likely landing fields. Many of the coast defences remain. Expansion of port and town has removed most of *Festung Le Havre*. South of Merville-Franceville-Plage, on the mouth of the Orne, is the Merville battery. This comprised three small artillery bunkers and one larger one, and was thought to house 150mm guns capable of engaging the British invasion beaches. It was attacked by Lieutenant-Colonel T. B. H. Otway's 9th Parachute Battalion on D-Day. The very complex operation had been meticulously rehearsed and although only 150 of the assault force of over 700 reached the battery, it was duly taken after a stiff battle. The attackers then found that the guns were only Skoda 100mms, without the range to affect the main battle. One of the bunkers now houses a museum with somewhat erratic opening times, but the site is well signposted and can be visited at all times.

Of the defences along the invasion beaches, the batteries at Longues-sur-Mer and the Pointe du Hoc are worth visiting despite the damage they received from bombing and shelling before the invasion. The former housed four 155mm guns, and despite having 1,500 metric tons of bombs dropped on it, engaged the invasion fleet until silenced by the fire of HMS *Ajax* and the Free French warship *Georges Leygues*: there is a lively controversy over whose fire was actually most effective. Granville, at the base of the Cotentin, has a cluster of intact bunkers up by its lighthouse.

The invasion of Europe began just after midnight on 6 June when Major John Howard's D Company, 2nd Oxfordshire and Buckinghamshire Light Infantry, arrived by glider at Pegasus Bridge over the Caen Canal and Horse Bridge over the nearby River Orne as part of the British 6th Airborne Division's descent on the Benouville–Ranville area to secure the left flank of the beachhead. US paratroopers dropped to the west, the 82nd Airborne Division around Ste-Mère Eglise and the 101st around Vierville, to secure the right flank. Both drop sites are well served by memorials and museums. Pegasus Bridge has its own museum and although the bridge itself has been replaced, the new structure looks very much like the old, which is to be found just behind the museum. The Pegasus Bridge café, owned by the wonderful Arlette Gondrée, has become a shrine for the dwindling number of veterans. The first officer across the bridge, Lieutenant Den Brotheridge, is buried in the churchyard at Ranville, next to Ranville War Cemetery, which contains over 2,000 British and fifty Canadians, most from 6th Airborne Division. There is a US Airborne Museum at Ste-Mère Eglise, a village made famous by the exploit of John Steele of 82nd Airborne's 505th Parachute Infantry Regiment. Steele landed on the church steeple and hung from it, his parachute snagged on the stonework, while the battle for the village hammered on below him: the episode figured prominently in the film *The Longest Day*.

Despite the confusion of the airborne drops, in which some unlucky soldiers fell into the sea or the flooded marshes of the Douve and Merderet, the first phase of the invasion was generally successful, with surprise being achieved and the flanks of the sea landings protected. The assault from the sea comprised two distinct thrusts. Dempsey's British 2nd Army would land between Arromanches and Ouistreham on Gold, Juno and Sword Beaches. Bradley's US 1st Army would go ashore on Omaha Beach, north of Colleville and St-Laurent, and Utah Beach, across the Vire estuary around La Madeleine.

The American landings enjoyed mixed fortunes. Accurate naval and air bombardment so hammered the defenders of

Utah Beach that they were able to offer little resistance to the US 4th Infantry Division, which hit the beach, about one kilometre south of the intended landing area, at 6.30 a.m., well supported by twenty-eight amphibious Sherman tanks. Contact was soon made with men of 101st Airborne who had secured exits from the beach and many of the defenders, from the German 709th Division, speedily surrendered. The story of events on Utah Beach that morning is well told in the museum on the beach, three kilometres east of Ste-Marie-du-Mont.

On Omaha Beach the story was very different. Here the bombardment had done little damage and the defending infantry of the decidedly average 716th Division had been reinforced by the rather better troops of the 352nd, on an anti-invasion exercise. The attacking infantry of the US V Corps found themselves pinned down on an open beach by savage small-arms and artillery fire. Most of the amphibious tanks, intended to help the attackers deal with just such an eventuality, had been launched too far from the coast and been swamped in the heavy sea. Landing craft, packed with drenched and seasick men, were riddled with fire and left rolling at the water's edge. Many of the engineers, on whose expertise hinged the clearing of the beach defences, were hit and only a handful of their armoured bulldozers survived.

Eventually small groups of brave and enterprising Americans forced their way off the beach: Brigadier-General Norman Cota of the 29th Division, on the right assault sector, played a notable part in getting men forward. The first half-hour of Steven Spielberg's *Saving Private Ryan* is set on Omaha Beach and comes perhaps as close as the cinema ever can to the reality of battle. By last light on D-Day the Americans held a beachhead at Omaha three kilometres deep and eight broad, but the effort had cost the two attacking divisions, 1st and 29th, well in excess of 2,000 men. Only German inability to launch a cohesive counter-attack had prevented a local reverse from becoming a serious disaster. Failure to get specialised armour ashore in the early stages ranks high amongst the causes of the costly setback at Omaha: not all the lessons of Dieppe had been uniformly learnt. Further to the west, three companies of 2nd Ranger Battalion under Lieutenant-Colonel James E. Rudder had landed at Pointe du Hoc, scaled the cliffs and taken the battery at their top. The 155mm guns it had contained were found further inland, and the Rangers destroyed these before determined infantry attacks brought them to a halt.

'Bloody Omaha' is the most striking of the Normandy beaches. It is approached down one of the gullies – the US *Official History* calls them 'draws' – through the cliffs. A sprawl of dune and scrub lies between the cliff foot and the low sea wall, and at low tide a broad band of sand slips down to the water's edge. Numerous monuments and memorial plaques commemorate the units that landed on D-Day and in Vierville-sur-Mer is a small museum, housed in a 1944 Nissen hut. One of the memorials, towards the centre of the landing area, marks the site of the first Second World War American military cemetery in France. After the war American dead were either repatriated or reinterred in large cemeteries, according to the wishes of relatives.

In an entirely appropriate location, on the cliffs above Omaha beach, is the American National Cemetery and Memorial at St-Laurent. A memorial colonnade encircling the bronze statue, Spirit of American Youth, looks out over the marble crosses and Stars of David marking the cemetery's 9,286 graves. Its restful elegance covers a deep well of human sorrow. Brigadier-General Theodore Roosevelt, assistant commander of the 4th Infantry Division, which landed on Utah Beach, was awarded the Congressional Medal of Honor for his bravery on D-Day. He died of a heart attack shortly afterwards, just before he heard that he had been appointed to command the 99th Division, and lies at St-Laurent, beside his brother Quentin, shot down over the Tardenois in 1918. There are thirty-two other pairs of brothers in the cemetery, as well as a father, Colonel Ollie Reed, lying beside his son, Ollie junior. An excellent new visitors centre was opened in 2007, although it is a sad comment on the way of the world that visitors are screened in an airport-style security check.

The assault on Sword Beach, at the eastern end of the

British line, was led by minesweeping flail tanks, which were first across the beach, followed by other specialist armour, like Crocodile flame-throwers tanks or Petard tanks whose heavy weapons destroyed bunkers. The defenders were already badly rocked by the time the landing craft carrying the first wave of infantry arrived and could do little to stem the tide. Juno Beach, to the west, was attacked by the 3rd Canadian Division whose infantry arrived on the beach slightly ahead of their supporting tanks and suffered more heavily than their comrades on Sword. There were similar difficulties on Gold, at the western end of the British sector, where the defences of Le Hamel, unsubdued by air and naval bombardment, caused problems for the 50th Division's right-hand battalion, 1st Hampshires. Very few of the Centaur tanks of the 1st Royal Marine Armoured Support Regiment managed to make their way through surf and beach defences: many of their hastily adapted landing craft foundered on their way in. Despite this, all 50th Division's infantry was ashore by early afternoon.

The British assault beaches are still strewn with the symbolism of D-Day. Ouistreham, just south-west of Sword Beach, was taken by Lord Lovat's 1st Special Service Brigade, which went on to link up with 6th Airborne Division, Lord Lovat crossing Pegasus Bridge accompanied by his piper, Bill Millin. A German flak tower in Ouistreham houses the Atlantic Wall Museum and opposite the casino, taken by No. 4 Commando, is the No. 4 Commando Museum. Memorials to the Commandos and to Commandant Philippe Keiffer, who led two French troops of No. 10 Inter-Allied Commando, stand nearby.

Through Colleville-Montgomery, renamed from Colleville-sur-Mer in honour of the British field marshal, is la Brèche d'Hermanville. A monument overlooking the beach pays tribute to the units of the British 3rd Division, which landed there, and a Churchill AVRE (Armoured Vehicle Royal Engineers) covers the centre of Sword Beach.

Although Luc-sur-Mer marks the boundary between Sword and Juno Beaches, the beaches at Luc and neighbouring Lagrune were unsuitable for landing, and it was at St-Aubin-sur-Mer that the attack on Juno began. Near a bunker on the beach are monuments to the North Shore Regiment from New Brunswick, 48 Royal Marine Commando, and a Canadian armoured regiment, the Fort Garry Horse. The Queen's Own Rifles of Canada landed at Bernières-sur-Mer, taking on a German strongpoint without tank support. They were followed by the *Régiment de la Chaudière*; both are commemorated on the beach. Courseulles, the next village to the west, was also a Canadian objective and amongst the memorials is a Canadian Sherman DD tank salvaged in 1970 from well out to sea where it had sunk on the run-in. At the foot of a very large Cross of Lorraine is a tank with an equally unusual history: this Churchill AVRE sank in a flooded culvert on D-Day and was recovered in 1970. Many of the Canadians killed on D-Day are buried in the Canadian War Cemetery at Beny-sur-Mer, on the high ground behind Juno Beach. The new *Centre Juno Beach* on the coast road just west of Courseulles is, I think, as good a museum as one might expect to find anywhere.

The 50th Division, liberators of Gold Beach, the westernmost British landing, are remembered by a memorial in Asnelles. German casemates still linger amongst the new building on Mont Fleury: it was in this area that Company Sergeant-Major Stan Hollis of 6th Green Howards won his VC, the only one earned on D-Day. The Green Howards disappeared in 2007 as a result of the restructuring of British infantry, but their reputation will endure as long as visitors pause at their memorial in Crépon, on the route of their advance inland. Just across the D176, north of the village of Creuilly, is the eighteenth-century château used as a headquarters by Montgomery, although in fact he lived in a caravan behind the building. It is not open to the public, but is easily visible through wrought-iron gates. Further towards Creully itself, just over the little Seulles, is the *Château-Fort*, a medieval fortress which housed the press corps.

While the D-Day landings proved easier than might have been expected – with the exception of the reverse at Omaha

– the advance inland was slower than had been hoped.
Although the Germans were thoroughly surprised and were
prevented from the rapid deployment of Panzer Group West
by Hitler's instructions, a combination of dogged defence
and brisk local counter-attacks helped slow down the Allies.

The British 3rd Division made good progress on the
morning of D-Day and by 11 a.m. its 185th Brigade, earmarked to seize Caen, was near Hermanville, ready to
advance. Unfortunately the tanks of the Staffordshire
Yeomanry, which were to have accompanied the three
infantry battalions, were trapped in a gigantic traffic jam on
the beach and the infantry set off without them. They then
discovered that 8th Brigade, ahead of them, had not yet
cleared two large German strongpoints, nicknamed *Morris*
and *Hillman*, whose fire covered the long ridge centred on
the village of Périers. *Morris* fell easily enough, but *Hillman*
was not taken until the evening (its commander, Colonel
Ludwig Krug, did not in fact surrender till the following
morning) and the delay proved serious. Although 2nd King's
Shropshire Light Infantry managed to push on down the
Caen road, flanking fire from *Hillman* hit 1st Royal Norfolk
as they followed up. The Shropshires' leading company
neared Lebisey Wood, then a mere five kilometres short of
Caen and now caught up in a *Zone Industrielle*, only to be
stopped short by the one major German counter-attack
launched that day.

General Feuchtinger of 21st Panzer Division committed
Colonel von Oppeln-Bronikowski's battle group into the gap
between the British and the Canadians. At Hill 61, where
the D60 and D60a cross between Beuville and Hermanville,
Oppeln lost thirteen tanks to tank and anti-tank gunfire, and
although some of his men managed to reach the 716th
Division's remaining strongpoints around Lion-sur-Mer, his
attack had too little weight to be really effective.

The area of the *Hillman* strongpoint has been cleared by
local enthusiasts and several of the bunkers can now be
entered at appropriate times: even when they are closed the
site offers an excellent view of the immediate hinterland of
Gold Beach. It lies on the narrow road south-east out of

Colleville-Montgomery, though getting a coach through the southern fringes of the village can be a nightmare.

By nightfall 3rd Division's leading elements were on the line Biéville-Benouville. The Canadians had pushed well inland, reaching the line Esquay–Coulombs–le Fresne-Camilly–Villons-les-Buissons. Of the troops landed on Gold Beach, 56th Brigade was approaching Bayeux and 47 Commando, tasked with joining the Americans from Omaha, was just south of Port-en-Bessin although it had not yet encountered its allies.

Failure to capture Caen on D-Day was an undoubted disappointment and it led, over the weeks that followed, to a series of British attempts to take the town. Historians and veterans alike still argue over the relative performance of British and German troops in this fighting, and many cast doubt on the wisdom of Montgomery's conduct of the battle on the eastern flank. It is perhaps safest to make two general points against which must be set what follows. The first is that the British army was already marked by five years of war – some of its veteran divisions were filled with soldiers who had been up the line too often in the past – and wherever possible British commanders let metal, not flesh, do their work. The second is that whatever Montgomery's failings, he had always intended that the British would 'write down' German armour on the eastern flank and his offensives, ill conducted though they often were, did indeed focus German interest on the British sector.

On 7 and 8 June direct attacks on Caen failed in the face of hardening resistance from 21st Panzer, opposite 3rd Division and Kurt 'Panzer' Meyer's 12th SS Panzer Division opposite the Canadians. On 9 June, after a painful approach march along roads harried by Allied aircraft, the redoubtable *Panzer Lehr* was in action against XXX Corps on the British right, and 7th Armoured Division was unable to make a clean break through well-handled German tanks and knots of infantry and self-propelled guns, which made good use of the hedge-bound *bocage* in which the XXX Corps was now becoming firmly enmeshed. There is abundant evidence of the character of the fighting inland. In Reviers a plaque commemorates the Regina Rifle Regiment, vigorously attacked by 12th SS Panzer on 8 June. Tanks got in amongst the Canadian positions and six Panthers were knocked out before Meyer's men pulled back.

Just off the small road entering Douvres-la-Deliverande from the west are the remains of a German radar station, which was one of the North Shore Regiment's D-Day objectives. The Canadians attacked towards it on 7 June but were fought to a standstill to its north, and responsibility for the attack then passed to 51st Highland Division, which launched another fruitless attack. There was an attempt on the station on 11 June and on the 17th a full battle group of 41 Royal Marine Commando with a squadron of the 22nd Dragoons and an engineer assault squadron mounted a set-piece attack from the village of Douvres on the heels of a heavy bombardment. Flail tanks ripped through mines and wire, AVREs lobbed explosive 'dustbins' at bunkers and the Commandos rushed in to take on the defenders, who surrendered after a sharp fight. It is now a radar museum, with the Second World War bunkers opened up and two post-war radars on display. The signs leading the visitor to the site take him round three sides of a square to minimise inconvenience to local residents.

Panzer Lehr's defence of the ground between the Aure and the Seulles encouraged Lieutenant-General Bucknall of XXX Corps to swing 7th Armoured Division west of the Aure, then hook down from Caumont to Villers-Bocage into the open German flank. While this move was in progress, 51st Highland Division would attack east of Caen. Both tines of this two-pronged advance were snapped off short. 51st Highland made little progress at heavy cost, while 6th Airborne's 12th Parachute Battalion lost 141 of the 160 men committed to its successful attack on Bréville, a sacrifice remembered by a memorial opposite the church.

Failure at Villers-Bocage was even more spectacular. 7th Armoured reached the Caumont area with little difficulty and on 13 June its leading element, 22nd Armoured Brigade Group, wheeled south-eastwards to take the Villers-Bocage Ridge. 4th County of London Yeomanry and a motor company of the Rifle Brigade entered the town, and Lieutenant-Colonel Lord Cranley led a Cromwell squadron and the Rifle Brigade company out along the Caen road, the N175, towards his objective, the high ground at Point 213, on the crest where the D217 joins the main road.

At this juncture *Hauptsturmführer* Michael Wittman of 501st SS Heavy Tank Battalion lunged into the Yeomanry from the south, his Tiger knocking out several tanks and even ramming a Cromwell before rattling through Villers-Bocage, destroying more tanks as it went. He refuelled and rearmed in time to join other Tigers in their attack on the leading squadron group, which was swamped some time later. Although his Tiger was knocked out as it moved back into Villers-Bocage, later in the day with a mixed force of *Panzer Lehr* and the newly arrived 2nd Panzer Division, Wittman and his crew escaped to fight again. The Germans destroyed twenty-five tanks, fourteen armoured trucks and fourteen Bren carriers in the engagement. 7th Armoured drew back to Tracy-Bocage and in the small hours of the 15th it withdrew to the line Livry-La Belle Epine.

The British *Official History* benevolently calls the Villers-Bocage battle 'disappointing' and observes that 7th Armoured had made its reputation in the desert not the *bocage*, and that the operation called for close infantry-tank co-operation, which was lacking. One might note that many German units also lacked experience of *bocage* fighting and that the recapture of Villers-Bocage was carried out by a scratch force from two different divisions including rear area troops. Both Bucknall of XXX Corps and Erskine of 7th Armoured were later relieved of their commands, and the commander of 51st Highland Division was also replaced.

With failure at Villers-Bocage the German line hardened across Normandy. Opportunities for exploiting gaps disappeared: room for manoeuvre would henceforth have to be created by hard slogging amongst the lanes and orchards. The British were planning their next attack on Caen when bad weather intervened to make their task even more difficult. Three days of severe but by no means unusual weather, beginning on 19 June, devastated the American

Mulberry harbour off St-Laurent and damaged the British *Mulberry* off Arromanches. Hundreds of vessels, from large landing craft to heavy Rhino ferries, were destroyed and the beaches were strewn with debris. The effect of the 'great storm' was less serious than it might have been, because there were already huge amounts of equipment ashore and it was relatively easy to land stores over open beaches. Despite its storm damage, the British *Mulberry* has withstood the passage of years remarkably well. Old ships, code-named *Gooseberries*, were sunk to form an outer breakwater and hollow concrete caissons were sunk on the landward side of them, forming a harbour wall. There were moorings inside the harbour, and piers ran out from the shore. Many of the caissons remain in a half-circle from Cap Maniveux round to Le Hamel, and further inshore are the remains of one of the piers. The view can be best appreciated from an orientation table on the coast road between Arromanches and La Fontaine St-Come. The panoramic film shown in the *Cinema Circulaire* nearby is very evocative, though it cannot be recommended to visitors with unsteady heads. There is an excellent museum, whose exhibits include a large model of the *Mulberry*, on the seafront in Arromanches.

The storm delayed the mounting of Operation *Epsom*. This employed both XXX Corps and the newly arrived VIII Corps: the former was to secure the area around Noyers, protecting the right flank of the attack, while the latter crossed the Odon and the Orne, outflanking Caen from the south. The battle began, in pouring rain, on 26 June and on the 29th a German armoured counter-attack hit the western flank of the salient, which now lapped out across the Odon south-west of Caen. Allied air power and artillery took the edge off the German armour long before the first Panzers engaged British tanks. However, the sheer vigour of the counter-attack persuaded General Dempsey of 2nd Army to withdraw behind the Odon and another opportunity had passed.

The Americans were busy in the west. On 8 June the US 1st Infantry Division crossed the Drome to meet the British

XXX Corps on its left and the fall of Isigny, on the night of 7–8 June, enabled the Americans to link up the V and VII Corps beachheads, but there could not be enduring contact until Carentan, the first proper crossing over the marshy Douves, was secured. The nakedly open causeway carrying the Valognes–Cherbourg road over the Douves and its water-meadows north-west of Carentan was crossed by Lieutenant-Colonel Robert G. Cole's 3rd Battalion, 502nd Parachute Infantry, which made its way as far as the outskirts of modern Carentan, where the D515 leaves the N13, before being forced to stop. The last leg of the battalion's advance was an old-style bayonet charge led by its commanding officer. A monument just on the Carentan side of the bridge over the Madeleine stream commemorates the action. 327th Glider Infantry, which had crossed the Douves at Le Moulin, advanced on Carentan from the east. The town fell on the 12th and the beachheads were securely linked the following day. The Germans immediately attacked up the Périers road with elements of 17th SS Panzer Grenadier Division, but the link held.

Over the next week, while Gerow's V Corps advanced southwards to Caumont, in step with the British on its left, Collins's VII Corps cut across the Cotentin, reaching Carteret, on its western coast, on the 18th. Some Germans managed to break through the corridor across the peninsula – 1,500 kicked their way past the 90th Division, which was making heavy weather of the advance and had its divisional commander and two regimental commanders replaced. General von Schlieben, commanding the Cotentin garrison, had the fragments of four divisions under his command, together with an assortment of flak gunners, naval personnel and Todt Organisation workers, up to 40,000 men in all. Once the peninsula was cut, Schlieben's men fell back quickly on Cherbourg and 20 June found them holding a perimeter on the hills around the town, from Vauville on the west coast, through Ste-Croix-Hague, Sideville, Hardinvast and Le Mesnil-au-Val north to the coast. On 22 June Cherbourg was heavily bombed and that afternoon VII Corps began its attack. Most defences held that day, but on the 23rd they began to crumble and on the 24th the Americans made deep penetrations against varying resistance.

On the 25th Bradley ordered a squadron of three battleships, four cruisers and escorting destroyers to give fire support to the final assault and air attacks continued. That day the Fort du Roule, a bastioned work built into a rocky outcrop and dominating Cherbourg from the south-east, was attacked by the 79th Division's 314th Infantry, and Corporal John D. Kelly and First Lieutenant Carlos D. Ogden were awarded the Congressional Medal of Honor for their part in the capture of its upper levels. It was not until the 26th that the lower levels were taken with the help of demolition charges and anti-tank fire.

On the 26th Schlieben was captured when tank destroyers put a couple of rounds into the entrances to his bunker in St-Sauveur, on the southern outskirts of Cherbourg. His deputy, Major-General Sattler, surrendered at the arsenal on the morning of the 27th and the last of the harbour forts fell two days later. The port facilities had been methodically destroyed. An American engineer officer called it 'the most complete, intensive and best-planned demolition in history', and Hitler awarded Admiral Hennecke, Naval Commander Normandy – captured with Schlieben – the Knight's Cross for doing such a thorough job.

The Fort du Roule now houses a museum and is an essential port of call. A memorial plaque on the Hôtel de Ville pays tribute to Sergeant William F. Finley of the 9th Division, the first US soldier to enter the town; he was killed in Germany on 1 April 1945 at the age of twenty. There is an exhibition dealing with Cherbourg's wartime experiences in the Fort de Roule. The decommissioned nuclear ballistic missile submarine *Le Redoutable*, built in Cherbourg and commissioned in 1971, can be visited, and there is a useful English-language commentary by its last commanding officer available on headphones.

With the Cotentin secure behind him, Bradley churned southwards into the *bocage*, but by the end of June he had stuck fast, and the Allied line ran from Portbail on the west coast, south of Carentan and down towards St-Lô, past

Caumont and up towards Tilly, bulging out over the Odon on the *Epsom* battlefield, then turning behind Caen to reach the coast at the Orne estuary. In early July the British took another bite at Caen with Operation *Charnwood*. On the night of the 7th the old city was heavily bombed and on the following morning I Corps attacked, only to discover that the ruins created by the bombing favoured the defender rather than the attacker and German will had not been broken by air power. The British and Canadians at last took most of Caen, though not its eastern industrial suburb of Colombelles. There was savage fighting in the triangle between Odon and Orne, with Hill 112, between the D36 and D8 north-east of Esquay-Notre-Dame, taken by 5th Duke of Cornwall's Light Infantry but then lost to a determined counter-attack. A memorial at its foot remembers the 43rd (Wessex) Division.

On 3 July Hitler relieved Field Marshal von Rundstedt of his command, replacing him with Field Marshal von Kluge, and General Eberbach took over Panzer Group West. The new commanders were given a firm directive to retain the ground they held and Kluge was warned that the Allies might attempt another landing in the Pas de Calais, so 15th Army could not be used to reinforce the Normandy front. Having reviewed the situation, Kluge decided to concentrate his tanks to attack the British. Rommel had little confidence in the project and said as much, but on 17 July he was on his way back from visiting Eberbach when his car was strafed by a British aircraft near the appropriately named Ste-Foye de Montgommery. The field marshal was badly wounded and was destined never to command troops again. Over the weeks that followed Eberbach was to be responsible for operations against the British on the eastern flank, while Hauser, newly appointed commander of 7th Army, faced the Americans in the west.

Bradley began serious attempts to move down to the line St-Lô–Marigny–Coutances on the 4th, in order to seize the St-Lô–Lessay road and break free of the difficult country on the Coutances front. VIII Corps began the attack on the western flank and VII Corps, redeployed from Cherbourg, soon joined in, attacking south from Carentan. The 83rd Infantry Division of VII Corps had a hard time of it, attacking on a very narrow front between the Gorges Marshes on its right and the River Taute on its left. Bradley's remaining Corps, XIX, extended the battle further east-wards, seizing crossings over the Vire at Airel. This encouraged Bradley to put 3rd Armoured Division at XIX Corps' disposal, and the corps commander ordered it to 'power-drive' through to the high ground west of St-Lô. The new division had to thread its way through congestion at Airel Bridge and its uncoordinated arrival in a congested battlefield produced no useful result.

With the Americans paying dearly for every inch of ground on the western flank, Montgomery prepared for another offensive, Operation *Goodwood*. He had established his tactical headquarters on the lawns behind the Château de Creullet, six kilometres inland from Gold Beach, whence he co-ordinated the activities of his two army commanders, and coped with the growing reservations of both Eisenhower and Churchill. This is not the place for a detailed study of the Eisenhower–Montgomery relationship, and in his three-volume biography Nigel Hamilton gives a scrupulously fair assessment of Montgomery's strengths and weaknesses. However, it is hard to resist the conclusion that by early July Montgomery was aware that he needed a victory, as much for political as for military reasons, and that his attempt to win one in Operation *Goodwood* bore all the hallmarks of the over-controlled battle, which some historians see as characteristic of large-scale British military operations.

As the British *Official History* observes, 'the conduct of the battle would be largely conditioned by geography.' The battlefield was bounded by Caen on its west and the close country running up from Troarn to the Bois de Bavent on its east. The embankment carrying the Caen–Troarn railway ran athwart the British line of advance, while further south the Caen–Vimont line cut obliquely across the field. The ground slopes up gently from the Orne valley, with the solid mass of the Bourgébus Ridge, crossed by the Caen–Falaise road, dotted by small, tightly nucleared villages with orchards and stone houses. Despite the standing corn, which gave some cover, fields of fire were long. Today, looking north from the lower slopes of the ridge, one has a front-row-of-the-stalls view of the opening stage of battle with the warehouses and chimneys of Colombelles as a backcloth.

Dempsey committed three corps to *Goodwood*. On his left, 1 Corps was to use its 3rd Infantry Division to buttress the Bavent flank, and on his right II Canadian Corps was to bridge the Orne and be prepared to advance on Fleury. The real weight of Dempsey's punch came in the centre, where O'Connor's VIII Corps would push its three armoured divisions – from the east Guards, 7th and 11th – on to the Bourgébus Ridge, keeping well together so that the Germans could not penetrate between them. Counter-attacks would be beaten off and the corps would exploit southwards. Doubt remains as to whether this exploitation was genuinely intended to be serious, or whether the attack was simply designed, as Montgomery later maintained, to attract and 'write down' German armour. At the time he wrote of getting a corps 'loose in the open country' and of O'Connor's 'plunge into his [the enemy's] vitals', and there is still no safe verdict on the question of the battle's real objectives.

The attack was prepared by a very heavy aerial bombardment. On the morning of 18 July British and American aircraft carpet-bombed large areas of the German front, and the guns of all three corps and those of the monitors HMS *Roberts*, *Mauritius* and *Enterprise* joined in. This storm of steel had the most profound effect on the defenders, and the 16th *Luftwaffe* Field Division, holding the front line, was very badly shaken. However, there was soon an immense traffic jam in VIII Corps' rear as units bunched on the Orne bridges and in minefield gaps, while further forward villages that had seemed to be bludgeoned into silence suddenly sprang to life as soon as the leading tanks had passed them. Major-General Roberts of 11th Armoured had been sceptical of the corps policy of giving infantry and armour separate objectives, and events proved him right, for patchy inter-arm co-operation was one of the features of the battle from the British side.

The leading elements of 11th Armoured Division, the first division through the gap, were across both railway lines by early afternoon, but the air was greasy with the smoke of burning Shermans as German 88mms reached out across the cornfields to take on British tanks at long range and the village garrisons joined in. Battle groups of 21st Panzer and 1st SS Panzer Divisions put in vicious little counter-attacks. On the British left, the Guards were held up by tanks and anti-tank guns around Emiéville and Cagny, while on the right 7th Armoured Division, stewed in the broth of vehicles on the Orne, only managed to get a single armoured regiment across the Caen–Vimont railway line by nightfall. The Canadians fought their way into Colombelles, and crossed both Odon and Orne south of Caen.

On the 19th, the second day of the battle, the 3rd Division was repulsed from Troarn but formed a solid defensive shoulder nonetheless. The Guards Armoured Division took Le Poirier and 7th Armoured, at last in action in strength, seized Four and came close to taking Bourgébus. 11th Armoured took both Bras and Hubert-Folie after heavy and costly fighting, and on its right the Canadians further improved their position. That afternoon Dempsey decided to secure the ground that had been taken and give his bruised armour a chance to refit. To achieve this II Canadian Corps would take over the line from Hubert-Folie to the Orne and work methodically forward.

On the 20th Bourgébus was occupied by the Guards, and the Canadians duly took over Bras and Hubert-Folie, and prepared to attack the Verrières Ridge feature across the Falaise road. The inevitable delays in relieving 7th Armoured meant that the Canadians were held up in their move on Verrières, and when their 2nd Division at last advanced it was counter-attacked very hard by elements of 1st SS and 2nd Panzer Divisions.

The Germans continued the pressure the next day and the battle focused on the D89, which runs from Hubert-Folie into the Orne valley through St-Martin-de-Fontenay. The Canadians had taken the road and the farms of Beauvoir and Troteval on the 20th, and although the Germans broke through to reach Ifs on the 21st, the Canadians had again secured the road by nightfall, despite confused fighting around the farms.

There was more fighting over the next few days, but *Goodwood* had effectively ended. The *Official History* admits that while it had achieved its main purpose by improving the Allied position around Caen and keeping German armour on the eastern flank, 'it had not attained all that was intended'. The battle cost 5,537 British and Canadian casualties, as well as 400 tanks. Such was Allied logistic mastery that most of the lost tanks had been replaced within thirty-six hours.

Colombelles has grown, many of the villages on the *Goodwood* battlefield have expanded markedly and the *Autoroute de Normandie* whistles across it almost on the line of the now dismantled Caen–Troarn railway. The Caen–Vimont line is no longer in use, but its embankment west of Soliers shows all too well the problems it posed to tanks. On that part of the field most were unable to cross it and instead had to use the cuttings that carry local roads beneath the railway. The villages, increasingly dominant as the ground rises towards the crest of the ridge, give a good feel for the strength of the German position. The 88mm gun, either on its ground mount or in the turret of a Tiger tank, could destroy a Sherman, the workhorse of allied armour, at a range of 2,000 yards and these nucleared villages put a premium on German firepower.

Cagny, on the Vimont road, provides a microcosm of the battle and events can best be appreciated from the area where buildings end on D225 on the north-eastern edge of the village. Major Hans von Luck, commander of a battle group in 21st Panzer, had reached his headquarters in Frénouville after three days' leave in Paris, to find British tanks already on the move. He set off for Cagny, where he saw a mass of Shermans heading for the gap between Cagny and Le Mesnil Frementel. Luck found a single Panzer Mark IV and an 88mm anti-tank gun that had escaped the bombing, as well as a battery of 88mm *Luftwaffe* anti-aircraft guns. The latter's commander explained that he was not in the business of taking on tanks, but Luck drew his pistol and explained, in a phrase incisively repeated on a score of Staff College battlefield tours, that he could 'either die now on my responsibility or win a decoration on his own'. The 88mms duly came into action on the edge of the orchard and within minutes Shermans of the Fife and Forfar Yeomanry were blazing in the cornfield between the villages. Recent research has thrown doubts on to the details of Luck's story – for instance, there is no sign of the 88mms in aerial photographs – but the principle remains instructive. Despite having had his battle group lacerated before it had fired a shot, Luck was able to generate purposeful activity, which trickled sand into the complex machinery of the British plan. For many years Pirbright in Surrey was the depot of the Household Division and the town was twinned with Cagny. There is a memorial to the Foot Guards regiments on the wall of the church, whose part-ancient and part-modern character reminds us of the effect of Allied bombing on these Norman villages.

After *Goodwood* the headquarters of Crerar's 1st Canadian Army arrived to take responsibility for the British left flank. American forces, too, were being reorganised and on 1 August Bradley became an army group commander, his 12th Army Group comprising Hodges's 1st Army and Patton's 3rd. Montgomery retained overall command of land forces for the time being. There was a last attempt to make fresh progress on the eastern flank on 25 July, when II Canadian Corps butted into stiff opposition on the Caen–Falaise road, and the emphasis thereafter shifted dramatically to the west as Bradley mounted Operation *Cobra*. On the morning of 25 July nearly 3,000 aircraft pounded German positions between St-Lô and Lessay. Although American units had been drawn back from the main road some bombs fell short, causing 600 casualties, amongst them Lieutenant-General Lesley J. McNair, commanding general of Army Ground Forces and the highest-ranking US officer killed in the war.

The first day of *Cobra* saw infantry at work, crumbling the defences on the line of the St-Lô–Périers road and on the next day Collins of VII Corps took what seemed the considerable risk of committing his armour. In fact, the bombing had done more damage than the first day's fighting had suggested and in places the German line was paper-thin. There was good progress on the 26th and on the 27th the advance built up real momentum: Avranches fell on the 30th.

History had unrolled through Avranches before: in 1172 Henry II of England had done penance, barefoot and wearing only a shirt, for his part in the death of Thomas à Becket. This took place at the entrance to the cathedral, which collapsed in 1794, but the site of Henry's humiliation is preserved as *la Plate-Forme* off the Place Daniel-Huet. At the beginning of August 1944 Avranches again saw history made. Patton squeezed his new 3rd Army through the Avranches corridor and across the single bridge over the Sée in twenty-four hours. He remains a controversial and bombastic figure, but in the late summer of 1944 he was the man of the hour, handling his armour with real flair and punching on down into Brittany in a campaign whose *blitzkrieg*-like panache left the misery of the *bocage* far behind. Patton is remembered by a monument where the D7 swings south out of Avranches. It stands in a square that is American territory, its soil and trees brought across from the United States. In Le Val St-Père, five kilometres to the south, is a museum which concentrates on the Avranches breakthrough. It is on a slip-road of the Avranches bypass, Junction 3 of the A84. .

The success of *Cobra* spread its ripples wide. On 27 July Montgomery issued orders for a redeployment to throw 21st Army Group's weight on to its right – Montgomery told Dempsey to 'step on the gas for Vire' – and on the 30th XXX and VIII Corps struck southwards from Caumont into the close, blind country between Orne and Vire.

Operation *Bluecoat* began with promise but was soon wallowing in the banked hedgerows and stifling orchards. 43rd Wessex Division took Mont Pincon, which dwarfs the surrounding countryside south of Aunay-sur-Odon. Its crest is best found by looking for the red-and-white TV mast next to the D54, north of Le Plessis Grimoult. The hill dominates not just the valleys of the Orne and the Vire, but also the ground out towards Argentan in the south. A 1992 memorial to the 13th/18th Royal Hussars is signposted from the TV mast and there is a memorial to 5th Duke of Cornwall's Light Infantry in Le Plessis itself.

On 7 August II Canadian Corps began Operation *Totalize* to the west of the *Bluecoat* battlefield. The story was a familiar one: initial progress was replaced by grim attrition as German reserves arrived. Once again losses were heavy. The British Columbia Regiment lost forty-seven tanks in a single day and the cemetery at Cintheaux shows something of the price Canada paid for her part in the liberation of Europe. The Polish cemetery at Grainville, further down the Falaise road, reminds us that Major-General Stanislas Maczek's 1st Polish Armoured Division was fighting under the Canadians, and many of its Lancers, Dragoons, Light Infantry and Mounted Rifles were destined to remain in France. It was also Michael Wittman's last battle. For many years it was believed that he perished when his unit was carpet-bombed, but it is now clear that this outstanding tank commander died under the guns of Shermans, although there is doubt as to whether these were Canadian or belonged to the Northamptonshire Yeomanry. Wittman and his crew are buried some distance away, in the German cemetery at La Cambe, just off the N13 between Isigny and Bayeux.

The Germans had mounted a major effort of their own the day before *Totalize* began. On the night of 6 August XLVII Panzer Corps, attacking on the direct orders of Hitler, jabbed into the US VII Corps at Mortain. The Americans had some advance warning from both ULTRA intercepts and air reconnaissance reports, and although the impact of the attack jarred some American units, when daylight came on the 7th Allied aircraft made good use of the superb flying weather, and of the seventy German tanks that made the initial penetration only thirty were in action at last light and the American position was entirely restored by 12 August.

Even before the Germans had been repulsed from

Mortain, Allied commanders had taken steps to capitalise on the opportunity that German persistence in retaining a hold on Lower Normandy now presented. On 8 August Bradley and Montgomery spoke on the telephone and agreed to a changed plan, which would swing Patton's 3rd Army up towards Argentan while the British pushed down to Falaise and Condé, encircling the bulk of the German army in a huge pocket. Montgomery issued a formal directive for this operation on 11 August and three days later the Canadians were approaching Falaise from the north with the US XV Corps near Argentan, some thirty kilometres away.

The difficulty was that XV Corps was stationary. It had halted on Bradley's order on 13 August. This decision was, like so much else in the Normandy campaign, to cause endless post-war recrimination, for it meant that the Falaise pocket was never fully watertight. Bradley feared that if American troops sealed it off they might be stamped flat by retreating Germans: he preferred 'a solid shoulder at Argentan to a broken neck at Falaise'. There was also concern about accidental clashes between US and Canadian troops, and the co-ordination of artillery fire and air strikes would be increasingly hard as the pincers closed.

In mid-August 1944 the shrinking Falaise pocket was the valley of tears of the German army in Normandy. Repeated air attacks, to which the Germans could make no effective response, destroyed tanks and half-tracks, trucks and staff cars alike. Much transport was still horse-drawn and the shocking memory of lanes choked with dead horses, bellies grotesquely swollen, still returns to many veterans of the fighting. The wounded Hauser was carried out of the Falaise pocket on the engine deck of a tank; 'Panzer' Meyer was guided out by a French civilian; Lieutenant-Colonel Heinz Guderian, Chief of Staff of 116 Panzer division and son of 'Hurrying Heinz', came out on one of the two assault guns his division could muster. In all, the Germans had lost 1,500 tanks and 450,000 men in the battle for Normandy. Allied losses were nearly 210,000 men, just under 84,000 of them from 21st Army Group.

Exploitation was rapid. On 25 August the military

governor of Paris surrendered to General Leclerc of French 2nd Armoured Division, the Canadians took Dieppe on 1 September and the British liberated Brussels on the 3rd. 1 September also saw Eisenhower assume personal command of Allied ground forces. Montgomery, in his headquarters in the Château de Dangu on the northern border of Normandy, received news that day of his promotion to field marshal. His conviction that the golden opportunities of the Normandy battle were about to be squandered by Eisenhower's policy of 'bulling ahead on all fronts' instead of mounting a single-minded thrust up the Channel coast took some of the pleasure from the occasion. It is unclear whether such a dramatic change of policy could have averted the operational pause, which enabled the Germans to draw breath in September–October 1944, and in any event the consequences of German recovery lie outside the compass of this chapter.

It took years for the debris of the German defeat to be cleared from southern Normandy. The tanks that remain, like the Tiger on the D979 about a kilometre south of Vimoutiers, have been heavily restored. A Polish Sherman (a 'Firefly', mounting a 17-pdr gun rather than the usual 75mm) looks out over the killing fields from the Polish memorial on Mont Ormel, north-west of Chambois, at the neck of the pocket. In Chambois itself, by the square keep whose flag bears the leopards of Normandy, a memorial stone recalls that it was there that Captain Waters of the US 358th Infantry met Colonel Zgorzelski of the Polish 10th Dragoons on 19 August 1944.

The battle of the pocket is well described in the August 1944 Museum in Falaise, though this clearly takes second place, in the eyes of the local tourist industry, to the William the Conqueror Museum in the castle. The Conqueror's birthplace was liberated by the 6th Canadian Infantry Brigade on 17 August 1944 and amongst the museum's exhibits is a diorama showing the state of the town when its population returned to the ruins. There are Second World War items on display in the Normandy Bocage Museum in the château at Flers, between Vire and Argentan. The fomer Rue du Banque is now named for the British 11th Armoured Division, which Miles Dempsey calls 'an outstandingly fine division. I have never met a better.' At St-Martin-des-Besaces on the N175 south-west of Villers-Bocage another museum, the *Musée de la Percée du Bocage*, open on summer weekends, recalls the *Bluecoat* offensive. At Tilly-sur-Seulles, on the D6 south of Bayeux, the beautiful white-stone chapel has been turned into a museum whose black-and-white photographs tell the story of the village's occupation and liberation.

It is in Bayeux that the knot linking Normandy to England is tied off tightly. Its suburbs were entered by patrols of 50th Northumbrian Division on the evening of D-Day and the town was liberated on 7 June. General de Gaulle visited it a week later, and the force of his enthusiastic reception helped convince the British and Americans they were right to recognise his National Liberation Committee as France's provisional government. It is a delightful little town, dominated by its Norman Gothic cathedral. The Bayeux Tapestry is sensitively displayed in its own museum: it is propaganda as well as history and much is made of Harold's 1064 oath to William. The excellent Battle of Normandy Museum stands in the Boulevard Fabian Ware, named after Major-General Sir Fabian Ware, guiding spirit of the Imperial (now Commonwealth) War Graves Commission and its first Vice-Chairman. Near the museum is the CWGC's Bayeux Cemetery. This, the largest British Second World War cemetery in France, contains 4,648 graves, including that of Corporal S. Bates, a Camberwell lad serving, ironically, in the Royal Norfolks, the only VC buried in Normandy, who was mortally wounded east of Vire on 6 July. Across the road from the cemetery is the Bayeux Memorial to the Missing, which bears the names of 1,537 British, 270 Canadians and one South African who fell in Normandy but have no known graves. A Latin inscription atop its colonnaded front remembers Britain's ambivalent relationship with Normandy. 'We, once conquered by William,' it proclaims, 'have now set free the Conqueror's native land.'

FORTIFICATION: THE BASTION SYSTEM

GROUND PLAN

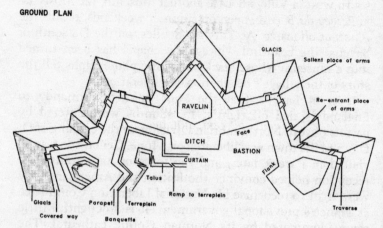

PROFILE

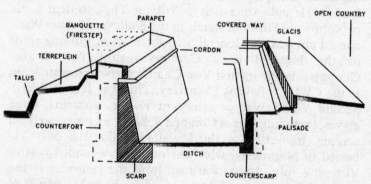

(NOT TO SCALE)

Index

Individuals are generally described by the military rank held when mentioned in the text, not necessarily the highest rank attained